D1766667

7 Day

University of Plymouth Library

Subject to status this item may be renewed
via your Voyager account

http://voyager.plymouth.ac.uk

Exeter tel: (01392) 475049
Exmouth tel: (01395) 255331
Plymouth tel: (01752) 232323

ITH

Item No.	200031
Class No.	388 . 094 EUR
Contl No.	0471957372

90 0283344 7

European Transport and Communications Networks

The **European Science Foundation** is an association of its 56 member research councils, academies, and institutions devoted to basic scientific research in 20 countries. The ESF assists its Member Organisations in two main ways: by bringing scientists together in its Scientific Programmes, Networks and European Research Conferences, to work on topics of common concern; and through the joint study of issues of strategic importance in European science policy.

The scientific work sponsored by ESF includes basic research in the natural and technical sciences, the medical and biosciences, the humanities and social sciences.

The ESF maintains close relations with other scientific institutions within and outside Europe. By its activities, ESF adds value by cooperation and coordination across national frontiers, offers expert scientific advice on strategic issues, and provides the European forum for fundamental science.

This volume arises from the work of the ESF Scientific Programme on Regional and Urban Restructuring in Europe (RURE).

Further information on ESF activities can be obtained from:

European Science Foundation
1, quai Lezay-Marnésia
F-67080 Strasbourg Cedex
France

Tel   (+33) 88 76 71 00
Fax   (+33) 88 37 05 32

UNIVERSITY OF PLYMOUTH
SEALE HAYNE
LIBRARY

European Transport and Communications Networks

POLICY EVOLUTION AND CHANGE

Edited by

DAVID BANISTER
University College London, UK

ROBERTA CAPELLO
Politecnico di Milano, Italy

WITHDRAWN
FROM
UNIVERSITY OF PLYMOUTH
LIBRARY SERVICES

PETER NIJKAMP
The Free University of Amsterdam, The Netherlands

UNIVERSITY OF PLYMOUTH
SEALE HAYNE
LIBRARY

JOHN WILEY & SONS
Chichester · New York · Brisbane · Toronto · Singapore

Copyright © 1995 by John Wiley & Sons Ltd,
Baffins Lane, Chichester,
West Sussex PO19 1UD, England

Telephone: National 1243 779777
International (+44) 1243 779777

All rights reserved.

No part of this book may be reproduced by any means,
or transmitted, or translated into a machine language
without the written permission of the publisher.

Other Wiley Editorial Offices

John Wiley & Sons, Inc., 605 Third Avenue,
New York, NY 10158-0012, USA

Jacaranda Wiley Ltd, 33 Park Road, Milton,
Queensland 4064, Australia

John Wiley & Sons (Canada) Ltd, 22 Worcester Road,
Rexdale, Ontario M9W 1L1, Canada

John Wiley & Sons (SEA) Pte Ltd, 37 Jalan Pemimpin #05-04,
Block B, Union Industrial Building, Singapore 2057

British Library Cataloguing in Publication Data

A catalogue record for this book is available from the British Library

ISBN 0-471-95737-2

Typeset in 10/12pt Times by Acorn Bookwork, Salisbury, Wilts
Printed and bound in Great Britain by
Biddles Ltd, Guildford and King's Lynn

Dedication

This book is dedicated to Bjørn Andersen who died in November 1994. Apart from contributing actively to all NECTAR activities over the last 10 years, Bjørn epitomized the spirit of international collaboration. He was always good fun to be with, he would always be willing to supply valuable data and information, he was always ready and keen to produce the next paper, and he was always a very generous host. He has also contributed two chapters for this book.

Contents

Contributors

The late Bjørn Andersen
Molde, Norway

David Banister
Planning and Development Research Centre, The Bartlett, University College London, 22 Gordon Street, London WC1H 0QB, UK

Sean Barrett
Department of Economics, Trinity College, University of Dublin, Dublin 2, Ireland

Ulrich Blum
Technical University of Dresden, Department of Economics, Mommsenstrasse 13, D-01062 Dresden, Germany

Kenneth Button
Department of Economics, Loughborough University, Loughborough LE11 3TU, UK

Roberta Capello
Department of Economics, Politecnico di Milan, Piazza Leonardo da Vinci 32, 20133 Milan, Italy

James Cornford
Centre for Urban and Regional Development Studies, University of Newcastle upon Tyne, Newcastle upon Tyne NE1 7RU, UK

Vlasta Dugonjic
United Nations, Economic Commission for Europe, Palais des Nations, CH 211 Geneva 10, Switzerland

Andrew Gillespie
Centre for Urban and Regional Development Studies, University of Newcastle upon Tyne, Newcastle upon Tyne NE1 7RU, UK

Veli Himanen
Technical Research Centre of Finland, Transport and Surveys, Sahkomichentie 3, SF – 02150 Espoo, Finland

Frank Leibbrand
Technical University of Dresden, Department of Economics, Mommsenstrasse 13, D-01062 Dresden, Germany

Kai Lemburg
City of Copenhagen, Sundvaengt 9, DK-2900 Hellerup, Denmark

Joost van Nierop
Department of Economics, Free University of Amsterdam, De Boelelaan 1105, 1081 – HV Amsterdam, The Netherlands

Peter Nijkamp
Department of Economics, Free University of Amsterdam, De Boelelaan 1105, 1081 – HV Amsterdam, The Netherlands

Juraj Padjen
Ekonomski Institut, Trg JF Kennedy-a 7, 41000 Zagreb, Croatia

Remigio Ratti
Institute of Economic Research, Stabile Torretta, CH-6500 Bellinzona, Switzerland

Piet Rietveld
Department of Economics, Free University of Amsterdam, De Boelelaan 1105, 1081 – HV Amsterdam, The Netherlands

Roberto Roson
Department of Economics, University of Venice, Ca'Foscari PD 3246, 1-30123 Venice, Italy

Fabio Rossera
Institute of Economic Research, Stabile Torretta, CH-6500 Bellinzona, Switzerland

Wladimir Segercrantz
Technical Research Centre of Finland Traffic, Itatuulenkuja 11A, SF – 02100 Espoo, Finland

José Viegas
Centro de Sistemas Urbanas e Regionas, Instituto Superior Tecnico, Avenida Rovisco Pais, 1000 Lisbon, Portugal

Jaap Vleugel
Department of Economics, Free University of Amsterdam, De Boelelaan 1105, 1081 – HV Amsterdam, The Netherlands

Eddy van de Voorde
Universiteit van Antwerpen (UESIA), Faculteit TEW, Valgroep Transport en Ruimte (SESO), Prinsstraat 13, B-2000 Antwerpen, Belgium

Stef Weijers
Policy Research Unit AVV, Ministry of Transport, PO Box 1031, 3000 BA Rotterdam, The Netherlands

Preface

Putting together a volume of papers such as this is a major enterprise, even in Europe where the transport and communications networks are of a high calibre. The original inspiration for this understanding was the Euro Conference held in Padua in December 1993 where several papers included here were first presented. The other papers come from a range of different sources, some invited and others again presented at major international seminars. There is a common denominator here—namely that all have a link with the European Science Foundation's Network of European Communications and Transport Activities Research (ESF-NECTAR).

The editors wish to thank the European Science Foundation for its continued support for the network, in particular John Smith for his continuous enthusiasm for NECTAR, and Caroline Grimont and Anne Guehl for their share in ensuring the success of the Padua Euro Conference. We would also like to thank the European Union Human Capital and Mobility Programme (EU-HCM) which has supported this conference as well.

The editors are particularly grateful to Prof. Roberto Camagni of Padua University, and his academic staff, for having so kindly and efficiently acted as the local hosts and organizers of the Euro Conference. Thanks to them, Padua and its university have provided the right pleasant and friendly environment where more than 50 experts in the field have been able to exchange scientific ideas and experience; this book is the major output from that event.

Moreover, financial support of the Italian Research Council (CNR) (project no. 93.01792.PF74 – Trasporti, directed by Prof. Roberto Camagni) is gratefully acknowledged, as well as support from local sponsors, namely the Fondazione Cassa di Risparmio di Padova e Rovigo, Urbania, Camera di Commercio di Padova and the Economics Department of the University of Padua.

In addition, the editors wish to thank all authors of individual chapters in this volume for meeting deadlines (some did better than others!), and to the numerous referees who were asked to comment on the chapters. This volume is a truly international venture with some 24 authors from approximately 12 different European countries collaborating on the major trans-European theme of transport and telecommunications networks. A sincere thank you goes also to Mrs Karen Woodword for her careful linguistic editorial work.

Finally, we would like to thank the directors and staff of the Tinbergen Institute in Amsterdam. Much of the final editing was carried out here together with the joint drafting of the Introduction and the concluding sections. This research centre is the ideal location to debate and write on European policy and economic issues.

David Banister
Roberta Capello
Peter Nijkamp

Tinbergen Institute, Amsterdam
January 1995

European Transport and Communications Networks: All Change

DAVID BANISTER, ROBERTA CAPELLO and PETER NIJKAMP

Europe is undergoing fundamental change as the twentieth century comes to an end. The European Union (EU) has grown in size, there are new agreements with the European Economic Area (EEA) and the countries of the old Eastern bloc. Internal barriers have been dismantled, many activities are now performed multinationally and internationally, decisions are increasingly being taken at the European level rather than at the national level. The net outcome of all these changes will be an increase in the demand for travel and an increased use of the communications networks. Business, leisure, work and even shopping will become international with the freedom of movement, the ability to work in any EU country and the expected growth in affluence.

The infrastructure is the key to an integrated Europe, yet much of the existing network is over one hundred years old and it is oriented towards local and national centres. Even the motorway network is feeling its age. Very substantial investment is required to replace existing roads and construct new ones, and these links should help integrate peripheral areas as well as open up new markets in Scandinavia and the old Eastern European countries. Road investment will be complemented by the new European high-speed rail networks and the telecommunications networks, including the value added networks (VANs) and the integrated digital services network (ISDN). It is this combination of networks which will facilitate the most fundamental changes brought about by knowledge advances and information technology. Included here are logistics planning, electronic data interchange, electronic route guidance, road pricing, emergency transport planning, information systems and databases for environmental monitoring.

As a consequence, the spatial imperative will no longer apply as cities will become much looser spatial organizations because the costs of urban centralization and high land prices will be balanced against the benefits of dispersal, with its high levels of mobility, better quality of life and flexibility. The movement out of cities will continue with only front office functions remaining. Growth will be concentrated in corridors of good communications and at peripheral urban locations where it is cost effective to link in with both the transport and information networks. Peripheral areas may still remain isolated and separate from the new infrastructure as access costs and capacity requirements may make the installation costs of the new networks uneconomic and the costs of using the system too high.

The most attractive locations in Europe are likely to be those where the transport and information networks link in with other factors such as a skilled labour force, a high-quality environment and the availability of low-cost land. Interchanges may provide particularly suitable locations for logistic platforms. International airports, high-speed rail stations, and major motorway intersections could all provide the sites of maximum accessibility which would minimize location and transport costs, and would also be on the international information network. Where more than two of these factors actually work together, then a major Euro-hub would develop. This is already becoming apparent at Charles de Gaulle airport at Roissy and at Schiphol airport in Amsterdam.

There is no doubt that very significant changes are taking place on both the political and technological fronts in Europe, and these profound changes are likely to affect all areas directly and indirectly. However, the exact nature and scale of that impact is less than clear, but it seems that the distribution of benefits will not be equal in all areas. Also the time scale of the changes is not clear, except to observe that the rate of change seems to be increasing rapidly. So the impact of these changes may be selective and take time for the full effects to become apparent. This is a foretaste of the exciting background against which this book is placed. This book has several objectives: to present the latest situation on the evolution of European transport and communications networks, as seen by some 24 experts; to integrate both policy and analysis issues in each of the chapters; to illustrate the arguments with case study material from a variety of national and international perspectives; and to contribute to the growing literature which is European and synoptic in its scale and coverage, thus moving away from purely national concerns and interests. The book has taken as its central theme one of the most important issues facing European transport policy makers in the next ten years, namely the quality and extent of the existing transport network, and the role that the new telecommunications and transport infrastructure will have on European integration and the new map of Europe at the end of the twentieth century.

This book has been divided into three main Parts. Part I concentrates on *Evolving European Networks* where the complicated interrelationships between the different modes of transport are introduced together with the key European problem of the well-integrated centre and the remote periphery. The debate here is over whether investment in networks at the centre to respond to existing capacity limitations will result in greater concentrations of economic growth at the centre, or whether there are regional development benefits which can be obtained through a greater investment in the peripheral regions. The balance of investment is a crucial issue as is the means by which the benefits from investments can be measured not just in direct terms, but in terms of the impacts on the local economy and the environment.

There is a new urgency about the role that European networks can have and should have as the membership of the European Union expands. Some means of communications are not restricted by physical and political boundaries, but other forms are severely constrained by the extent of the network, its quality and capacity. In designing systems for the next century, politicians, planners and economists must look towards standardization of the technology, common operating systems, the maximum efficiency of the network, and value from the very substantial investment costs required. But there is also a responsibility to ensure that macro economic objectives and technological sophistication are matched by a concern over the distributional and equity implications from the decisions made, and the longer-term environmental and ecological consequences of decisions in the transport and communications sectors.

In the first chapter Peter Nijkamp and Jaap Vleugel focus on the critical role that the transport infrastructure has on the spatial–economic evolution of national and European economies. Although the arguments are illustrated only with examples from the international freight transport and high-speed rail sectors, the messages are true for all modes of transport. True European integration is dependent upon completing the Pan-European network of missing links and networks. If this objective is not achieved, then the competitive position of the core regions will continue to be strengthened at the expense of the weaker peripheral regions. The argument leads to the conclusion that at present too much emphasis is placed on priorities related to national interests. This national focus must be replaced by European objectives. To some extent this has been achieved in international freight and passenger movement, particularly with the advent of the high-speed rail network, but this is only a beginning. The Maastricht Treaty was complex and dense, but the need for transport and communications systems which are interoperable, interconnected and intermodal are all clear requirements for European networks.

The land-based arguments are complemented by sea-based arguments in the chapter by Eddy van de Voorde and José Viegas. Long-distance travel will increase substantially, both in the passenger and freight sectors. This increase in demand will be matched by the globalization and internationalization of companies, new logistic concepts, more efficient information and communications systems, and a greater importance being given to the total costs of transport and distribution. Inefficiencies in the infrastructure act as a severe constraint on progress. One means to gain the benefits of these changes, but at a relatively low cost, is to make much more use of the sea network which effectively has an unlimited capacity. It would also assist in the development of the peripheral countries within Europe, as all of these are accessible to the sea. At present, the sea acts as a barrier, but it can and should be utilized as part of the new European networks. Short-sea shipping is cost effective as it requires limited additional infrastructure and it is environmentally friendly as the costs of moving freight by sea is very energy efficient. A case study is used in this paper to present the relative competitive positions of road and sea routes, yet despite clear cost advantages for the shipping alternative, the road is still used. To achieve the competitive advantages of short sea shipping, its potential must be recognized, and it must become an integral part of a European transport policy.

This theme of transport networks in the peripheral regions of Europe is taken up by Veli Himanen, Juraj Padjen, Vlasta Dugonjic and Wladimir Segercrantz in their contribution. Although it is now much easier to get from one part of Europe to another, new factors have emerged—congestion, pollution, social costs of transport—and there is now a greater dependence on the car. The barrier effects of national borders within the European Union have been eliminated, and the formation of the European Economic Area (EEA) has also made movement between EFTA and EU countries very easy. However, substantial barriers still remain between East and West and between South and North. This collaborative chapter is illustrated with examples from the Nordic countries and the links between the EEA and Russia.

Barriers still remain, even though border delays have been reduced. The distances are long, the frequency of air services is low, and many movements are dependent upon sea crossings. Consequently, costs are high and the competitive position of the peripheral nations in the EU and the EEA is weakened.

Even though barriers still remain, they are being reduced and Remigio Ratti analyses the effects and consequences of the gradual abolition of institutional barriers. This change is illustrated by the impact that the new logistics networks have had on forwarding agents and on small and medium-sized transport enterprises in a border region. A 'functional space' approach is developed incorporating production space, market space and supporting space, so that the traditional operations within a border region can be compared with the new situation with no borders. The regional impact depends on the use made of logistics networks, a substantial capital investment for many forwarding agents. It will also depend on the organizational dynamics and the operational strategies at the level of the functional space. Location at borders will no longer be a key factor and it is likely that a hierarchy will develop with centres of continental logistics (Rotterdam, Frankfurt, London) or at the level of large metropolitan centres. Other logistics platforms may develop at the major airports or at particular nodes near the barriers (Verona, Lyon, Basle). In addition to this spatial hierarchy, a specialization hierarchy may also develop as the market readjusts to changes in prices, and the role of subcontractors and niche markets may become dominant. New complex patterns of operation will emerge as the institutional barriers are removed.

The final chapter in Part I examines the existing literature on network externalities and then presents a behavioural analysis of the importance of network externalities on diffusion processes for two contrasting economic environments in which these technologies have been developed, namely North and South Italy. Roberta Capello and Peter Nijkamp develop a taxonomy of network externalities in the telecommunications industry, which links together production and use of telecommunications with concepts of productivity (intermediate use) and utility (final use). Their conclusion is that network externalities have been expanded to embrace a much wider range of factors, in particular manufacturing firms and service providers. They then proceed to examine consumption network externalities through the behavioural analysis of the links between telecommunications and regional development in Italy. The aim is to test the role that network externalities play in the decision to adopt the technologies. The analysis is both descriptive and interpretative. The contingency table analysis strengthens the role of the number of existing subscribers as the main reason for adoption. The logit model demonstrates the importance of the pricing regime and the supply side variables. In both cases there were regional variations between North and South Italy, and this has led the authors to conclude that the STAR programme[1] has been successful in stimulating the willingness to adopt in the near future—there are consumption network externalities. But it is still unclear whether production network externalities are exploited by firms and regions.

Part II develops the theme of evolving European networks by suggesting appropriate methods for the analysis of networks. There are four chapters on Applied Network Analysis, and each of them takes a different approach to analysis, but all also illustrate the method with case studies.

Piet Rietveld and Joost van Nierop review the different contributions to network location modelling. These include cost minimization algorithms, demand-oriented formulations such as maximizing the total use of the network, and maximization of the rate of return on investment in the infrastructure. The practice always differs from the theory, as decisions are influenced by political factors, regional policy objectives, strategic reasons, cross-border considerations and even status factors. Rationality or economic efficiency arguments may only feature at a lower order of importance. The basic aim of the chapter is to establish whether the development of the Dutch railway system in the last century can be analysed through network location models. The first line was constructed in 1839 and the network was fully

---

[1]This research is part of the European Union's Special Telecommunications Action for Regional Development (STAR) programme which commenced in 1987.

connected by 1875. The modelling results were mixed with only 40 per cent of the first 20 links in the railway networks being accurately predicted. But the approach does offer opportunities for further extensions to the model as the simple demand functions can be refined through modal competition, the development impacts of the railways and the introduction of several competing companies rather than the single railway company assumed here.

One of the major barriers within Europe is the Alps, and its presence has resulted in strange routes being used by freight traffic. This is partly because of the limited number of crossing points, but it also results from the political pressures in Switzerland to reduce the adverse environmental effects of heavy transit traffic. Fabio Rossera explores the choices of mode and routes which have resulted as a consequence of the physical barriers of the Alps, and then also investigates the impact of the restrictions imposed by the Swiss government. A database has been established for eleven European regions, and the flows by mode have been modelled through a singly constrained gravity model. The results are mixed and it is difficult to draw clear overall conclusions, so further more focused analyses have been carried out. The impact of restrictions imposed by the Swiss government has resulted in a doubling of transit traffic in Austria and a quadrupling of traffic through France. Diversion of traffic back to Switzerland would result in a formidable increase in transit volumes. The action taken by one government has therefore significantly impacted upon the mode and the route choice of freight hauliers across the Alps, resulting in longer routes and substantially increased volumes of transit traffic in neighbouring countries.

A very different modelling approach is presented in the next chapter by Roberto Roson. He has taken the so-called MITER model and a regional econometric model generating an exogenous scenario of economic growth for the Italian economy, implying an increased demand for transport services. Supply constraints on transport infrastructures were simulated by holding the infrastructure stock constant at 1988 levels—the calibration year for the model—so that the congestion costs on interregional links were generated by the model. The assignment of freight flows on the interregional road network was done under two alternative hypotheses: the system optimum where overall transport costs were minimized and user optimum where a non-cooperative Nash equilibrium was achieved in path choices. The second approach seems to be more suited for modelling flow distribution in decentralized markets and the first approach gives the bench-mark for the determination of a Pigovian tax system, correcting system inefficiencies due to congestion externalities. The numerical case studies have provided some insights into policies which can internalize the social costs of transport.

The final chapter in Part II examines the border and barriers in European transport networks by analysing business trip distribution in Europe. Ulrich Blum and Frank Leibbrand formulate and estimate a trip distribution model centred on the Federal Republic of Germany which can also include different types of barrier effects. The model is estimated using an algorithm that simultaneously allows the parameters to be estimated, the functional form to be determined and the error term to be specified. The conclusions reached indicate that strong barrier effects exist, particularly in the Alps and between Germany and its European Union neighbours. Smaller barrier effects exist between East and West Germany, between other East and West countries, and between countries separated by the sea (e.g. Great Britain, Ireland and Scandinavia).

Part III moves away from the primary concern over changes in the structure of the transport and communications networks and the methods available for their analysis towards the policy implications. Policy Responses and Issues examines the problems from a variety of political and economic perspectives, arguing again that new methods of analysis are required as the scale of the problems faced by decision makers are of an order greater than those in the past. Chapters cover finance, regulation, competition, decision-making, environment, policy and fragmentation.

In the first chapter, David Banister, Bjørn Andersen and Sean Barrett argue for new methods of finance for the transport infrastructure. In the past, most infrastructure has been funded in the public sector as these projects are large scale, involving high risk and long payback periods. But with the limitations on public budgets and the ability to charge users directly for the use of infrastructure (through tolls and road pricing), there may now be a case for much greater private sector involvement, either through separately funded projects or through joint projects between the public and private sectors. The difficulties of attracting private sector investment into transport are well known, and in the past interest has only really been shown in transport terminals and interchanges. The authors present some of the basic economic arguments for the greater involvement of the private sector in transport, but the conclusions reached leave many questions open. In a competitive market, privately owned enterprises perform better than similar publicly owned enterprises, at least in economic terms. But although there is substantial involvement of the private sector in certain activities, there is still a reluctance in other activities. It is here that the possibilities of partnership are developed as this would allow the risks and return to be shared between the public and private sectors. The scale of investment required would indicate that joint ventures between the two sectors offer the greatest potential for actions.

Bjørn Andersen examines different regulatory regimes in the transport sector, focusing on franchising systems which have been used in Scandinavia and France. This issue has become of importance to all European countries as a result of the agreement on procurement between the European Union and the European Free Trade Association countries. There should be no discrimination between countries on the basis of nationality with respect to public procurement. This means that there has to be transparency in the procurement of services, including transport, with clear criteria for selection including tendering and franchising. This paper explores the current situation on transport service procurement in France and Scandinavia, including new forms of public/private partnership in the financing and operations of public transport services. It is argued that franchising has clear advantages over other forms of regulation as substantial efficiency gains can be achieved, greater flexibility in investment in rolling stock is possible, and there are greater opportunities for joint public/private ventures.

Similar conclusions are reached in the freight sector, namely that new approaches are required to regulation, finance and organization to meet the European challenges. Three key questions posed in Stef Weijers' chapter relate to the nature of European integration and the disappearance of frontiers, the opportunities presented by new logistics systems, and the development of complex European transport networks. Many large shippers in Europe are reconsidering their procurement and distribution structures as well as their production methods. There is also a concern over matching their inward and outward bound transport flows to achieve maximum efficiency. All of these questions are discussed with evidence from Dutch shippers and carriers. With the trend towards liberalization of markets, new strategies are required for survival and hauliers must become more service-oriented towards shippers than has been the case in the past. Quality of service and the use of logistics are seen as the key developments in this field, and this in turn means that new management and operational procedures are needed. However, it seems from the case study material that the response of many hauliers has not been as active as anticipated, as expansion into new management and operating practices have been modified by recession and uncertainty over the future. Although there is a keen awareness of the important changes taking place through integration, logistics and new transport networks, the hauliers have been conservative rather than dynamic in changing their current practices.

The approach taken by Kai Lemberg in his chapter is both sociological and political as it discusses the decision-making processes leading up to the decision to build fixed transport links across the Great Belt (within Denmark), the Fehmarn Belt (between Denmark and Germany), and the Sound (between Denmark and Sweden). Important investment decisions have been made concerning major new links in the national and international transport network, yet by following the actual sequence of events it is possible to understand why particular outcomes have occurred. It is possible to identify which economic and political actors have been most influential in this process. It seems that certain vested interests have manipulated the debate so that the preferred outcomes for these interests have been achieved. A more objective analysis seems to suggest that there is no clear case in economic terms for the Great Belt Bridge and the Sound Bridge, and the assumed economic development benefits have not been demonstrated. There has been no clear and systematic comparison between the alternatives, and prestige projects have been favoured. The real decisions have not been made by the Danish Parliament, but by powerful Inter-European companies, combined with the influential European Round Table of Industrialists and the Danish motor lobby, together with Danish trade union interests.

Underlying much of the current political and economic debates on European networks is the desirability for increased levels of travel and communication as there are substantial environmental costs. However, the popular notion of sustainable development does not offer quantitative criteria to help balance priorities for the environment against those of the economy. The chapter by Peter Nijkamp and Jaap Vleugel develops some ideas on identifying maximum allowable pollution levels in the transport sector, assuming a critical level of maximum resource use, a maximum carrying capacity, a maximum environmental utilization space, and a maximum sustainable yield or some other critical threshold level for environmental decay. They develop the concept of maximum environmental capacity use to indicate the maximum resource use of a given environmental capital stock which is compatible with socio-economic and environmental conditions. The conceptual arguments may be clear, but it appears to be very difficult to operationalize such a measure in the transport sector, so that pollution quota systems are difficult to implement. Therefore, they also examine the possibilities of using insurance strategies where premiums would be related to the total social costs imposed by the different sectors, including transport.

Radical policy alternatives have also been introduced in the telecommunications sector and these are discussed in Ken Button's contribution. As with the transport sector, telecommunications has been characterized by state-owned monopolies, but these are now being liberalized and privatized. The paper provides a wealth of background material on the nature and scale of the European telecommunications industry and highlights some of the national differences. It then introduces some of the economic arguments for change including intervention failures, asymmetric information theory, theories of contestability and competition in the market. The experience in the USA is informative here as it may give some indications as to how the liberalization will develop in Europe. However, even though liberalization of European telecommunications has been discussed for over ten years, it was only in 1990 that any action took place, but even this was a compromise. The Services and Open Network Provision Directives confirmed the intention to open the value added services markets for competition, but there were several possibilities for opting out by Member States. National strategies have tended to follow different paths, even though the European Union wishes to separate the operations and regulatory functions of telecommunications administrations. Most Member States have kept these functions within the government ministry, and only in the UK has an independent agency been set up. The opportunities for the telecommunications industry are substantial and the rate of change is phenomenal, yet regulatory reforms have only influenced certain parts of the supply of services. The more complex issues of infrastructure development have not been tackled as outcomes of change (e.g. in network externalities) are less certain.

The last chapter examines the arguments for and against network diversi-fication and fragmentation, and how these forces will affect the evolution of European telecommunications in a competitive market. Andy Gillespie and James Cornford identify two different options for the future development of telecommunications networks. One is the public monopoly provision of a unitary network and the second is the multiple, competing networks. At present the former is in decline and the latter is in the ascendancy. The arguments are moving in favour of network diversity which can now be achieved through the new technology, but there is still a strong possibility of network fragmentation, particularly if the market is allowed complete freedom. Their review takes the form of a spatial analysis by contrasting the idealistic interconnected network, mainly provided by the public sector, with the more realistic fragmented network. The balance, at least in the UK, which has gone furthest towards the multiple network, has been towards fragmentation. The authors conclude that further regulation is required in Europe to ensure that there is not a sharp variation in the range and quality of services provided. However, as new technologies emerge, it becomes increasingly difficult to maintain or provide a service which is available to all, so this differentiation will increasingly impact on competitiveness and location decisions.

From this review of the contents and structure of this book, certain general conclusions can be drawn. There needs to be a European strategy for transport and communications infrastructure investment which can be agreed by all Member States and those neighbouring countries in the wider Europe. This strategy needs to be followed over a period of time. In the past the primary concern has been over increasing the physical capacity and extent of the network to meet the continuous growth in demand. It is now time to look at a much wider range of options including:

1. the means to limit the growth in demand and to make the best use of the existing infrastructure (through pricing, regulation, management and the use of telematics and other technologies);
2. to fully cost the externalities created by transport and attribute them to the user;
3. to examine the possibilities of substitution of travel by telecommunica-tions;
4. to put environmental and quality of life objectives higher up the political agenda;

5. to examine the means by which new regulatory regimes can be introduced to maintain fair competition between and within the different modes of transport;
6. to emphasize the new opportunities offered to private network operators in transborder and multi-modal transport;
7. to establish organizational structures appropriate for the new internal market and the freedom of movement within the European Union;
8. to examine the nature of logistics and management systems to optimize the use of the freight fleet;
9. to explore the role that the private sector should play in the development of new transport and telecommunications systems.

The transport and telecommunications sectors are dynamic and expanding, and there is considerable potential to raise substantial amounts of revenue from the user, yet it is unclear as to how these monies should be used. Underlying all of the debates and uncertainties is the fundamental question of the vision of how the European transport and communications networks should develop over the next twenty years. Once that vision has been set at the European level, then many of the questions listed above will fall into place. It is hoped that in this book both the vision and the debate have been advanced.

Part I

EVOLVING EUROPEAN NETWORKS

1 Transport Infrastructure and European Union Developments

PETER NIJKAMP and JAAP VLEUGEL
Free University of Amsterdam, The Netherlands

1.1 THE CRITICAL ROLE OF TRANSPORT

Transportation plays a critical role in the spatial–economic evolution of our economies. The development of the transport sector—and spatial interaction in general—mirrors the socio-economic, spatial and political dynamics of our societies. Research in transportation planning has in past decades devoted much attention to demand analysis, e.g. mode choice, route choice etc. In particular, the behavioural models in transport research dealt mainly with the demand side. The *supply* side has received far less attention, especially in a modelling context.

In recent years the profound changes in economic and spatial policy have brought about a *re-orientation* in transportation with a clear focus on supply-driven mechanisms, in which the role of the public sector is increasingly at stake.

This trend is reinforced by *various force-fields* emerging in many countries, such as public budget deficits, the need for more competitiveness in (semi-) public goods delivery in order to enhance efficiency, the need for more customized service supply at a local (decentralized) level, and the drastic re-orientation in former centrally planned economies where privatization is a *sine qua non* for bureaucratic inefficiencies, insufficient fiscal revenues and new equity and ownership considerations.

In view of the skyrocketing mobility at the demand side and the strategic role of transport infrastructure as a critical success factor for competitive advantage and internationalization at the supply side, transport policy deserves full-scale attention. The positive externalities of transport networks and operations run the risk of being offset by negative externalities in the form of pollution, congestion and lack of safety. As a result, *various types of government interventions* (initiating, regulatory, financial or market-oriented) have emerged. However, the high costs of modern transport infrastructure in all modes have at the same time put an unprecedented burden on the government budget, so that in recent years the debate has started on *private financing* of infrastructure, based e.g. on 'user charge' principles. Thus, in our era

European Transport and Communications Networks: Policy Evolution and Change. Edited by David Banister, Roberta Capello and Peter Nijkamp. © 1995 John Wiley & Sons Ltd.

transportation planning requires a balanced implementation of actions which ensure a consideration of both private and social costs and benefits, and a network orientation which exceeds local or single-modal policy interests.

The present chapter aims to offer a sketch of modern transportation policy issues, taking Europe as a frame of reference. In recent years, the completion of the European market has provoked much interest in transportation and infrastructure as an integrative strategy for socio-economic cohesion. And therefore, much attention will be given to the structuring role of infrastructure networks, using, *inter alia*, the notions of missing links and missing networks. Particular emphasis will be placed on the identification of critical success factors, using the so-called Pentagon model. The arguments will be illustrated by considering two important European transport issues, viz. international freight transport and high-speed trains.

1.2   EUROPEAN INTEGRATION, MISSING LINKS AND MISSING NETWORKS

1.2.1   INTRODUCTION

The completion of the internal EC market, the increasing linkages with EFTA countries followed by the extension of the EC, and the socio-political and socio-economic accessibility of East-European countries have drastically changed the face of European cities, regions and countries. Openness, competitiveness, innovation, infrastructure connections and private–public initiatives have become new magical words in economic development strategies at all spatial levels in Europe. All these terms suggest that regional development policy is entering a new stage in which the *indigenous potential* of regions based on self-reliance strategies will come to the fore (cf. Suarez-Villa and Cuadrado-Roura 1993).

The new situation in Europe also provokes new policy questions which are directly linked to the three-tier structure of the new Europe, viz. the *competence of various actors involved in regional development policy*: European (i.e. supranational), national (i.e. supraregional) and regional policy-makers. Especially after the Maastricht Treaty and the Danish referendum, this question of institutional competence has played an important role in many countries. The fear of new supranational and bureaucratic authority in Brussels which would take over many responsibilities of lower-level actors has prompted many Europeans to resist the glamour of a new Europe, and as a reaction against the widespread 'Europhoria' many have recently called for sound policy principles based on bottom-up initiatives and decentralization. The *subsidiarity principle* has become an important institutional paradigm which suggests that the responsibility for policy initiatives should rest with authorities at the lowest possible decision level, while reasons of efficiency, coherence, equity and standardization may necessitate policy coordination at a higher level.

Another problem concerns spatial–economic equity in Europe. The completion of the internal EC market will likely aggravate the problem of *socio-economic disparities between regions* in the EC. It is generally expected that the relatively weak competitive position of peripheral regions will prevent them from a full participation in the process of European integration, so that the integration gains will most likely show up in the central regions in Europe (cf. Gaudard 1971; Ratti and Alberton 1993). For example, Quévit (1991) states: 'The main effect of the attainment of the European market will be the concentration of economic activity in a limited number of locations' (p. 34).

A situation of *interregional convergence and divergence* after a market integration will depend on:

1. the degree of cost reduction in each region as a result of economies of scale and market expansion;
2. the efficiency rise in firms as a result of rationalization and of a price policy that is more in accordance with production costs in a competitive market;
3. the degree of industrial restructuring and specialization as a result of more pronounced comparative advantages in an integrated market;
4. the degree of product and process innovation following investments in research and development (R&D) as a competitive tool in an integrated market.

The above observations point at serious questions on the *interregional distribution of benefits* of a unified Europe, as it seems plausible that the strong, central and highly competitive regions will become the winners in the new Europe, absorbing the lion's share of the economic activity at the expense of the peripheral, weaker regions. This issue of efficiency versus equity is even more important, as the efficiency–equity dilemma is likely to generate a competition which may be at odds with environmental quality. Therefore, it is likely that the following questions will emerge in the European setting of the 1990s:

1. the distribution of integration benefits between nations and regions;
2. the degree of socio-economic disparity between central and peripheral areas;
3. the threat to ecologically sustainable economic development at both regional and national scales;
4. the development of proper policy strategies at supranational, supra-regional and regional levels which alleviate the conflicts between economic efficiency, social equity and environmental conservation.

It goes without saying that transport is the blood circulation in an economy and hence infrastructure plays a critical role in performing the multiplicity of functions in an economy. In this section infrastructure policy in Europe will be discussed in the context of economic integration in Europe. It starts with a brief overview of conventional policy-making commonly found in most European countries until recently (Section 1.2.2), followed by an introduction to more recent and new ways of infrastructure policy-making (Section 1.2.3).

1.2.2 NATURE AND BACKGROUND OF MISSING LINKS

In the literature a distinction is usually made between *demand*-oriented policies on the one hand and *supply*-oriented policies on the other. Demand-oriented policies are reactive in response to transport demand, meaning that an increase in mobility is followed by expansion of physical infrastructure. Supply-oriented policies on the other hand are proactive, since supply of infrastructure is used as a tool to manage and influence transport demand.

Infrastructure policy-making in Europe (and elsewhere) has mainly been demand-oriented. In light of the vast problems related to growing infrastructure networks and use—notably congestion and deterioration of the environment and living conditions in general—one would expect a bias in current transport policy-making towards supply-oriented policies. In reality, an emphasis on supply policies is not very common in most countries.

There are various reasons why this change in policy-making has not happened. The first reason is *inertia*: in most European countries the political will for such a policy change is lacking, for both political (pressure groups) and financial–economical reasons (excise-duty income). The position of transport policy in the context of a so-called political life-cycle leads to short-sighted transport policies. A second reason can be found in the fact that the transportation sector has become one of the vital parts of the modern service and information economy; economic development in most countries has become transport- and distribution-oriented. Changing this situation is both very difficult and expensive. A third reason is that it is very hard to change old habits and behaviour (inertia), not only in policy-making, but also in user behaviour; car use is more highly valued (in terms of travel time, flexibility and quality) than the use of other modes of transport, in both passenger and goods transport.

A second feature of European transport policy-making is its rather *nationalistic* character. In general, policies are developed and implemented in a segmented way, each country seeking its own solution for each transport mode without keeping an eye on the synergetic effects of a coordinated design and use of advanced infrastructures. National infrastructure building companies, vehicle producers and transportation companies are often given a competitive advantage at the cost of their foreign counterparts. As other countries will use the same tactics, however, in most cases all parties will be losers in this way, since efficient scale is not reached and large sums of public investments are lost. A consequence of such inefficient behaviour may be that external competitors (e.g. Far Eastern or American companies)—while having large home markets—may outperform European companies. The problems created by the reduced efficiency of infrastructure have often been viewed as pure infrastructure bottlenecks with only two dimensions, viz. physical infrastructure and financial funding, while neglecting demand management and institutional and organizational aspects as well as ecological implications. Examples of the effects of neglecting these important infrastructure development criteria include, *inter alia*, inefficient use of vehicles because of cabotage restrictions, protection of national carriers, segmented European railway companies and lack of multi-modal transport facilities. In fact, efficient border crossing is a test case for most infrastructure networks.

As a result, the European transport scene is characterized by many bottlenecks; so-called 'missing links'. They are partly caused by the absence of efficient border-crossing operations and regulations. Since 1993 a gradual removal of such impediments is taking place, but there is still a long way to go.

Until recently, transportation policies in European countries have in most cases been short-term-oriented and demand-following, since they did not take into account that expanding physical infrastructure in most cases means attracting additional traffic and transport; the removal of missing links often created new missing links elsewhere. However, such policies do not deal with the real causes of mobility growth and tend to cope with symptoms rather than with underlying structural processes.

1.2.3 THE NOTION OF MISSING NETWORKS

Crossing geographical borders in Europe is not (made) easy, even nowadays. Long queues of freight trucks at border crossings have become normal phenomena. Cross-border infrastructure has always been an underdeveloped area. There are various reasons for this situation (see Giaoutzi and Nijkamp 1993). The first reason relates to the fact that cross-border traffic and transport used to be a minor part of total traffic; national borders appear to act as barriers to transport (see also Bruinsma and Rietveld 1993). Since investment costs in border infrastructure were only partly met by additional traffic and since the neighbouring country would also benefit from such investments, a lot of cross-border links were of poor quality. The second reason is that the planning of cross-border infrastructure incorporates international planning, whereas infrastructure planning used to be based on national, segmented policy-making. European integration will remove some of these problems. It will also lead to additional international transport. International planning of railway lines—but not of networks—is also beginning to emerge.

In strong contrast to the planning world, the economic and political arena in Europe has changed dramatically, as was shown in Section 1.1. Globalization is the keyword in this context, stressing the importance of linkages between economic and political developments and decision-making in all regions of the world. As economic and political borders are reduced and finally removed—a 'Europe without internal borders' is one of the goals—increasingly, however, the need for multinationally planned and implemented *networks* will arise, instead of conventional policies based on piecemeal and *ad hoc* linking of national networks. In some, especially southern, countries this will lead to major expansions of existing networks, which in turn will lead to important quality gains in transport, at least in the medium term, given the vast growth of transport and vehicles. In most countries in Europe, especially in the northern part with its relatively high quality and dense networks, this might mean a shift from quantity to quality in network planning and use.

As transport planning by firms becomes more and more a process of planning of transport chains, each link of this chain is critically dependent on the other edges. This holds especially true for areas where major harbours or airports are located and for areas crossed by through traffic. Their local decisions will then increasingly obtain a higher-level importance.

Since economic growth in Europe tends to be concentrated along one or more axes or corridors with high concentrations of population (witness, e.g., the well-known 'blue banana' concept), it is foreseeable that network and corridor planning becomes a new field of infrastructure planning (see Vickerman 1992). In this context the notion of missing networks has emerged. 'Missing networks'—as a general term for a substandard functioning of infrastructure—refers to the absence of various strategic components and necessary conditions of transportation and/or communication infrastructure. Missing networks may refer to both *single-mode* networks and *multi-mode* networks. The concept may concern (1) the physical absence of a given infrastructure (e.g. inland waterways), (2) the lack of strategic linkages between various modes of transportation (e.g. absence of combined transport for road and rail), (3) absence of critical success factors for the proper functioning of a given network (e.g. absence of a sophisticated logistic system for international container transport). Central to the concept of missing networks is the idea that the performance of infrastructure is far below the maximum potential in terms of services delivered. The way in which a network can be designed and developed can be assessed and evaluated by means of five types of success factors, which may be summarized in a so-called *pentagon prism* (see Figure 1.1). In the context of missing networks this pentagon prism can be interpreted as follows.

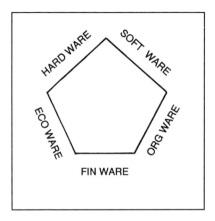

Figure 1.1. Pentagon of long-term infrastructure development criteria

1. *Hard ware* refers to the tangible material aspects of transportation infrastructure (e.g. technical equipment, terminals, railways, road networks or harbours). They serve to physically facilitate transport services or flows generated by consumers or firms. A great many problems in the hard ware of transport systems can be seen in Europe. For example:

(a) a lack of a network of standard quality international transport connections between European countries (the so-called Trans European Networks);
(b) many missing links in major European networks (e.g. north–south connections, sea crossings, Alpine crossings);
(c) a lack of intermodal transport opportunities on both short and long distances;
(d) an absence of mechanisms promoting advanced new infrastructure systems (e.g. Maglev, high-speed tube networks).

2. *Soft ware* refers both to computer soft ware used to control sophisticated hard ware facilities and the services (viz. information systems, communication facilities, data services/banks, route guidance systems) offered to the user of infrastructure. In the area of soft ware various problematic cases can be seen in Europe today:

(a) unequal speeds of introduction of new logistic systems (e.g. route planning) in European countries;
(b) lack of coordinated and standardized information systems for both users and operators of transport infrastructure (the so-called electronic data interchange (EDI));

(c) insufficient use of new tools for transport planning (e.g. geographic information systems).

3. *Org ware* comprises all regulatory, administrative, legal, management, and coordination activities and structures regarding both the demand and the supply side of transport (in terms of legislation, regulations, fares, procedures etc.) which form the private and public institutional framework of the transport system. Features of devolution (e.g. liberalization, decentralization, deregulation or privatization) are also important in Europe at present.

Problems related to org ware include:

(a) lack of political will and non-engagement in decision making;
(b) lack of coordination between regional, national and international infrastructure plans in all modes;
(c) lack of institutional frameworks for designing, using, operating and managing transport infrastructure, with a special view on saturation trends in various infrastructures;
(d) lack of a European ('holistic') view on market entry and transport behaviour (especially in a transnational setting);
(e) lack of a proper international organization to support combined transport, including rules for standardization and harmonization of weight and size limits;
(f) overemphasis on technological characteristics of advanced transport systems to the detriment of the services to be rendered.

4. *Fin ware* refers not only to the socio-economic cost–benefit aspects of new investments, but also to the ways of financing and maintaining new infrastructures, to fare structures, to state contracts for guaranteed finances for public transport deficits, etc. Projects which cross national borders are of course very relevant here.

Examples of problems in a European context include:

(a) lack of private initiatives for financing major (transnational) transport infrastructure;
(b) lack of a uniform European system for user charges (e.g. road pricing, tolls);
(c) insufficient use of the potential offered by market contestability, for instance, making a distinction between the operational and capital costs of transport infrastructure;
(d) lack of an evaluation framework for infrastructure appraisal seen from a European perspective.

5. *Eco ware* refers to environmental and ecological concerns (including safety and energy questions) in transport systems, as well as to abatement

measures for environmental degradation (e.g. user charge principles). It concerns both the infrastructure owner (e.g. landscape deterioration caused by visual pollution) and the infrastructure user (e.g. emission of exhaust fumes of cars). Problems here include:

(a) lack of a European view and planning system regarding the environmental deterioration caused by transportation;
(b) lack of standard rules and control regarding the transport of hazardous materials;
(c) lack of European-wide incentives for the development and use of more environment-friendly transport systems;
(d) lack of coordination between physical planning, land-use planning and environmental policy (including separation of work and home places);
(e) unsatisfactory application of 'polluter pays' principles in European transport systems;
(f) insufficient use of market incentives for improving current vehicle technology.

It is clear that using the pentagon as a practical evaluation tool for transport planning is not an easy task, since fulfilling all five criteria *simultaneously* seems possible only when using compromise solutions. Another important caveat is the need for quantification (e.g. via indicators) and weighting of these five criteria. The pentagon prism will be used in this study to evaluate major problems in European transport infrastructure, with a particular view to identifying critical success factors and policy lessons to be learned. In the sequel of this study the attention will be focused on freight transport (by road and rail) and on the European high-speed trains in order to test the above Pentagon model from the viewpoint of failure and success stories in European transport development.

1.3 EUROPEAN FREIGHT TRANSPORT (ROAD AND RAIL)

1.3.1 GENERAL TRENDS

The steady increase in the level of economic activity within and between the EC countries has been accompanied by the ever-increasing need to haul industrial goods and agricultural products over long distances, mainly across national borders of EC or non-EC countries. For various reasons, over two-thirds of this traffic is being carried by heavy trailer-trucks while only 20–25 per cent of all freight is carried by rail transport and this share is still declining. In practical terms this pattern of transport implies under-utilization of available rail capacity, despite its seemingly economic and other advantages, and over-utilization of roads and highways.

The problem of under-utilization of rail capacity is a complicated one, however, since in many cases the lack of sufficient supporting facilities is a major barrier to further growth of rail activities. In the case of Germany, for example, the expected growth of 1.8 billion tonne-kilometres of freight volume over the next ten years will be feasible only if the capacity of goods storage facilities and container terminals is increased drastically and the concept of block-trains is further developed.

Heavy trailer freight trucks travel over surface local roads and major highways and concentrate in large numbers in trailer parks and other parking areas. Given the limited available capacity of these infrastructure facilities, *congestion* due to truck movements within general motorized traffic is rapidly becoming the number one traffic problem in all EC (and adjacent) countries. In order to alleviate this traffic problem various countries are finding it increasingly necessary to invest heavily in highways and other supporting facilities. Moreover, heavy trucks with many axles are known to be the prime cause for highway damage and deterioration, which in turn, calls for additional investments in maintenance and traffic control activities. In general, the position of freight operators in the whole of European commodity transport deserves careful attention.

However, for various reasons such as lack of financial resources, planning delays and geographical and topographical barriers, the expansion of the road capacity to accommodate this growth in truck freight movement is lagging behind the needs at an increasing rate. Consequently, over time, the real cost of hauling a tonne of freight between any given points of origin and destination in the EC markets is continuously rising. The immediate results are additional costs of labour, capital, inventory and spoilage, which eventually are reflected in the final prices of goods borne by the consumers.

Another problem which stems from this pattern of freight transport is that of distribution of equity. The transport of freight between countries at the fringe of the EC market (e.g. Greece to Scandinavia) must pass through the territory of many EC and non-EC countries which bear most of the traffic and environmental costs involved, without directly benefiting from this transport activity. Needless to say, this creates economic and political friction. In some cases, e.g. that of Switzerland, through truck traffic has created unbearable environmental problems threatening the survival of forests and the countryside. The reaction was to ban most of the traffic from Swiss and Austrian highways, diverting it to neighbouring countries with all the economic and political ramifications.

Switzerland has imposed a maximum weight limit of 28 tonnes for trucks as opposed to 40 tonnes in the EC at large. This protective regulation has initially been introduced not because of environmental concerns but in order to protect rail freight traffic. Switzerland and the EC have now agreed on a quota of 50 trucks of 40 tonnes per day which can be surpassed only if all rail capacity has been used.

The type of goods transported is also a major problem relevant to the overall issue of rail–truck competition and complementarity. To illustrate, in Germany, the share of rail in total haulage, as a function of freight type, varies from 75% for fuel to only 7% for food products. Therefore, changes in regulatory policies will influence modal split between rail and truck in very specific ways, depending on the goods transported and the effective available network structure.

In examining rail networks in Europe it is evident that there is no global European rail network but individual national networks. This reality, in turn, puts rail in a major disadvantaged position next to truck hauling. As a result, international rail shipments in Europe, being affected by operational conditions, level of service and management attitude on individual national rail networks, can hardly provide a reliable and flexible service. The truck, which is in contrast largely a door-to-door type service, does not suffer from such adversities.

Since many studies suggest a continuation of the growth of freight transportation—especially by road—a process amplified by the liberalization process of trade in Europe, intensive policy actions in this area are needed in order to alleviate or to remove the existing bottlenecks in road and rail transport (especially the lack of a commercial attitude of railway companies and the current miscoordination in the area of international commodity transport). These problems also create new opportunities for combined transport by means of road–rail combinations. This type of combined transport may become one of the solutions to tackle current and foreseeable bottlenecks in freight transportation in Europe.

Several attempts to assess the current volume and future trends in goods transport in Europe have been undertaken in the eighties. The picture of the future of goods transport given in different forecasts is rather diverse. There is a certain agreement on income elasticities of goods transport. For domestic goods transport this elasticity is generally found to be below 1 (indicating a less than proportional rise of domestic freight transport as compared to gross national product). For transit transport the elasticity is found to fall somewhere between 1.5 and 2 for the past decades. This finding is accepted in most forecasts, so that in general the growth of freight transport at a European level is expected to be twice as high as general economic growth.

There is a general agreement on several general economic and behavioural trends which determine freight transport demand in Europe.

A first trend that can be identified is the *functional/spatial division of labour* which results in an increasingly complex European network of production and service units with a growing share of intra-industry trade. It can be expected that this trend is significantly reinforced by European integration. The goods produced and transported in Europe gain in value per weight. Hence the trend towards transport of *high-value goods* will persist. Another characteristic of transport that will be increasingly requested is flexibility. The demand in freight transport will most probably be for *small deliveries at irregular intervals.*

Besides these demand side trends, some other developments clearly show up in goods transport. With the spread of *just in time production* (JIT) there is an increase in on-line calls for goods kept in stock somewhere on a rolling transport vehicle. An increasing demand for informatics services in the transport sector will be the consequence. There will be a specific need in road hauling, rail freight and combined transport where the introduction of Electronic Data Interchange (EDI) services seems to be more difficult (due to the small-scale and fragmented organizational structure of the branch) than in the case of water transport where most ports seem to have more incentives to supply these services. Another problem linked to just in time production is the problem of partly empty trucks which have to make a trip just because a specific load is being ordered 'just in time'. As a result of modal split, the railway companies will be asked more and more to act as *European-wide suppliers of services* which in addition are not restricted to their own mode but also comprise door-to-door services. As mentioned before, a shift from a competitors view of the road haulage sector to a partnership view is therefore needed. The *collector's and distributor's function* of the suppliers in the transport market will gain in importance over the pure transport view.

Finally, various problems need to be identified which are specifically related to the *European integration.* In an integrated Europe, a simplification of procedures and regulations in crossing the borders concerning transport—specific obstacles which have not been automatically abolished by the end of 1992—is urgent. This relates first of all to technical problems like railway voltage, standardization of wagons, trucks, combined transport infrastructure etc. Besides these technical problems, the logistics will have to develop from paper to screen logistics if the advantages of the open frontiers have to be exploited (see below).

As far as missing European networks in transport are concerned, it will not be sufficient to take a pure EC perspective on Europe. The problem of the Alpine countries in relation to north–south traffic must be solved in order to prevent a successful integration of Europe being hindered by transport problems elsewhere. And the problem of transport infrastructure in Eastern Europe might become an even more important issue in the years to come. For the sake of the economic development of the EC after 1993 it will be vital to solve the transport problems in Europe in general from the viewpoint of *European benefits* (e.g. a win–win situation for all partners involved).

Finally, institutional issues appear to be very important for the development of transportation. It is clear that economic integration—changes in regulation—will lead to economic restructuring and to changes in transport flows in general. This has various implications for transportation by road and rail and for combined transportation. *Removal of barriers* means an increase in competition, especially between former more or less monopolistic railway operators. But also in road transportation, various forms of regulatory and pricing barriers will be removed. Since transportation is in general a business with high costs and low or even negative returns on investment, economic integration may certainly lead to a 'shake out' process in the transport business. This process may become even stronger, if external costs are to be internalized.

Perhaps the main problem regarding road freight transportation lies in the lack of freedom for cabotage; international road freight traffic is still restricted by a system of permits. Since there are far fewer permits than needed in most countries, the EC has adopted a policy of a step-wise increase of permits until 1995, when cabotage should become unrestricted; only a qualitative regime will then be used to regulate the market (Commission of the European Communities 1992a,b).

3.2 BOTTLENECKS AND MISSING NETWORKS

In view of the difficult problems confronting road and rail transport (not to mention the opposition to road traffic caused by its environmental impacts (*eco ware*)), the issue of *combined transport* has become an important argument in recent transport discussions. Above all, the railway companies and the combined transport firms which are more or less closely linked to the railways push this 'alternative'. In February 1990 the European railways and the combined transport operators issued their common 'Brussels declaration' where they presented a common strategy on design, operation and marketing of combined transport.

A.T. Kearney & Co. demonstrate in their report (1989) that there are important problems which limit the capacity of this network and which relate to things like terminal capacity and loading profiles (*hard ware*). Besides these problems of terminal capacity and profile differences there are others which hinder a European-wide performance of combined transport. They relate to organizational issues (*org ware*), logistics (*soft ware*) and technical coordination of the wagons and the loading techniques.

It is clear that missing networks in the field of combined transport cannot be identified simply on the basis of maps of infrastructure networks. First, a strategic view on networks should be *intermodal*. Except for road haulage, the goods are generally not transported from door to door by the same vehicle. Therefore, if some modal parts of the network are weaker than others (or even absent), this influences the performance of the whole network for a certain type of transport. A second dimension which has to be recognized is the *multi-layer view*. Transport networks do not only consist of *hard ware* (infrastructure). They are also a function of the organizational structure of the service suppliers (*org ware*) and of the logistics that guarantee a smooth functioning of the operations (*soft ware*). In addition, the financial implications (*fin ware*) and the ecological impact (*eco ware*) of the networks have to be taken into account.

The intermodal and multi-layer view can best be illustrated if we start by looking for missing networks in the domain of combined transport. A good impression of the elements that are missing on a European level in combined transport is given by the elements of the Brussels Declaration of the European railways and combined transport operators. In the framework of a common strategy of design, operation and marketing they propose the following fields of action:

1. block trains
2. service contracts and guarantees
3. equal treatment regarding rates
4. tariff contracts
5. European network
6. terminals
7. wagons
8. new techniques.

Among these fields we find infrastructural domains (*hard ware*) like the foreseen investments in new railway infrastructure and the adaptation of existing installations in order to cope with additional traffic as well as the realization of new terminals and the expansion of existing ones. Another infrastructure issue is the intention to coordinate future innovations in combined transport technologies. Organizational issues (*org ware*; until recently unsolved and now given priority in this Declaration) are the predicted Europe-wide tariff structure, the equal treatment of railways and operators regarding rates, and the coordination of the selection of wagons.

On the *soft ware* level, the engaged parties judge several elements of a European network as being absent. These relate to the promotion of block trains and service contracts and guarantees. It is interesting to note that there is no mention of the logistics needed for the European control over the wagon fleet, which is a basic requirement for the operation of a European combined transport network. This regards the monitoring of the units (wagons, containers etc.), their routing and their repair. Thus there is not only the technical and organizational lack of a network that would provide a universal wagon and container type for combined transport, but there is also a missing network to be diagnosed in the logistic control of combined transport movements. In order to solve these problems it will be vital to consider all elements making up combined transport, viz. the traction, the carrying, the handling and the containers. There is a strong feeling that a *hub and spoke* system—consisting of performing links with unified rolling stock and nodes which are multimodal multiservice freight terminals—is what is needed in Europe. Only if this missing network is realized is there a fair chance that centralized logistic solutions will be feasible.

While the realization of a true European combined transport network might solve some problems of freight transport in Europe, other problems will remain. One of these is the *limited additional capacity* that can be provided by such a network. A.T. Kearney & Co. foresee a tripling of combined transport until the end of the century with a doubling of the market share from 4 to 8 per cent. Hence, the combined transport network will not be able to solve the problems in the short run. Important infrastructure investments, like new Alpine rail axes, investment in countries with limited free profiles like Italy and the adaptation of the Spanish gauge to the normal European standard, will significantly raise the capacity in combined transport.

It follows that the missing networks in combined transport with a short-term character are to be found on the *org ware* and *soft ware* level. Creating these networks would help to solve some problems. This means that national railway companies have to learn to behave like market-oriented enterprises, instead of being controlled to a large extent by their nationalistic bias.

1.3.3 POLICY ISSUES

From the foregoing observations various policy questions arise, which are of interest in an analysis of missing networks. In general, the major bottleneck in road transport does not merely lie in crossing the borders, but in a general lack of capacity due to a rising demand for transportation and the combination of a comparatively low user cost of road transport and difficulties in extending the networks (Maggi 1992). A closer look at congestion reveals that, although the number of freight vehicles is growing at a rapid pace, the dominant and still growing use and ownership of private cars is the main cause of congestion. So there is a case for governments to reduce (the growth of) the number of private car owners and users. The problem, however, is that fewer private cars on the roads lead to more freight vehicles instead, increasing the imbalance of the modal split and, in a few years, leading to new congestion.

Another question is related to investments in road networks, especially in the south and the east of Europe. These investments—although reasonable seen from a short-term economic point of view—are likely to favour the ownership and use of road vehicles. In a longer-term perspective, large-scale investments in rail, water (ports) and combined transportation are likely to offer a far better option.

There is a case for deregulation/reregulation and privatization in rail and combined transport. The main questions in this field relate to the new organization, the logistics, the financing and regulatory regimes by governments.

Seen from the viewpoint of interoperability, an important policy issue is the organization of the railways on a European level. In this regard the liberalization of rail markets to private firms along with the separation of rail operations and infrastructure are relevant policy alternatives. However, such solutions are likely to hinder the stability of existing national railway monopolies. Therefore, a global European view should be considered as superseding national interests. These policies would have to be defined in terms of capital investment, fiscal measures, regulatory changes and organizational changes that countries should undertake to achieve the desired goals. The fragile divide between public actors and private actors is again at stake here.

Regarding the necessity of combined transportation, one should keep in mind that train transportation has its main potential over distances of more than 500 km. In fact, at present 90% of freight transport is on distances of 200 km maximum (Commission of the European Communities 1992c). It is clear that the EU leads to more long distance transport. However, it is also clear that the market share of combined transportation will not grow considerably, especially if more congestion on rails and at terminals is to be expected, thereby raising costs (European Round Table of Industrialists 1992). Thus, this raises intriguing questions on the real potential of combined transportation.

Another problem is the lack of coordination between projects in different countries. Winkelbauer (1992) shows that the foreseen new railway tunnel projects (the 'Neue Eisenbahn Alpen Transversalen'; NEAT) in Austria and Switzerland are in heavy competition with one another. This means that the cost–benefit ratio of these projects may even be negative, because none of them is able to reach a minimum level of traffic. A related problem concerns the split in monetary costs and benefits between countries; according to EC law each country has to pay the costs of its own infrastructure even if it is mainly used by other countries. For instance, the recent German proposal to levy road taxes on foreign truck owners using its roads has been cancelled by the EC for this reason. However, it should be admitted that more tunnels offer far more capacity, so that they are able to deal with a strong growth of traffic and give truck owners more flexibility.

In European freight transport trucks and delivery vans are the dominant means of transport, leaving very little room for other modes of transport. It is foreseen in various reports that this imbalance in modal split will continue to grow, as will total freight transport. Since it is clear that such a development will lead to an increase of the already high social cost of transport and the cost of infrastructure—to mention only two of the most important problems—there is a potential market (niche) for combined transportation. Nowadays, combined transportation is not favoured by most shippers because of various bottlenecks, most profoundly in the rail part of combined transportation. At present, growing congestion on the road network is not a sufficient condition to increase the demand for combined transport. Only in combination with a removal of the vast amount of bottlenecks (viz. regarding tariffs, reliability and flexibility), internalization of social costs and in some cases explicit regulation and (preliminary) subsidization, will road congestion act as an incentive for combined transportation.

1.4   EUROPEAN RAPID TRAIN NETWORKS

1.4.1   GENERAL TRENDS

The term 'high-speed rail' covers both conventional wheel-on-rail systems with significantly increased maximum (160–200 km/h) and average speeds, the advanced TGV/ICE systems (above 200 km/h) and Maglev systems (500 km/h or more). They open up the possibility of services competing with air transport in terms of journey times, frequency, comfort, reliability and safety. High-speed travel seems to be an excellent solution for Europe because of the relatively short distances between its capitals (from 200 to 1000 km, equivalent to a maximum travel time of only 4–5 h during the day per TGV or ICE). Night-time services in just 8–12 h could be the best solution for longer journeys involving distances of up to 2500 km. Whereas the maximum technical speed of the TGV wheel-on-rail system is some 500 km/h, the maximum economical speed will be around 300 km/h, because of disproportionate maintenance costs for trackage and wheelsets, geographical conditions and the spatial pattern of towns in a country (Massoni, 1988). Running on a magnetic cushion, the TRANSRAPID system—after a period of testing since 1974—is also ready for commercial use, especially for transportation on medium and long distances. Its major drawbacks lie, however, in its fully incompatible infrastructure (forcing the construction of a completely new infrastructure), its noise and its visual intrusion on the landscape.

Only a very small part of the conventional railway network in Europe can stand maximum speeds higher than 140 km/h, as higher speeds lead to much higher construction and maintenance costs, and safety and signalling systems are still inadequate. In densely populated countries, average speeds on longer distances usually do not even exceed 80–90 km/h. Sometimes even lower average speeds are necessary, most notably when lines are operated with single track and diesel units. This is one of the many reasons why railways are in most cases not a real threat to transportation by other modes of transport, especially by road. Railways are in a better position to compete in long-distance travelling, at least when they do not have to cross national borders. But there are only few large enough countries in Europe.

Aware of this challenge, the community of European railway companies of the twelve EC members plus Austria and Switzerland presented in January 1989 a project for a European high-speed network. This project covers infrastructures, rolling-stock and the timing of future connections, thereby redrawing the European railway network map.

At this moment many technically—and also physically—distinct, high-speed railway networks or connections are being developed in many European countries (for a snapshot, see e.g. Massoni, 1988, Espieussas, 1989); well-known projects are those in France (TGV), Germany (ICE), Italy (Diretissima; Roma–Firenze) and Spain (AVE; Sevilla–Madrid). All these systems are excellent examples of nationalistic transportation policies, as the following two examples from France and Germany convincingly show.

The first French main line, from Paris to Lyon, is an unrivalled success for a railway line, since trains on this main line were able to compete with domestic air transport, both in terms of price and travelling time. Investments in track and equipment have been paid back in 10 years, a situation never previously found in railway history. The demand for TGV services has been far greater than expected and still continues to grow strongly. The success of the TGV is to a large extent due to the favourable spatial pattern of cities and towns in France; major towns and cities lie in the 300–800 km range, which is best served by the TGV. This success induced the French government to plan and execute extensions of its network to all parts of the country, while international extensions are also foreseen before the turn of this century; so supply is also growing strongly—Say's law is very vivid in railways.

The Paris–Bordeaux and Paris–Geneva lines, although still under construction, are, however, not as successful as the Paris–Lyon line. This is partly due to the fact that TGV trains on those lines have to use parts of conventional track with high traffic density as well, so reductions of travelling time are limited. The importance of the TGV lies not only in the improved travelling quality of railways (equal or better than found in air transportation, including telephone and fax services), but also in its capacity effect. The use of parallel TGV sections has freed capacity which can be used for high-speed freight trains (160 km/h) and regional trains.

The German ICE high-speed train service started in the beginning of 1991 with the operation of the Mannheim–Stuttgart and Hannover–Würzburg lines, lagging more than 10 years behind the French TGV. The ICE and TGV differ in many respects, the first one being conceptual, as the ICE network will be used for both passenger and freight transport, where the TGV is used only for passengers and postal transport. The second major difference lies in the technical characteristics; the ICE has a higher axle load than common in France and other countries and the train body is wider than the French TGV. As their electrical power supply systems are also incompatible, the ICE and TGV trains are not yet able to run on each other's networks. Germany is, however, developing a multi-circuit-power unit, the ICE-M, for different voltages and frequencies, so one of these problems may be solved in due time. A third difference between the TGV and the ICE is found in the comfort and the service level; the ICE comfort and service level is far higher than that currently found in the TGV-Lyon type (stewards, bidirectional telephone and fax services etc.). However, the latest (third generation) TGV types also offer much more comfort.

The plans for the German railways also include the upgrading of existing tracks for conventional traffic up to 160–200 km/h. Where this is impossible, reconstruction of tracks (*Neubaustrecken*) is necessary. Some new lines are developed for passenger transport only, whereas other—mainly junctions—are for dual purposes (freight trains up to 120 km/h) (Massoni 1988).

One of the main reasons for the independent development of the high-speed systems in France and Germany is found in government protection of firms producing national railway equipment. In fact, Germany refused to accept the French offer to cooperate in favour of developing its own (yet incompatible) ICE systems (cf. Gérardin 1990). This means that high-speed trains still face the same problems as conventional ones when crossing borders between various countries.

Suppliers of high-speed trains are in close competition to attract customers, as was shown recently with the Spanish purchase of high-speed trains. As this first high-speed project outside both supplying countries could mean a breakthrough in selling high-speed trains and thereby setting a European or world-wide standard, governments were strongly involved in bidding and selling of high-speed trains to other countries. At first the Spanish government opted for Japanese train systems, because these were less expensive. Strong political pressure led to the purchase of the AVE, in fact an excellent mixture of the TGV and ICE state-of-the-art technology.

Planning and running railways in Europe has always been a task of national governments. This explains the large, bureaucratic, non-market-oriented railway organizations in most countries. It also explains to a large extent the lack of private sector involvement in the financing, management and operation of railways. The picture of road transportation is very different. This explains the dominant use of road vehicles in both freight and passenger transportation.

1.4.2 BOTTLENECKS AND MISSING NETWORKS

When dealing with missing networks in European railway systems, a distinction should be made between bottlenecks which have hampered the development and use of conventional railways (traditional problems) and those which are specific for high-speed railways (new problems). Solving both types of problems is necessary, because the development of high-speed railways cannot be isolated from the development of conventional railways; they should be viewed as complementary, not as substitutes. Specific *high-speed* rail problems include the following.

At the *hard ware* level most of the purely technical difficulties of driving at high speeds have been solved. Both the TGV and ICE use sophisticated hardware and the latest electronic and telecommunication technologies (*soft ware*). The basic hard ware problem is that Europe has two more or less incompatible high-speed train standards. Since none of the two industrial conglomerates is eager to accept the other one's standards, given their own efforts in terms of time and money spent in developing their own high-speed train systems, only a compromise would help to overcome this problem. This is the case for the Paris–Cologne–Brussels–Amsterdam TGV, for which new multi-current locomotives have to be developed to ensure hard ware compatibility, at least for trans-border connections (Community of European Railways 1993). Another important problem is whether the high-speed network should be opened to passenger traffic and postal services only (as is the case in France) or should it also be used for freight trains (as in Germany)? Freight transport demands much higher standards for railway construction and maintenance, and therefore much higher investment.

A further problem is related to the fact that the TGV has a lower axle-load than the ICE. As the French railway infrastructure is built to the TGV axle-load, the standard ICE is at present unable to use the French high-speed rail network. A related question is the use of high-speed tracks for conventional use too; is the high-speed network an extension of the conventional network? The use of conventional trains on high-speed networks has both benefits (higher average speeds, more efficient use of capacity) and costs (lower average speeds of high-speed trains). In the opposite case, high-speed trains may be as slow as conventional trains and sometimes even slower because of track geometry. High-speed trains compete with conventional trains in terms of costs and revenues. This means that the use of high-speed trains may lead to a reduction of conventional train services (lower frequency, fewer passengers, fewer stops).

Conventional train technology is not invariable. The Italian 'tilting' train technology for example—which enables driving on existing curved tracks with higher maximum speeds without increasing the risk of derailment—is quite successful and has recently also been put in use in Germany. These and other techniques make maximum speeds of up to 200 km/h or more already possible on a number of tracks in Europe. As soon as the physical (including control) infrastructure is capable of handling these speeds, conventional trains ('Super-Eurocities') may even compete with high-speed trains, especially in terms of economic efficiency. More homogeneity of train flows via separation of local, regional and long-distance trains—as proposed in the Dutch National Railways plan Rail 21—may be part of such a scheme.

Org ware problems seem to be the most difficult question in the European context. The identification of a solution for institutional and organizational issues needs a lot of diplomacy and political intuition, because there are some 30 parties involved (and none of them are future users; Sundberg 1990). Three main problems should be mentioned in this connection:

1. The tendency of national governments to protect national producers of both infrastructure and trains, by means of exclusive government contracts incorporating national standards.
2. The tendency of national governments to upgrade their existing networks according to national demands, instead of European ones. In the past this was quite logical, given the profitability and efficiency of domestic investments (Espieussas 1989). Nowadays this is no longer the case, as international traffic and its profitability will rise significantly.
3. Strong differences in time loss because of incompatible procedures between countries. In France, for instance, once a project is under construction, objections of people involved do not stop actual construction, whereas in Germany construction is stopped each time.

Although the first effect may become less important as frontiers vanish in the European integration process and transnational mergers grow in importance, the second effect is harder to overcome, given the degree at which frontiers reduce international transport, particularly for business travel. There is a real chicken–egg problem in this field, as planning and investment are based on actual needs for transnational transportation, which will not develop until rail transportation reaches a higher quality level. Actual construction will therefore show a very diverse pattern between different European countries, thereby hindering the emergence of a transnational network.

One of the major bottlenecks in *fin ware* is the shortage of public funding, while the list of priority projects grows longer. This problem will become even greater, however, since in most countries stabilization or even reducing income tax and social security contributions is favoured. Private funding has already gone some way to solving this problem. The supporters of these projects urge public loan guarantees from the states and the Community, especially when the revenues from transportation are expected to be insignificant on some links. This effect is especially significant in densely populated, yet small, countries such as the Netherlands. Should revenues from more profitable parts of the network be used to finance less profitable parts and to what extent?

Another problem related to budget problems is that they may force decision-makers to choose between investment in conventional trains and infrastructure, and investment in high-speed trains. Since the latter need vast amounts of money, but cover only a small percentage of traffic, it is an important question whether these investments are worth while, or whether the whole train network should be upgraded.

From an environmental and ecological viewpoint (*eco ware*) high-speed trains are a better solution compared to air and road transport, most notably in the use of energy and the air pollution. Land use per passenger–kilometre is also smaller than what is needed for road transport. High-speed trains have an environmental and ecological impact and this has a strong effect on recent plans for new high-speed connections in Europe (e.g. in Kent, Flanders and Holland). The *local* environmental concerns arising from the building of these high-speed links and of the retooling of existing links are immense. Those advantaged are usually different from those who have to accept local environmental deterioration from the building of rail tracks and cargo, especially container, facilities, etc. However, the proposed use of conventional routes will not lead to significant increases in performance, as only significant increases in speed will lead to important savings in travelling time.

1.4.3 POLICY ISSUES

High-speed railways have considerable advantages over conventional ones for customers living in one of the larger towns and one of the richer parts of the country. French experience shows that most small towns and villages are not offered a better service than before. On the contrary, when high-speed trains use conventional track their operation leads to a lower quality of conventional train services. High-speed trains also favour the richer parts of Europe (Gérardin and Viegas 1992). So further economic polarization between urban and rural parts of a country because of the high-speed trains is foreseeable. The split between national (= feeder and regional) and international services can only be justified and implemented against the political opposition of those potentially disadvantaged if compensatory measures are given, most notably fast feeder connections (Commission of the European Communities, 1991).

What are the long-term economic effects of high-speed trains? This question is especially relevant, since French experience suggests that high-speed train usage mainly stems from existing train travellers, who travel more and over longer distances, and from air traffic, not from car traffic. In small countries, the impact of the high-speed train on domestic modal split is (therefore) forecast to be minor, whereas investment outlays are very large. Some argue that it might be a good option to improve the quality of the whole conventional network instead of constructing one single TGV-line.

Although economic integration is a term used by many politicians in

Europe, transport practice shows various signs of nationalism. The future European high-speed train network is a good example. It shows that planning without establishing independent multinational, centralized political decision-making units and centralized funding is a waste of time. A TGV-North is planned to run between Paris-Nord and Amsterdam. The French part of this line is ready for operation, whereas in Belgium and the Netherlands the trajectory, due to local, environmentally induced, resistance, the time-path and the Belgian funding are as yet uncertain. If finished, this TGV will be unable to run at full speed in both countries, thereby reducing its market potential.

Another, but not final, question concerns the deficits of railway companies, which provoke important political questions. Is it reasonable to keep on subsidizing railway companies for services which too few people want to use, just for social reasons? Or should those subsidies be directed towards services with a higher market potential, such as high-speed ones or conventional ones on busier lines?

In this regard it is interesting to note that the management of the German Railways, in the face of its future privatization, is planning to sell the regional train part of its network to the *Länder* (provinces) and to concentrate on the higher-level services. Such an option has many advantages, including a smaller, more market-oriented organization. It could also help to establish better regional and interregional economic cooperation, instead of competition. On the other hand, there is the fear that many branch lines will soon disappear, since the local and regional communities have no money to provide regional passenger and/or freight services and to invest in maintenance or even new tracks.

The 'explosion' of mobility makes it clear that high-speed trains are only one of the means necessary to cope with mobility growth; overall modal split will not change very much. One of the most important causes of this situation is the fact that most countries have not developed some kind of master plan for public transport. France is still the only country working with these plans, thereby enlarging the potential of both high-speed train and public transport (Rijck et al 1991).

High-speed railways, together with the upgrading of conventional ones, may be one way to attract new customers and to ease the ongoing shift in modal split towards road and air transportation. If competing modes are not able to digest all of the forecast transport growth, railways may be able to attract part of this growth. When all transport modes have to pay the real cost of transportation their market potential may become even larger.

To arrive at such a situation various bottlenecks have to be removed. The central issue is that high-speed rail transport is at its best at long distances. It is precisely long distance international transport that is the weak spot of railway companies, since cross-border cooperation has never been a favoured issue within the railway companies and their national governments.

Integration is an important *supply* factor for improving the performance of train systems. It can only exploit its full potential once speed is added to the system.

1.5 CONCLUSIONS AND REFLECTIONS

The context and nature of European trade and transport is entering a new era. In recent years, there have been some dramatic changes: integration of the EC market, disintegration of various nation states, and more openness between all countries and regions in Europe. The full benefits of the foreseen internal European Market will only be reaped by means of effective (physical and non-physical) infrastructural adjustments in Europe. What is needed in this context is European, and *not* national, thinking and action in infrastructural policy, based on the knowledge of past successes and failures in infrastructural planning and the future needs of the economy. The fast emergence of rapid trains as a success story and the slow introduction of intermodal transport as a partial success story illustrate that a focused effective European transport policy is needed.

Although at both national and European levels attention is increasingly focused on Trans-European Networks, for the time being interest in Europe is mainly addressed towards separate, i.e. single-mode, transport solutions. Only recently has it been realized that interconnected networks supported by modern telecommunications technology may offer a realized high value added. *Interoperability* between different modes to use the transport capacity as efficiently as possible appears to be very difficult to achieve. Two factors of strategic importance have to be considered in this context:

1. *complementarity* between different modes in order to benefit—in terms of added value networks—from synergy (e.g. rails and waterways, roads and airports etc.);
2. *competition* between different modes in order to operate under the most cost-efficient conditions (e.g. common carriage).

In view of the role change of actors in the transport sector, complementarity (by means of interconnectivity, intermodality and interoperability) is a *sine qua non* for competition in the sector.

It has been claimed above that the future of a unified Europe will be critically dependent on the functioning of strategic infrastructure networks which are *interconnected* in terms of (1) *integration* between different layers of network (e.g. coordination of high-speed/long-distance networks such as TGV or aircraft and lower-speed local networks such as light rail or roads), and (2) *intermodality* between different competing or complementary network modalities. In this respect also the quality of *nodal centres* (terminals, stations, urban centres) plays an important role, as well as the frequencies of different types of transport (or carriers) in Europe.

The notions of *interoperability and interconnectivity* of networks, as advocated in the Maastricht Treaty, generate a series of important local points which deserve thorough attention from the side of policy-makers and the research community:

1. the operation of transnational networks, seen from the viewpoint of European cohesion and East-European (re)integration;
2. the close connection between the development of transport networks and (tele)communications networks (including new logistical systems) and their potential implications for the European space (e.g. polarization tendencies towards larger metropolitan areas);
3. the new regulations for competitive behaviour for all actors and modes with a view of upgrading and increasing the entire European network performance;
4. the new roles of public and private decision-makers, where a creative division of tasks has to be found between public authorities (urban/ regional, national, European) and private actors (transport operators and logistic suppliers) in order to generate value added networks;
5. the interconnectivity of high-speed long-distance networks and new regional–local infrastructures in central nodes of the European network;
6. the role of physical barriers (and organizational impediments) which reduce the benefits of economic integration in Europe (including the connections with Eastern Europe);
7. the emerging conflict between environmental sustainability, infrastructure expansion and completing networks (notably competing transport modes);
8. the fiscal or regulatory ways of charging to the user all external (social) costs of transport infrastructures and network use in Europe, not only locally but also European-wide;
9. the impact of new transportation, logistics and (tele)communication technologies on both mobility behaviour (demand) and infrastructure life cycles in the European space (supply);
10 the lack of standardization of transport systems technologies (and uniformity in transport systems in general) in Europe, which hamper the full benefits of an interoperable European network (especially in relation to intermodal transport);
11. the rather different financing regimes of states and network operators for European transport modes, which prevent a fair competition;
12. the lack of a strategic viewpoint on the linkage between European networks and global networks developed in other regions outside Europe.

Consequently, the policy agenda for interoperable and interconnected European networks is vast and deserves much attention in the near future.

REFERENCES

A. T. Kearney & Co. (in Zusammenarbeit mit Logitech) (1989) *Gemeinschaft der Europäischen Bahnen: Studie über die Perspektiven eines Europäischen Netzes des kombinierten Verkehrs, Abschlussbericht.*

Bruinsma, F. R. and P. Rietveld (1993) Urban agglomerations in European infrastructure networks, *Urban Studies*, **30**(6), 919–34.

Commission of the European Communities (1991) *Europe 2000, Outlook for the Development of the Community's Territory.* Communication from the Commission to the Council and the European Parliament, DG for Regional Policy, Brussels/Luxembourg.

Commission of the European Communities (1992a) *Verordening (EEG) van de Raad betreffende de Toegang tot de Markt van het Goederenvervoer over de Weg in de Gemeenschap (etc.)*, COM (92) 104 def., Brussels.

Commission of the European Communities (1992b) *Verordening (EEG) van de Raad tot Vaststelling van de Definitieve Regeling voor de Toelating van niet in een Lid-Staat (etc.)*, COM (92) 283 def., Brussels.

Commission of the European Communities (1992c) *Communication from the Commission on the Creation of a European Combined Transport Network*, COM (92) 230 final, Brussels.

Community of European Railways (CER) (1993) *European Rail*, Newsletter No. 15, February.

Espieussas, J-L. (1989) Proposals for a European high-speed network, in: H.G. Smit (ed.), *European Transport in 1992 and Beyond*, pp. 22–35, Infotrans Foundation/ETC, Delft.

European Round Table of Industrialists (ERT) (1992) The renewal of Europe's infrastructure, *Conference Proceedings*, Lisbon 1–2 June.

Gaudard, G. (1971) *Le Problèm des Régions—Frontières Suisses*, Cahiers de l'ISEA, Paris.

Gérardin, B. (1990) *Missing Infrastructure Networks in Europe, A Case Study: The European Rapid Train Network* (case study report NECTAR), INRETS, France.

Gérardin, B. and J. Viegas (1992) European transport infrastructure and networks: current policies and trends, paper presented at the *NECTAR International Symposium*, Amsterdam, 18–21 March.

Giaoutzi, M. and Nijkamp, P. (1993) Barriers and missing networks in European infrastructure: inland waterways and coastal transport, in: P. Nijkamp (ed.), *New Borders and Old Barriers in Spatial Development*, Gordon & Breach, London.

Maggi, R. (1992) *Borders and Barriers: Road-Rail Networks*, Research Paper, NIAS, Wassenaar.

Massoni, M. (1988) European Network, New Lines, Peripheral and Transit Countries, report presented at the *11th International Symposium on Theory and Practice in Transport Economics*, ECMT, Brussels, 12–14 September.

PROGNOS (Cerwenka, P. et al) (1988) *Gemeinschaftsuntersuchung Güterverkehrsmarkt Europa, Zusammenfassungen und wichtige Ergebnisse in Schaubildern*, Prognos AG, Basel.

Quévit, M. (ed.) (1991) *Regional Development Trajectories and the Attainment of the European Internal Market*, Research Paper, Louvain-La-Neuve.

Ratti, R. and S. Alberton (eds.) (1993) *Theory and Strategy of Border Area Development*, IRE, Bellinzona.

Rijck, L. de., H. Baeyens and D. Lauwers (1991) *Cities as Loci of Applied Technology, Impact of the High Speed Train*, FAST-Prospective Dossier No. 4 on the Future of European Cities, Draft Report, Mens en Ruimte, Brussels.

Suarez-Villa, L. and J. R. Cuadrado-Roura (1993) Regional Economic Integration and the Evolution of Disparities, *Papers in Regional Science*, **72**(4), October.

Sundberg, L. (1990) Aspects of advancing the European transport system, in: G. R. M. Jansen (ed.), *Future European Travel Demand and Infrastructure*, pp. 3–22, Infotrans Foundation/ETC, Delft.

Vickerman, R. W. (1992) *Changing European Transport Infrastructure and their Regional Implications*, Research Paper, University of Kent at Canterbury.

Winkelbauer, S. (1992) Regional cost–benefit analysis of a new rail transversal through the Alps, Paper presented at the *IVth World Congress of the RSAI*, 26–29 May, Palma de Mallorca.

2 Trans-European Networks: Short-Sea Shipping

EDDY VAN DE VOORDE
University of Antwerp, Belgium

and

JOSÉ VIEGAS
Instituto Superior Tecnico, Lisbon, Portugal

2.1 INTRODUCTION

All recent analyses forecast an enormous growth of long-distance European transport in the years to come, both for passenger and commodity transport. Some examples to illustrate this: the port of Rotterdam expects in the best of four scenarios an increase in throughput (and hinterland transport!) from approximately 290 million tonnes now to 394 million tonnes in 2010. The port of Antwerp expects from now to 2010, depending on the scenario, a growth of 30–50 per cent (Meersman and Van de Voorde, 1993).

Together with this growth some important market developments will occur (Van Willigenburg and Hollander 1993, p. 184): globalization and internationalization; new emerging logistic concepts such as rolling stock and just-in-time; the fact that much (semi-) manufactured cargo requires speed and security in delivery; better-organized information and communication systems; a higher importance of the total (integrated) cost of transport and distribution. In particular, the congestion phenomenon will create serious service problems for commodity transport: the just-in-time transport becomes disrupted, with consequences for production facilities; disrupted hinterland transport to and from ports creates higher costs due to higher time losses.

All this will have consequences for the existing transport infrastructure. More congestion and environmental problems in road haulage seem to be unacceptable both to those responsible and to society. The intention is to solve current transport problems by influencing the growth of road transport in favour of cleaner and cheaper modes of transport. A traditional sectorial or even national policy will not be sufficient. We need an integrated policy, developed in the context of European integration.

This offers new opportunities for short-sea shipping. Short-sea shipping

European Transport and Communications Networks: Policy Evolution and Change. Edited by David Banister, Roberta Capello and Peter Nijkamp. © 1995 John Wiley & Sons Ltd.

seems to be an environmentally friendly alternative to inland transport on long-distance haulage networks. The particular characteristics of this mode of transport make it suitable for certain market segments, fairly impossible for others. The information we need now is which segments, and why?

The strength of Europe is that it has many ports and a pronounced peninsular shape, especially in the extreme western and northern parts. It should be possible to resolve some of the problems of congestion on inland transport routes by means of short-sea shipping (ECMT 1993, p. 105). But, as matters now stand, information and studies in this connection are lacking. An even bigger problem could be that we lack a clear definition of short-sea shipping.

In the short term, there should be a clarification of the major European networks of long-distance freight transport in which short-sea shipping is, or could be, involved. Here the key concept will be the systematic integration between parties involved in sea transport and hinterland transport. All this information has to be combined with an investment policy in order to avoid bottlenecks.

2.2 SHORT-SEA SHIPPING: HOW TO DEFINE IT?

There exists no common definition of short-sea shipping. Short-sea ships have often been defined as sea-going cargo-carrying ships (including passenger carriers) of less than 5000 GT. Ships of less than 100 GT, non-propelled vessels, and harbour or inland waterway service vessels are not included (Crilley and Dean 1993, p. 1). Bjornland (1993, p. 63) defines short-sea shipping as seaborne goods transport (tonnes) which does not cross oceans.

However, defining short-sea shipping in terms of trading patterns seems much easier than defining it in terms of ship characteristics, e.g. size. Short-sea shipping, as a global phenomenon, means to a very large degree European short-sea shipping in particular; in other words, coastal shipping in the larger European area, from the Barents Sea and Scandinavia, via the North Sea and Baltic Sea, down to the Mediterranean and the Black Sea, and managed by European shipping companies (Linde 1993, p. 107). A hundred per cent exclusive use of ships for short-sea or deep-sea shipping appears not to be practicable.

The significance of short-sea ships can clearly be shown in terms of the number of ships and the market in which they operate. Table 2.1 gives data concerning the major European-owned short-sea fleets and the ship type composition.

Table 2.1 only gives static material, indicating nothing about the utilization of the fleet. But even then we can state that European shipping is to a high degree dry-cargo shipping, appearing as a variety of dry-bulks, neo-bulks (such as forest and steel/metal products), and containerized cargoes.

Table 2.1. Major European-owned short-sea fleets (data end 1991)

	Short-sea fleet		% Composition		
	Number	Percentage	Liquid cargo	Dry cargo	Other cargo
ex-USSR	1735	17.1	14	61	25
Greece	1206	11.9	23	45	32
Germany	1086	10.7	10	71	19
Norway	1076	10.6	9	54	37
Italy	698	6.9	32	29	39
UK	688	6.8	26	49	25
Turkey	642	6.3	14	66	20
Denmark	544	5.4	17	60	23
Netherlands	482	4.8	10	70	20
Spain	360	3.6	10	53	37
Sweden	341	3.4	18	47	35
Yugoslavia	203	2.0	7	43	50
France	150	1.5	23	37	40
Finland	131	1.3	8	58	34
Romania	116	1.1	7	84	9
other EU	194	1.9	18	52	30
other Europe	481	4.7	20	71	9
European totals	10133	100.0	16	56	28

Source: Crilley and Dean (1993), p. 17.

Nevertheless, liquid cargo (especially mineral oil products and chemicals) cannot be totally ignored.

Hoogerbeets and Melissen (1993, p. 347) state that short-sea shipping within Europe can be divided into a number of categories: the traditional single-deck bulk carriers, the container–feeder vessels and the ferries. We can think of the ro–ro concept, to speed up the import and export of goods by trailer and container.

But a number of bulk commodities transported within Europe are carried by large bulk carriers and tankers, which can hardly be considered as short-sea traders in the traditional sense. According to these authors this category accounts for over 50 per cent of the total sea-going transport within the geographically traditional coaster area. Deep-sea container services, although multiporting within the European area, may be excluded here, as these ships in general will not carry intra-European cargo.

Apart from all the difficulties in defining short-sea shipping, one thing is clear: short-sea trades have undergone change during the last decade. This can be considered partly as a function of the incentive to achieve economies of scale, provoking a reduction of the number of port calls. Accordingly, short-sea trades had to react positively to feeder services.

2.3 STATISTICAL PROBLEMS

At the First European Research Conference on Short-sea Shipping (Delft 1992) a number of policy recommendations were made to help to create opportunities for short-sea shipping and improve the possibilities for further development. One of these recommendations concerned the development of a database containing detailed statistics on short-sea shipping.

Indeed, analysing existing cargo flows and assessing the market potential of short-sea shipping is hampered by the lack of reliability and consistency in the available statistical material. The comparison of short-sea shipping flows derived from different databases is a delicate matter, for several reasons (Wijnolst et al 1993, p. XXXVI):

1. Since short-sea shipping is not defined in an unambiguous fashion, flows may vary considerably, depending upon the definition.
2. The definition of import and export may vary considerably depending upon the reporting statistical agency.
3. Cargo flows based upon different databases using different goods classification are hardly comparable.

The need for detailed short-sea shipping statistics has already been expressed several times by both researchers and policymakers. As a beginning, one should try to create origin/destination matrices on a port-to-port basis limited to the most important ports. Such an analysis is necessary to facilitate planning of future long-distance freight networks. It will indicate where volumes of freight flows are sufficient for investment in freight logistic systems and pinpoint where good integration with inland transportation is required.

At present an attempt is being made to use Eurostat data, providing information on external trade of the EU Member States. However, this procedure has several drawbacks. The Eurostat database is restricted to national statistics on the 15 EU Member States, creating limitations for e.g. interregional studies. No port or regional statistics are supplied. Also, since the database is limited to import and export data, transit is either lost or has to be compiled from other databases. However, transit flows are very important for Belgian and Dutch ports.

Those data are necessary and even indispensable for two reasons. First, we need to have insight into the current market: the tonnage, the parcel size, the freight commodities, the geographical links. The second reason concerns the potential market: which potential freight flows can be transported from an economic point of view by the short-sea shipping mode, and what are the economic consequences for the transportation market?

One has to recognize that the compilation of a complete and consistent database will be a difficult undertaking. Awaiting such a database, we have to accept second best procedures to get an idea of the importance of short-sea shipping in the European context.

Stuchtey and Zachcial (1993, p. 36) indicate the importance of short-sea shipping by its share of total goods volume in European long-distance transportation. It has a 34 per cent share of all intra-European EC cross-border traffic. This is about 240 million tonnes out of a total 700 million tonnes in 1990. An additional 235 million tonnes come from EC-cabotage.

The importance of coastal trade in intra-European transport can be estimated as about 35 per cent (Bagchus and Kuipers 1993, p. 52). Coastal shipping then is defined as all forms of maritime transport within Europe and between Europe and adjacent regions, irrespective of whether it involves small ocean-going vessels, large ocean-going vessels or coasters.

Awaiting such a complete and detailed database we can use an interesting report by Bjornland (1993), presenting some information on the absolute and relative volume of European short-sea shipping. As shown in Table 2.2, intra-ECMT short-sea shipping transported in 1988 was slightly more than 400 million tonnes, being 38 per cent of all intra-ECMT goods transport. At the time of the research (1991) the following countries were members of the ECMT (European Conference of Ministers of Transport): Austria, Belgium, Denmark, Finland, France, Germany, Greece, Ireland, Italy, Luxemburg, The Netherlands, Norway, Portugal, Spain, Sweden, Switzerland, Turkey, United Kingdom and Yugoslavia. Approximately 56 million tonnes concerned container transport, of which approximately one-third was feeder transport.

For that year (1988) a simplified 8 × 8 origin–destination (O–D) matrix

Table 2.2. Intra-ECMT short-sea shipping, in 100 000 tonnes (1988)

Countries /Regions	A	B	C	D	E	F	G	H	Total
A	279	248	249	75	260	15	18	7	1151
B	74	—	22	8	64	10	—	6	184
C	113	37	8	66	200	29	32	37	522
D	16	11	42	—	67	19	66	23	244
E	116	161	440	170	74	62	52	10	1085
F	25	11	85	30	55	13	31	10	260
G	5	3	19	43	19	18	—	43	150
H	4	9	66	109	46	13	193	58	498
Total	632	480	931	501	785	179	392	194	4094

Source: Bjornland (1993), p. 96.
Country codes: A: Denmark, Finland, Norway, Sweden
                B: Germany
                C: Belgium, Luxemburg, the Netherlands
                D: France
                E: Ireland, United Kingdom
                F: Spain, Portugal
                G: Italy
                H: Greece, Yugoslavia, Turkey

has been drawn up, limited to freight statistics since no adequate statistics on total intra-European passenger transport were available.

A number of important conclusions can be drawn from Table 2.2:

1. The four Nordic countries (Denmark, Finland, Norway and Sweden) together with Ireland and the United Kingdom accounted for 55 per cent of total intra-ECMT seaborne exports and 35 per cent of imports.
2. The largest importing block of countries is the Benelux.
3. The largest single flows can be found between the British Isles and the Benelux countries, and between the Nordic countries themselves.
4. There is a pronounced imbalance between incoming and outgoing trade volumes for the ECMT-countries, with effects on transport costs.

In 1987 short-sea shipping accounted for about 60 per cent of the total volume loaded for international transport in the ports (Bjornland 1993, p. 71). This means that European short-sea shipping has a major share in the international outgoing transport from the ports.

Road transport and short-sea shipping have been the expansive transport modes in intra-ECMT transport. From 1980 to 1988 intra-ECMT short-sea shipping increased on average by 4.7 per cent per year. However, the situation differs country by country. For the Nordic countries shipping is much more important than the average ECMT figures indicate.

One must realize that intra-ECMT short-sea shipping does not cover all European short-sea shipping, since it includes neither shipping between eastern and western Europe, nor shipping between the countries of eastern Europe. Bjornland (1993, p. 66) estimated that we will have to add some 10–15 per cent on top of their intra-ECMT short-sea shipping to get a more correct figure of European short-sea shipping.

Bjornland (1993, p. 96–97) also published tables on container trade (1988). The European container trade increased rapidly during the 1980s, with an annual average growth rate of approximately 9 per cent.

2.4   THE COMPETITIVE POSITION OF SHORT-SEA SHIPPING

Short-sea shipping has to compete with inland transport, mainly in the form of road haulage. Short-sea shipping must be considered as a viable alternative, especially within a combined transport system. However, some literature claims that road haulage is too cheap, i.e. does not pay for all the costs it generates (e.g. external costs).

Road transport, as a single-mode alternative, has some characteristics that are very close to the requirements of the current transport market demand, cf. the just-in-time concept and the door-to-door concept. Moreover,

European road transport has been stimulated by enormous investments in road infrastructure.

Only in recent years has some opposition arisen, based on congestion and environmental problems. For those geographical links and those commodities where road transportation is not absolutely necessary, one could aim at the use of environmentally friendly modes, but this requires a commercially attractive alternative.

In this context some so-called pull factors can be considered, directed towards the improvement of the competitive position of other modes, or push factors directed towards slowing down the growth of road transport.

The competitiveness debate has to be concentrated on the competition between combined transport, including short-sea transport, and single-mode transport such as road haulage. The entire transport chain has to be studied, since short-sea shipping is only one part of a transport and logistics chain, and must be considered in connection with other transport modes and including elements such as storage, production and consumption, risks of loss, damage or delays. Indeed, competition is a function of relatively generalized costs, including the out-of-pocket costs but also including other elements indicating the quality of service a carrier can provide.

Important logistical factors at the present time are transport costs, service frequency and reliability, avoidance of cargo transfers and the capacity provided. We know the impact of each of these components taken individually but not their combined effect, so that it is difficult to make any estimate of the market shares which technological innovation could win for short-sea shipping (ECMT 1993, p. 106).

Within such a transport chain approach, large carriers and forwarders could risk acquiring a dominant position. Competition will be oligopolistic even in highly segmented markets (ECMT 1993, p. 107). Looking at the sub-markets separately, a type of market such as short-sea shipping is contestable since maritime transport operators are large international companies which can move into or withdraw from a market and, in any event, take a dominant position on it. 'In such a context, government intervention would become necessary to prevent the formation of an oligopolistic market.' (ECMT 1993, p. 108)

2.4.1 PORTS AND NETWORKS

Ports can be considered as an essential link in a complex chain, since port costs are a decisive aspect in the competitiveness of maritime transport. The port is expected to provide a service in which reliability and on-time deliveries are essential. Port productivity is a very important factor, bearing in mind that port expenses may account for up to 70 per cent of the total costs of short-sea shipping services (ECMT 1993, p. 109).

The competitive position of short-sea transport will depend upon the itiveness of the network configuration. Therefore, we also need an ⌐... into the transhipment facilities in ports, including the smaller European ports. This can give us an idea of 'missing links' and 'bottlenecks'. We have to realise that especially in the smaller ports of southern and eastern Europe the past investments remain very low, both for transhipments operations in ports and hinterland connections to and from ports.

Within the European Union the possibility exists of obtaining infrastructure subsidies within the Trans-European networks, i.e. networks contributing to the integration of peripheral areas. Ports and hinterland connections become part of such a network. In addition, funds are available within the framework of the cohesion fund, for infrastructure investments in four European countries where the GNP is much lower than the average GNP of the other Member States.

2.4.2 POLITICAL FACTORS

Some institutional factors have to be considered: the completion of the internal market, the continuation of the European Economic Area (EEA), and the opening of the east European market. In this domain, the deeper relations of the European Union with the Scandinavian countries will stimulate short-sea shipping, whereas this mode will play a very minor role in East Europe due to its very continental shape.

However, the most significant political factor for short-sea shipping will probably be the evolution of local and regional politics with respect to road transport. If there is any reinforcement of opposition to through heavy traffic on the roads, modal shifts must occur to the benefit of rail and shipping. Short-sea shipping will probably be able to respond much quicker than the railway, since there is substantially less new infrastructure to build.

2.4.3 NATURAL ADVANTAGES AND DISADVANTAGES OF SEA TRANSPORT

Here we can stress the natural advantages of sea transport:

1. Sea transport seems to be cost effective.
2. An increase of short-sea shipping does not require too much additional infrastructure.
3. Sea transport is environmentally friendly, and an economical consumer of energy.

Here, on the other hand, are its disadvantages:

1. Lower frequency of service for any pair of points, because of larger unit capacity.

2. Lower reliability of departure and arrival times due to variability of weather conditions.
3. Higher risk of damage to transported goods.
4. More companies participating in the service supply.

2.4.4 INCREASING COMPETITIVENESS OF SHORT-SEA SHIPPING

To stand a better chance in the competition against road transport, three attributes are essential besides the price for the client: frequency of service, reliability of times, and simplicity of administrative procedures.

The simplification of procedures could be strongly favoured if intra-European Union transport is allowed to proceed in separate areas of the ports, with no need for customs, and also if vertical integration of the transport chain suppliers is achieved.

The need for high frequency of service creates a problem of start-up critical mass. The volume of investments in ships, trucks, fixed facilities and organization is probably too big to allow market forces alone to create a strong and stable supply of intra-European short-sea transport. Incentives are probably needed, such as innovative measures in the European registration of trailers with fiscal benefits, or even the widening of the EUROS register concept for the whole combined transport fleet.

2.5 MODAL SHIFTS AND COST DIFFERENCES: A CASE APPROACH

On the basis of detailed cost estimates, qualitative assessments of the political and macro-economic settings, and perspectives for freight traffic growth, it is possible to derive estimates of the traffic volumes which are transferable from land to sea transport. Stuchtey and Zachcial (1993, p. 41) reported about such a detailed study. The most appropriate traffic flows between origin and destination were chosen, including flows with both sea traffic and at least one participating land carrier dealing with the traffic. The research team also included some flows which currently make little or no use of the seagoing alternative.

The flows were selected for transport cost comparisons. The main conclusions can be summarized as follows (Stuchtey and Zachcial 1993, pp. 41–2):

1. The price elasticity of demand is generally low for the choice of a carrier, and only becomes significant for a very limited range of goods and transport flows, e.g. containers.
2. The overwhelming majority of traffic is largely fixed by combined aspects like type of goods, consignment size and route, not easily attracted by the sea transport alternative.

3. Moderate differences of transport price between sea and land transport thus have no noteworthy influence on the modal split.

The Dutch shipowners' association (KVNR 1993) refers to a study carried out by the Dutch consultants MERC-NEA, with the aim of making an inventory of the possibilities short-sea shipping could offer to slow down the growth of road transport, by stimulating a shift from road to short-sea shipping. In practice the work could be divided into two steps: to analyse the commodity flows in and around Europe; to analyse and compare the door-to-door tariffs between road and short-sea shipping.

What are the results for The Netherlands? Based on data of 1986 it was found that 31.5 million tonnes are potential maritime traffic. At the same time the road transport to and from The Netherlands amounted to 86 million tonnes.

In a next step tariffs were compared, including some drawbacks of sea transport like extra transhipment, longer transit times, extra hinterland transport. This comparison resulted in a potential cargo shift of 4.5 tonnes. In the next step, the hypothesis was formulated that a cost advantage of 35% would be needed, i.e. short-sea transport has to be 35% cheaper than road transport before the cargo shift would take place. At that moment the potential cargo to be transferred decreases to 1.37 million tonnes, with the main destinations in the Mediterranean Sea and in Eastern Europe. This 1.37 million tonnes corresponds with a potential decrease of 4.2 billion tonne km of European road transport.

Doing the same exercise with another scenario, including harmonization of petrol taxes and vehicle taxes, and also including environmental and infrastructure taxes, the potential cargo shift could increase to 4.3 million tonnes, being 5% of the Dutch international road transport.

Bagchus and Kuipers (1993) carried out a case study concerning a shift from road transport to coastal shipping on the link between the Netherlands and Portugal. They looked at the following two questions:

1. What impact does a shift from road transport to coastal shipping have on the turnover of these two sectors?
2. What is the effect if the operations concerned are carried out by a foreign shipowner?

Their answer is based on a containerized cargo of 1000 tonnes on the above route between The Netherlands and Portugal, being an average coastal shipping route. The case, using turnover as an indicator, was based on several assumptions: door-to-door transport is assumed to be possible; the cargo is containerized; a distinction should be made between turnover and costs.

Table 2.3. Turnover in a road transport and coastal shipping chain (data of 1991)

	Road (1)	Coastal (2)	Coastal (3)
Transport			
Distance (km)	2250	1900	50/50
Cargo (tonnes)	1000	1000	1000
Tonne km (mill.)	2.25	1.9	0.1
Cargo/journey (tonnes)	20	1000	20
No. of journeys	50	1	100
Tonne km/journey	45 000	1 900 000	1000
Turnover-ct/km (0.01 fl)	10	2	50
Turnover/journey (fl)	4500	38 000	500
Transport turnover (fl)	225 000	38 000	50 000
	(50 × 4500)	(1 × 38 000)	(100 × 500)
Transhipment			
No. of transhipment operations	100	100	200
Charge/transhipment operation (fl)	60	120	60
Transhipment costs (fl)	6000	12 000	12 000
Transport + transhipment (fl)	231 000	50 000	62 000
Coastal shipping total (fl)			112 000
Per container	4620		2240

Source: Bagchus and Kuipers (1993), pp. 63–4.
(1) Road transport
(2) Coastal shipping, Main route
(3) Coastal shipping, Supply routes
fl = guilders

The empirical results are summarized in Tables 2.3 and 2.4.

If 1000 tonnes of cargo are involved, the turnover in the Dutch goods transport sector as a whole falls by at least 119 000 guilders in the case of Dutch coastal shipping, and by a maximum of 194 000 guilders in the case of foreign coastal shipping. This latter figure corresponds with 3880 guilders per container.

What are the consequences for consignors? Based on the figures of the example above, consignors can save 2380 guilders per container by using Dutch coastal shipping. The authors conclude that this is an attractive saving for consignors, amounting to over half the calculated turnover of a road container, which is 4620 guilders for the route in question.

Referring to another study (NEI, NEA, 1990) we know that consignors only switch from road transport to rail and inland navigation if the transport costs are 30–50 per cent lower. For coastal shipping, the percentage is put at 35 per cent (MERC, NEA, 1991). The next question should then be: why do consignors ignore the above-mentioned price difference and continue to use road transport on route between The Netherlands and Portugal?

From an economic point of view the only reason could be the fact that

Table 2.4. Total Dutch road transport chain versus coastal shipping provided by a foreign shipowner and an onward journey provided by a foreign carrier (in guilders)

	(a) Dutch supply route	(b) Foreign coastal shipping and onward journey	(c) Total (a) + (b)
Coastal shipping		38 000	38 000
Supply route	25 000		25 000
Onward route		25 000	25 000
Transhipment			
Port	6000	6000	12 000
Supply and onward routes	6000	6000	12 000
Total:	37 000	75 000	112 000
Per container:	740	1500	2240

Source: Bagchus and Kuipers (1993), p. 64.

Explanatory notes
Column (a) concerns solely the activities carried out in The Netherlands. The total gives the contribution of the Dutch transport sector.
Column (b) concerns the activities carried out by a foreign shipowner and a local Portuguese carrier.

transport costs are only a relatively small part of the overall logistical costs. For the same route Bagchus and Kuipers (1993) considered differences in logistical costs. Only if the overall logistical costs are lower will a consignor opt for another mode of transport.

A modal shift has several effects on logistics: higher interest charges, because the goods take longer to complete their journey by sea than by road (for a study concerning the value of time in freight transport, see Blauwens and Van de Voorde, 1988); a lower frequency, resulting in higher storage costs for coastal shipping.

We saw earlier that coastal shipping costs 2380 guilders less per container than road transport, a sum that could be available for logistical costs. In this case the latter amounted to 781 guilders. So the conclusion could be: by using coastal shipping, a consignor can save 1600 guilders per container in overall logistical costs.

For a 1000 tonne cargo (i.e. 50 containers carrying 20 tonnes each) this corresponds to 80 000 guilders. So, if interest charges are included, the difference between coastal shipping and road is still 30 per cent, so that a shift makes sense in terms of overall logistical costs.

But in practice even with this difference road transport is still used. That means that costs alone are clearly not the decisive factor in provoking a modal shift. Bagchus and Kuipers (1933, p. 61) argues that for consignors some additional factors remain important in their modal choice: the image of the transport mode; the market approach of a transport sector; the commercial organization of that mode; etc.

2.6   WHAT DO WE KNOW ABOUT THE POTENTIAL MARKET?

Research activities like those of Stuchtey and Zachcial (1993) and Bagchus and Kuipers (1993) have to be continued and even enlarged to a much larger scale. The market situation and the market perspective have to be investigated from a double point of view: What will be the growth of sea transport in general? What do we know about the modal shift from land transport towards short-sea transport?

Concerning sea transport in general, we have to start from the principle of derived demand. The shipping sector follows the cyclical movement from the economy, mostly with a time lag. Moreover, sea transport is dominated by low-value goods, while the future growth in world trade will be dominated by high-value goods. Some OECD forecasts for sea transport expect an average growth percentage of approximately 2 per cent. The main growth markets will be the transportation of liquid gas and containers.

The modal shift will be, among other things, a function of a number of factors that make inland transport difficult, e.g. congestion and capacity problems (e.g. some railway links and terminals), border procedures and problems in transit transport. These problems create changes with total costs, and especially with relative total costs, and can provoke modal shifts. These modal shifts can be stimulated or opposed by EU policy.

Stuchtey and Zachcial (1993, p. 38) argue that in Europe ocean traffic, short-sea container operations and ro–ro traffic will form the core of future growth dynamics. On the basis of their own forecasts, they expect an annual average growth rate of more than 7% in origin/destination container traffic by sea within Europe. Converting this into capacity requirements this means that we will see more than 110 000 TEU in 1995 and 150 000 TEU by the year 2000 compared with the 1988 capacity of about 69 000 TEU. For bulk traffic European coastal shipping is likely to see only moderate growth rates. Ferry traffic will show an important and continual growth rate, especially in the Baltic region and in the traffic between the UK and Ireland. But the Channel Tunnel will certainly divert some of the traffic previously carried by the ferries, leading to levels of intensity of use of short-sea shipping from Great Britain more similar to those of the continent.

In assessing the future role of European coastal shipping, Stuchtey and Zachcial (1993, p. 37) think that the following five aspects and tendencies must be taken into account:

1. realization of the European domestic market with its growing exchange of goods and services;
2. high growth of land transport following German unification;
3. the increasing liberalization of trade between eastern and western Europe;

4. deeper economic relations among the Baltic states, Scandinavia, eastern and western Europe;
5. the growing division of labour, connected with better logistical solutions, such as the just-in-time concept.

But forecasting future traffic growth involves uncertainty. In a continually changing European situation, market perspectives can change quickly and dramatically. An example could be the possible organization of long-distance traffic with points of origin and destination in Scandinavia, the former Soviet Union (including Lithuania, Estonia and Latvia) and Poland, as well as in the Baltic part of Germany (e.g. Rostock). The success of such an operation depends on the prerequisite of a fast improvement of the hinterland connections.

2.7 CONCLUSIONS

A competitive system, in which the terms of competition are fair, would be to the advantage of short-sea shipping. Indeed, the costs of the various transport modes should be transparent, which means, among other things, the incorporation of environmental and infrastructural costs in the tariffs paid by the users.

Further (quantitative) research has to be carried out to identify the development potential of short-sea shipping clearly. The topics to be investigated concern the volume and geographical split of freight and passenger flows, the cost structure, the market organization and the competitiveness of port infrastructure. More detailed analysis is needed of the modal shift of freight from road to sea, and of the effects on the different sectors. More research is needed into the trade-off between land and sea transport.

Other research themes concern strategies for achieving integration in European transport using short-sea shipping as a link, the effects of technical and operational innovation in the logistical chains, the effects of the European maritime policy and the consequences of regulation and deregulation.

Starting from the idea that future competition will be between transport chains, short-sea shipping must be recognized as an integrated part of a coherent European transport policy.

REFERENCES

Bagchus, R. C. and B. Kuipers (1993) Autostrada del mare, in N. Wijnolst, C. Peeters and P. Liebman (eds.), *European Shortsea Shipping*, Lloyd's of London Press, London, pp. 52–65.

Bjornland, D. (1993) The importance of short sea shipping in European transport, in *ECMT, Short Sea Shipping*, Economic Research Centre, Paris, pp. 59–93.

Blauwens, G. and E. Van de Voorde (1988) The valuation of time savings in commodity transport, *International Journal of Transport Economics*, 15, 78–87.

Crilley, J. and C. J. Dean (1993) Shortsea shipping and the world cargo carrying fleet—a statistical summary, in N. Wijnolst, C. Peeters and P. Liebman (eds.), *European Shortsea Shipping*, Lloyd's of London Press, London, pp. 1–21.

ECMT (1993) *Short Sea Shipping*, Round Table 89, European Conference of Ministers of Transport, Paris.

Hoogerbeets, J. and P. Melissen (1993) Facilitation of shortsea shipping: improvement in the sea/land interface (the Dutch case), in N. Wijnolst, C. Peeters and P. Liebman (eds.), *European Shortsea Shipping*, Lloyd's of London Press, London, pp. 346–350.

KVNR (1993) *Maritiem Keerpunt. Strategienota van de Koninklijke Vereniging van Nederlandse Reders*, KVNR, Rotterdam.

Linde (1993) Status and perspectives of technological development in European shortsea shipping, in N. Wijnolst, C. Peeters and P. Liebman (eds.), *European Shortsea Shipping*, Lloyd's of London Press, London, pp. 107–124.

Meersman, H. and E. Van de Voorde (1993) De toekomstige Antwerpse haventrafiek, *Het Tijdschrift van het Gemeentekrediet*, 73–85.

MERC, NEA (1991) *Mariniseerbare ladingen*, MERC-NEA, Rotterdam-Rijswijk.

NEI, NEA (1990) Vervoerwijzekeuze in het goederenvervoer, Rotterdam/Rijswijk.

Stuchtey, R. W. and M. Zachcial (1993) Perspectives and trends of the European short sea shipping industry, in K. M. Gwilliam (ed.), *Current Issues in Maritime Economics*, Kluwer Academic Publishers, Dordrecht, pp. 36–44.

Van Willigenburg, J. R. and S. Hollander (1993) Coastal shipping, opportunities in a changing market, in N. Wijnolst, C. Peeters and P. Liebman (eds.), *European Shortsea Shipping*, Lloyd's of London Press, London, pp. 184–192.

Wijnolst, N., C. Peeters and P. Liebman (eds.) (1993) *European Shortsea Shipping*, Lloyd's of London Press, London.

3 Evolution of Transport Networks Around the Core Area of Europe

VELI HIMANEN
Technical Research Centre of Finland, Espoo, Finland

JURAJ PADJEN
Ekonomski Institut, Zagreb, Croatia

VLASTA DUGONJIC
UN-ECE, Geneva, Switzerland

WLADIMIR SEGERCRANTZ
Technical Research Centre of Finland, Espoo, Finland

3.1 INTRODUCTION

The current trends in the transport geography of Europe can be expressed by two words: *integration* and *expansion*. The European Community established a single market in January 1993 to promote the integration of western Europe. This integration is one of the most important policy principles of the Community and it is assumed to be related to economic progress. Integration expanded in the beginning of 1994 when the EFTA countries (except Switzerland) formed with the Community the European Economic Area (EEA). The EFTA-countries—Austria, Finland and Sweden—have also negotiated membership of the EU commencing in 1995. European integration is still expanding because the iron curtain has fallen down and ex-socialist countries are renewing their relationships with other European nations.

In this process the integrated Europe, which previously was shown on maps to end at the eastern border of the former Federal Republic of Germany and in the north around Copenhagen, is now moving some 2000 km northwards to the Arctic Ocean·and some 3000 km eastwards to the Ural Mountains. A new transport geography is shaped by the Community and three different groups of actors are still outside the Community: Nordic Countries (Norway), former socialist countries between EEA and Russia, and Russia itself (see Figure 3.1). In this connection it is necessary to remember that central goals of the Community's transport policy (Anon.

European Transport and Communications Networks: Policy Evolution and Change. Edited by David Banister, Roberta Capello and Peter Nijkamp. © 1995 John Wiley & Sons Ltd.

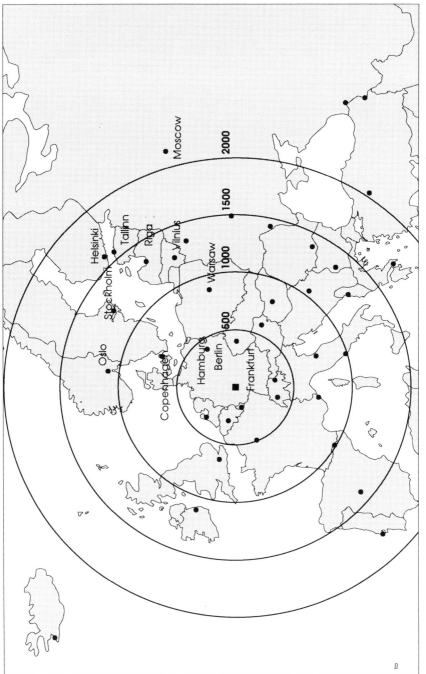

Figure 3.1. New actors in the transport geography of Europe, the Nordic Countries, former socialist countries between the EAA and Russia, and Russia.

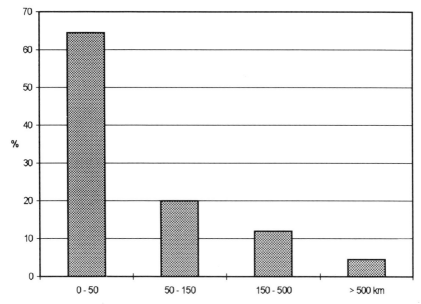

Figure 3.2. Average haulage length in road transport inside the Community (data gathered by NEA Transport Research and Training)

1992) include the linking of peripheral countries with the central regions of the Community and also the development of appropriate relations with third countries.

Inside the Community the distances covered in the transport of goods are usually short; in road transport journey lengths over 500 km are very rare (see Figure 3.2). However, 500 km is just the starting point in the emerging transport map when considering distances from the peripheral countries into the Core Area (see Figure 3.3). In addition, the abolition of border controls inside the EEA will emphasize the influence of borders outside the EEA and together with long distances and sea crossings will determine the level of interaction costs (see Figure 3.3).

The connections between the Nordic Countries and the Core Area have been discussed recently by Kristiansen (1993). Transport on the transition period in Central and Eastern Europe has been lately described by Button (1993) and Hall (1993) and from the German viewpoint by Blum and Leibbrand (1993). The transport situation in Russia has been less discussed in current international literature. Martellaro (1992) has provided an overview which was rather meagre compared to the vast task. In this chapter new types of accessibility problems are discussed. Then some major aspects of transport networks in the above-mentioned countries are discussed with the aim of presenting facts and trends not sufficiently covered by the recent literature.

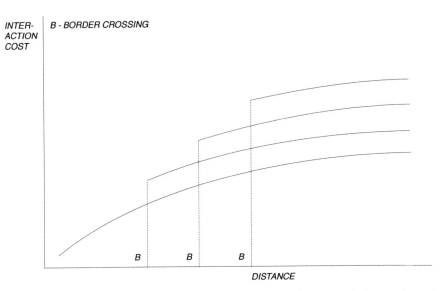

Figure 3.3. Interaction costs as a function of transport distance and the number of border crossings (inspired by a lecture of Börje Johansson in the seminar *Future of the Baltic Rim*, Turku, 25 August 1993)

3.2 ACCESSIBILITY

3.2.1 TRANSPORT PROBLEMS

Citizens in developed countries have achieved an unforeseen mobility. However, new problems seem to arise immediately after the old ones have been solved. A modern transport problem—accidents and pollution—is the result of solving the historical problem: how to get from one place to another. Our solution involves a tremendous use of material and energy. After the modern transport problem we have also got a post-modern transport problem. Traffic volumes in major cities and on main roads of the European Core Area have reached the available road capacity, resulting in severe congestion. It is obvious that it is no longer possible to expand road infrastructure or urban parking space to meet forecast demand (Goodwin et al 1991). Congestion has reached such proportions that some are waiting for the final gridlock (Banister 1989). This kind of congestion is, however, restricted to large built-up areas. The Nordic Countries and especially the former socialist countries are more concerned with historical and modern transport problems (see Table 3.1).

Table 3.1. A comparison of the Nordic countries (Finland, Sweden and Norway), the core area of Europe and former socialist countries in the context of historical, modern and postmodern transport problems

Transport problem	Nordic countries	Core area of Europe	Former socialist countries
Historical			
domestic	Well-developed transport networks	Well-developed transport networks	Inadequate transport networks
international	Sea crossings and long distances except from southern Sweden	Short connections by road and rail	Long distances and many border crossings except for the most western ones
Modern			
safety	Well-developed road safety	Varying road safety also security problems	Poor road safety
air pollution	Mainly sparsely populated regions with modest concentrations of pollutants	Densely populated regions with high concentrations of pollutants	High emission rates because of old, poorly maintained vehicles
Post-modern			
domestic	No severe congestion, room for construction	Severe congestion, almost no room for construction	Some congestion because of poor networks, room for construction
international	Congestion around Hamburg	Congestion in main corridors	Congestion at borders and in Austria and Germany

3.2.2 TRANSIT PROBLEMS

When the intensity of through traffic was still relatively low and when it mainly relied upon railways and inland waterways, the issues regarding through traffic were not a serious concern for anybody. A strong increase in the volumes of road transport and correspondingly large investment requirements in transport infrastructure as well as negative impacts on environmental quality have induced numerous difficulties in coordinating the interests of countries that supply and those that demand through transport services. The latter are countries situated on the periphery of Europe, whilst the former are situated in the central regions of Europe. However, there are many countries that are simultaneously both suppliers and users of through traffic services.

Development of through traffic has its technical–exploitation and transport–economic constraints. The former constraint is determined by the relationship between the volume of installed transport capacity and the volume of provided transport services and its structure (quantity of goods, number of passengers, or number and types of vehicles). The latter constraint is determined by the relationship between total revenues and total costs generated in transport activity.

Until recently, the relationship between revenues and costs focused exclusively on direct costs. However, the influences of increasing environmental consciousness intensify the necessity to add also the indirect costs to this relationship. Consequently, such an approach substantially changes the framework of through traffic development. Instead of attracting any type of transport at any cost, there is an increasing tendency for selective attraction of through traffic. In this case the focus is set on quality and profitability of transport. In other words, the objective of this approach is to maximize revenues from provided transport services, and to minimize expenditures at any level of revenue.

Such policy is overwhelmingly gaining ground in some central European countries, particularly in Austria, Switzerland and Slovenia. In Austria, for example, the load on through traffic routes has reached the unacceptable level which calls for destimulation of road cargo by different measures and the redirection of through transport to railways and inland waterways.

The profitability of transit transport from the viewpoint of transit countries depends very much on the modal split. In the case of cargo changing mode—from ship to truck or train—the possibilities for economic gains are much better than in the case where just trucks are driving straight through the country. This explains different attitudes among transit countries. Compared to the negative opinions in Central European countries, the Netherlands, Finland and the Baltic Countries have much more positive attitudes. In any case it is obvious that in the expanding Europe the most easterly EEA countries will be transit countries for their eastern neighbours outside the EEA (see Table 3.2).

Table 3.2. EEA countries linking their eastern neighbours

Finland	Russia
Germany	Poland
	Czech Republic
Austria	Slovakia
	Hungary
Italy	Slovenia
	Albania
Greece	Macedonia
	Bulgaria

3.2.3   RAILWAY INCOMPATIBILITIES

Because of long distances and a high share of railway transport in former socialist countries (cf. Button 1993) railways form an important element in the integration of peripheral countries to the European Core Area. There are already many problems in the cooperation and coordination of railways inside the Community and these are manifold in some eastern countries.

For historical reasons, technical disparities have developed and in the absence of common requirements they now constitute obstacles to border crossings. Examples are as follows:

1. Differences in electrical supply which require change-overs of hauling vehicles, thus wasting time (at least 30 min).
2. Gauge differences (Iberian Peninsula, Finland, the Baltic countries, CIS-countries) requiring change of bogies at the borders. However, this can be done speedily with proper technology as is the case in Hanko harbour in Finland.
3. Differences in rules relating to train formation which gives rise to an additional formality at the border involving the preparation of a formation notice. The length over buffers and the maximum train length are factors which vary from one network to another. For example, the transit trains from Russia must be cut down in Byelorussia because the length of railway yards is too short (Anon. 1993).
4. Differences in the rules for calculating train braking (brake/speed ratio) which means that a brake label has to be prepared, not to mention the additional complication of braking systems which differ from one network to another.

3.2.4   BORDER CROSSINGS

Inside the Community border checks are no longer a major obstacle to transit although they could be improved. The situation in the common borders of EFTA and the EU countries has improved dramatically since the beginning of 1994 with the formation of the European Economic Area (EEA). However, it will be difficult if not impossible for former socialist countries to give up all forms of border checks given their economic situation of transition towards a market economy. Generally speaking, border crossing difficulties diminish when travelling from south to north and from east to west.

When considering difficulties in border crossing, two types of difficulties have been distinguished: (a) difficulties that fall within the competence of State bodies, and (b) those falling within the competence of transport operators. When considering the first category, a distinction should be made between fiscal measures and measures intended to ensure State security.

3.2.4.1   Difficulties falling within the competence of State bodies

Passenger traffic may be subjected to checks at fixed points and also on board trains. Border police checks are of two kinds, (i) trans-border traffic, which could be relaxed or abolished; and (ii) police anti-criminal operations, which cannot be abolished. It may be noted that in all countries, these checks have been increased as a result of migratory pressures from eastern Europe and the Mediterranean basin. There exist many possibilities for the improvement of current procedures.

Goods transport may be subjected to several kinds of controls: (1) custom checks, (2) inspection of dangerous wastes, (3) phytosanitary and veterinary inspections, (4) checks for tax reasons, (5) reasons of civil and labour law.

1. Regarding customs checks, since 1 January 1993, the movement of goods has been completely unrestricted within the European Community and is subject neither to inspection nor to formalities, except where a fraud is suspected. In practice, customs checks inside the Community now concern only goods from third countries. This kind of checking system has been expanded also to the EFTA countries instead of the former simplified procedure which enabled goods to be carried without border stops between these countries. Countries that remain outside the EEA are put in a difficult situation.
2. Transport of dangerous goods has become a very complex issue especially because there is no legal liability to ensure safety standards. The transport of these goods is currently affected on the basis of the consignor's declaration. Therefore, inspection of dangerous wastes has become an acute necessity.
3. Phytosanitary and/or veterinary inspections are still carried out at borders (particularly in Austria, Italy, Poland and the Czech Republic). Whilst the volume of goods involved in such checks is relatively small, the length of the delay which they cause at borders may be prejudicial— 24 h at the Czech and Polish borders, and from 3 to 15 h elsewhere.

    Negotiations initiated between the Community and the third countries which are their main trading partners are aimed at reducing the level of inspections according to the health risk of products imported into the Community. Member States may also make the procedure more flexible through bilateral agreements which allow for document checks at the border and the other checks at the destination. In any case, it is difficult to envisage the complete elimination of veterinary inspections at borders, in view of the risks incurred. In all cases, there will remain at least a document check.
4. The differences in tax regulations (particularly between the Common Market and non-member countries) may induce forwarding agents to resort to re-consignment (with the resulting border delays) in order to

avoid certain taxes, which produce stops at borders for tax purposes. All the routes which cross the Italian frontier present special difficulties in connection with the Guardia di Finanza.

5. *Liability law–labour law in railways.* It has been pointed out that the crossover of personnel and equipment would be facilitated if the problem of the international liability of the railway companies with respect to their own crews and equipment (particularly *vis-à-vis* third persons) could be settled. There is, for example, a bilateral agreement between Norway and Sweden which reciprocally allows trains from either country to continue their journey beyond the border with the same crew. In Europe generally, Directive EECV/91/440 has already addressed this problem by creating international groupings which will enable crews to continue to perform their duties beyond national borders, within the Community.

It should be recalled that the trans-border use of rolling stock is outside the scope of the *CIM international transport contract*, and that two consecutive technical inspections are therefore necessary—one by the outgoing network and one by the incoming network. Even if the 'honour' procedure (practised by a small number of networks) were to become the general rule, the consequences which could arise from the use of the rolling stock would come under the national civil or criminal liability law of the user network. Naturally, the legal regulations applicable may vary from one State to another. It would therefore be appreciated if Governments could consider this issue and, in the medium term, plan standard international rules governing this type of liability.

3.2.4.2 Difficulties falling within the competence of transport operators

The forwarding of the commercial and technical data appearing on the transport documents constitutes the main source of difficulties. It is therefore essential to avoid the development of national systems which will lead to further international incompatibilities. Although the HERMES system (comprising 11 networks) enables an interface to be established between various systems, the universal view is that the quantity, quality and reliability of the data transmitted should be improved. The current inadequacies often necessitate a complete check of trains at borders in order to verify the commercial data.

A draft for the criteria to be met in a pan-European context states as an aim that waiting times at borders should not exceed:

- 15 min for passenger cars;
- 20 min for coaches and trains;
- 60 min for freight vehicles.

Table 3.3. Average waiting times for lorries on some borders during August/September 1993 (*Transport ja Logistika* **1**/94, 6). Reproduced with permission

Border crossings	Direction	Hours or km
Schwedt	D – PL	4 h
Frankfurt/Oder	D – PL	23 h
Guben	D – PL	14 h
Forst	D – PL	11 h
Neugersdorf	CZ – D	2 h
Zinmvald	PL – D	1.6 km (length of queue)
Zinmvald	D – CZ	1.5 km (length of queue)
Schönberg	CZ – D	1.9 km (length of queue)
Narva	EST – RUS	24 h
Rajka	D – H	4 h
Nagylak	H – RO	10 h
Gyula	H – RO	9 h
Biharkeresztes	H – RO	3 h

These criteria are in sharp contrast with the actual practice. An examination of borders of former socialist countries revealed that the target for freight vehicles was never obtained during summer 1993 according to an Estonian source (see Table 3.3).

In addition, according to the expedition firm of Lithuanian Railways (interviewed 6.5.1994 by Wladimir Segercrantz), waiting times can be rather long. They gave following waiting times which in normal conditions will not be exceeded by lorries:

- Lithuania–Russia (towards Kaliningrad) 12 h
- Lithuania–Latvia 16 h
- Lithuania–Belarus 24 h
- Belarus–Russia (Smolensk) 48 h.

Border crossings between Finland and Russia are operating sufficiently. The waiting times for lorries in May 1994 were normally less than one hour, and in exceptional cases a couple of hours (*source*: the Ministry of Transport and Communications in Finland).

3.3 NORDIC COUNTRIES

In this context only three countries—Finland, Sweden and Norway—two of which have joined the EEA will be considered. Kristiansen (1993) has given an exhaustive description of the complex issues related to the connections between the continent and the Scandinavian peninsula. This is why the Finnish viewpoint—which is lacking in Kristiansen's paper—will be our main concern here. It must be noted that the geographical situation of Finland differs much from that of Sweden and Norway. Finland is situated much more towards the northeast. Even though the Nordic capitals of Helsinki, Stockholm and Oslo are located near the 60th northern meridian, a major part of Sweden's population and industry is situated south of Stockholm. Similarly, a significant part of Norway's territory is south of Oslo, but Helsinki is located in the southern end of Finland. Finland is also separated—except in the far north—from its western and southern neighbours by the Baltic Sea.

For the time being the Baltic Sea is full of short- or long-distance ferry lines (Figure 3.4). A special aspect of transport geography is that almost all connections from Scandinavia to the Core Area and beyond are directed through the Hamburg region. This configuration will remain in the future. Four different plans for the improvement of the connections between the Nordic Countries and central Europe have recently been emerging and three of them are also directed through the Hamburg region (see Figure 3.5). All these links already exist and the current plans aim only to improve the quality of transport services.

The major impact will occur with the improvement of ScanLink (Figure 3.5) already heavily used, with fixed connections between South Sweden and Germany through Denmark. NorLink is a concept aimed to improve logistical services on the already existing route. The improvement of FinLink is based on an advanced ship concept (see Table 3.4).

BaltLink includes different actions to improve the road/rail connections through the Baltic states. Many border crossings with long delays induce major obstacles on this route.

3.4 COUNTRIES BETWEEN THE EEA AND RUSSIA

Because of the thorough discussions about the impact of economic and political transition on transport in central and eastern Europe by Button (1993) and Hall (1993), only a few aspects will be considered here.

The countries between the EEA and Russia differ from each other in regard to transport geography (Figure 3.1). Poland, the Czech Republic, Slovakia, Hungary and Slovenia are situated very near to the European Core Area and they all have only one border to cross to get into the EEA. According to the latest forecasts (sources: Reuters, *Financial Times* and the *Economist*), these countries, except Slovakia, are also the first to move from the transition period towards economic growth. All other countries are less favourably situated when considering connections with the EEA. However, most of them also have a sea route possibility.

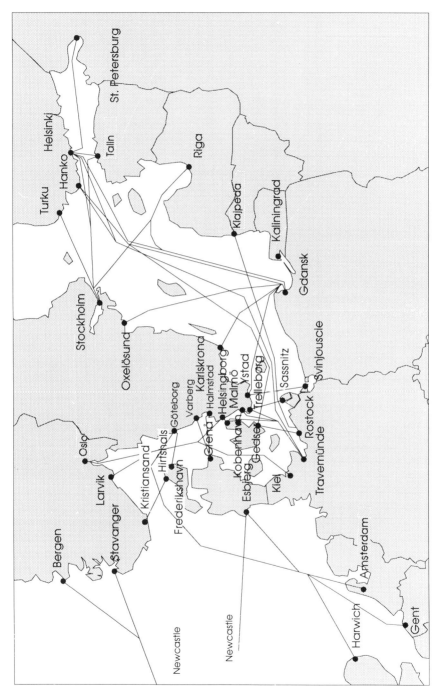

Figure 3.4. Ferry connections across the Baltic Sea

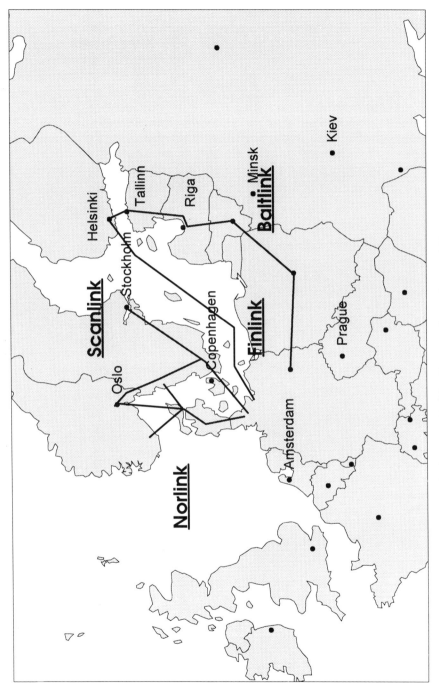

Figure 3.5. The improvement of four links between the Nordic countries and central Europe

Table 3.4. Ferry services between Helsinki and Travemünde (from Levander 1992, with permission)

	Today	With a fast ship
At sea	36 h	18 h
In port	12 h	6 h
Total hours	48 h	24 h
Average speed	17 knots	34 knots

Belarus is perhaps less favoured in transport geography. It does not have a common border with an EEA country nor a sea route connection. On the other hand, Croatia and Estonia are better situated. Even though they do not have a common land border with any EEA country they have very short sea crossings; Croatia towards Italy and Estonia towards Finland. Croatia is also situated very near to the European Core Area.

In line with their autarchic policy, the Council for Mutual Economic Assistance (CMEA) nations had conducted a policy of redirecting the flows of goods and passengers to the transport systems of CMEA countries only. Apart from economic measures, the strength of the barriers to transport movement had been amplified by formation of political, institutional, social, ideological and other similar barriers and obstacles which created discontinuity in the normal process of communication and diffusion of people, goods, information and knowledge, described for example by Nijkamp et al (1990).

With the dissolution of the Soviet bloc such policy simply vanished and radical changes in the socio-economic relationships and the position of individual countries in the interregional cooperation are emerging. Such movements are reflected in the rapid decline in trade among the former CMEA countries. According to the data presented in a study by OECD (1992, p. 85), trade decline among these countries since 1989 has amounted to 50 per cent in volume terms in 1991. Hungary, Poland and the former Czechoslovakia have been able to compensate this loss by increased trade with the Community and with other OECD countries and even with non-OECD countries.

In addition, according to Richter and Toth (1993, p. 3) one may expect a dramatic reorientation towards the Community and towards Germany in particular. The final conclusion is that the natural trade partners of former CMEA countries are the advanced industrial countries of Europe. By contrast, the share of trade between these countries is expected to decline drastically, to less than 25 per cent of the whole of foreign trade. However, trade with Russia will continue to have some importance in the future.

Political and economic changes in Europe will radically change the established economic and social relations and the position of particular regions regarding interregional cooperation, and substantially influence the redirection of transport flows. Furthermore, it will induce the process of restoration of business connections that had been forcefully suspended for the past 50 years, as well as the renewal of legal, institutional and other means of inter-regional cooperation. The modernization of transport facilities which are presently in poor technical condition and the improvement of transport infrastructure between East and West are needed. However, it is no use improving transport infrastructure without improving border-crossing procedures. Because of huge financing demand the improvement of the transport network will be a slow process.

3.5   RUSSIA

3.5.1   FEW FOREIGN CONTACTS WITH RUSSIA

In 1991 Russia had 148 million inhabitants. Even though the country is vast, the majority of people live in the south-west. The biggest cities, Moscow and St Petersburg, are near the western border (see Figure 3.6). However, major natural resources such as oil, natural gas and minerals are situated in the northern and eastern parts of the country.

Because of statistical deficiencies the estimate of Russia's GNP is uncertain. Figures just over US $3000 per capita have been presented by various bodies. In 1987 import and export in the Soviet Union was US $720 per capita, in Finland $8077 and in the USA $2784 (Himanen et al 1993). In big countries the need for foreign exchange is less than in small ones. Also the very small GNP diminishes export and import volumes in Russia. Few foreign contacts are highlighted in the number of air passengers. In 1989 Moscow had only 5 million international air passengers; the same amount as Helsinki airport (Himanen 1993). It must also be remembered that the economy in the former Soviet Union was very much involved with the military industry (Gaponenko 1993). The question remains to be answered: how much will Russia trade with Central Europe in the future?

3.5.2   HUGE DOMESTIC TRANSPORT VOLUMES

Transported goods in tonne/km per dollar of GDP in the former Soviet Union was six times that of the United States and over four times that of China (Holt 1993). This was not solely caused by the vastness of the country but was also the result of the centrally planned way of industrial and social sector organization. Costs of energy, transportation and protection of the environment were abnormally low and made it possible to transport by railways huge volumes of bulk over long distances between western and eastern parts of the country. With increasing transport costs this kind of hauling cannot continue.

Figure 3.6. The biggest cities and railways in Russia and the neighbouring countries

Table 3.5. Main roads and railways (Anon. 1993)

Country	Public roads		Roads (km/km^2)	Railways (km)	Railways (km/km^2)
	Primary	Total			
Russia (study area)	54 000:	150 000	0.09	51 000	0.023
Ukraine	estimate:	180 000	0.30	18 100	0.030
Belarus		48 902	0.24	9 800	0.047
Lithuania	20 900	40 565	0.62	2 672	0.041
Latvia	8 368	20 688	0.32	2 397	0.037
Estonia	6 453	14 811	0.33	1 026	0.023
Poland	46 600	231 700	0.74	26 550	0.085
Norway	26 100	88 174	0.27	4 180	0.013
Finland	11 500	76 717	0.23	5 890	0.017
Total		1 034 700	0.25	121 615	0.030

3.5.3   INFRASTRUCTURE

The importance of railways and the lack of roads is highlighted in a recent study about North-eastern Europe (see Table 3.5). The road network is fairly well developed only around some metropolises. Otherwise a local road network is almost non-existent and pavements are in poor condition along existing main roads. The transport of goods relies heavily on railroads which connect the biggest cities with each other (see Figure 3.6). Russia has also many inland waterways but for the time being they are not open for international transport.

In Russia various plans for infrastructure improvements have been presented. One obvious corridor for improvement is that between Moscow and St Petersburg. Plans for a motorway and a high-speed railway are ongoing. In addition, a special project aimed at the settlement of officers transferred to the reserve under the reduction of the army has recently started. The project envisages the formation of special development zones along the corridor between Moscow and St Petersburg.

3.5.4   WATERWAYS

The break-up of the Soviet Union and the birth of independent states on its former territory has produced a totally new geopolitical situation. The best ports in the previous Soviet Union in the Black and Baltic Seas are now in the hands of foreign countries. From the Russian standpoint it is problematic that there are also political, territorial, and racial difficulties with these countries. In this situation it is amazing that the freight through St

Petersburg harbour has about halved since the late 1980s. However, the operating and freight-handling times in St Petersburg have been increasing and are now measured in weeks and even months rather than in days. The deterioration of handling capacity in the harbour is obvious. In Russia there also exists interest in building new port capacity in St Petersburg and in Novorossijsk. A major part of Russian trade in the Baltic Sea is transported through the harbours of the Baltic states (Figure 3.7). Russia's total foreign trade in 1992 was 165 million tonnes, some 30 million tonnes up since 1991.

3.5.5 AVIATION

Air transport was heavily subsidized in the former Soviet Union. Its share of all domestic passenger mileage was 19 per cent in 1988. In 1987, Soviet aircraft carried 120 million passengers, of which 115 million were in domestic routes. In the Soviet Union there were 6950 airports, of which 1050 are operational and have permanently surfaced runways (Martellaro, 1992).

The most frequent flight connections to Moscow are from Frankfurt and to St Petersburg from Helsinki (Table 3.6).

Moscow has daily flight connections to major metropolises in Europe. It is probable that this pattern will remain in the future. International air traffic to St Petersburg is still very small. In the future it is probable that St Petersburg will have daily non-stop connections to major cities in Europe as well as Moscow. However, for the time being Helsinki acts as a gateway for St Petersburg. In particular tourists come by air to Helsinki and then continue by train or by bus to St Petersburg.

Aeroflot had a monopoly in the former Soviet Union. The difficulty in obtaining reliable data and/or the speed of development are highlighted by the unknown number of new airlines replacing Aeroflot, which was 20 according to Martellaro (1992), almost 70 according to Barrett (1993) and 92 according to Hall (1993). The reliability of the new network has been rather poor. International companies operating in Russia and in other CIS countries have been obliged to find new more reliable ways of transport. Privately chartered air flights from Helsinki are nowadays widely used for these purposes.

Helsinki is particularly well situated for flights to oil and gas fields in the northern part of Russia. Half of the flight hours for some Finnish business aviation companies are already made up by flights to CIS countries. As mentioned above, in CIS countries there are thousands of airports, but the use of them needs special skills. Outside a few international routes, aviation maps are in the Russian language and local air traffic controllers speak only Russian. The only practical solution is to use Russian navigators (article published in Kauppalehti, 6 October, 1993).

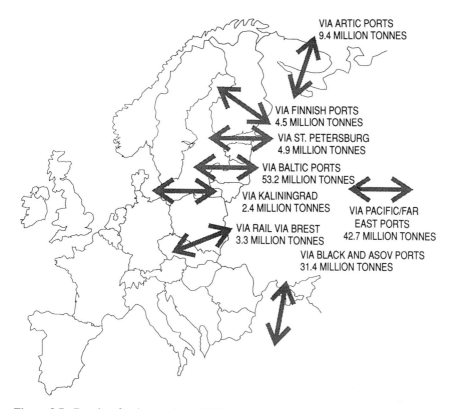

Figure 3.7. Russian foreign trade in 1992

3.6   CONCLUSIONS

The major trends in the European transport market are integration and expansion. In this expansion new groups of actors are emerging with new kinds of transport problems. Major transport problems between the Nordic Countries and the European Core Area include long distances—except for Southern Sweden—and sea crossings. Former socialist countries between the EEA countries and Russia can be divided into those with a common border with the EEA and those without. The latter are plagued with many border crossings and long distances to the Core Area of Europe. Transport between Russia and the Core Area of Europe suffers from many border crossings— except through Finland—and long distances.

Because of long transport distances, short-sea shipping, railways, and aviation have potential. However, railways are plagued with many border crossings and even include the necessity to change bogies because of differ-

Table 3.6. Daily non-stop one-way flights to Russia in summer 1993 (airline timetables and Amadeus data bank)

From	to Moscow	to St Petersburg
Helsinki	2–4	2–3
Stockholm	1	1
Copenhagen	1	1 on Sundays
Frankfurt	4–5	4 per week
Berlin	3–5	–
Düsseldorf	1–2	1 on Sundays
Munich	1	1 on Sundays
Vienna	2–3	0–2
Paris	2–4	1–2
London	2	0–2
Amsterdam	1–3	0–1
Brussels	0–1	1 on Saturdays

ent gauges. The improvement of border-crossing procedures is extremely important in order to cut down extremely long delays in the borders. Because of huge financing demands the improvement of transport networks will be slow.

REFERENCES

Anon. (1992) *The Future Development of the Common Transport Policy; A global approach to the construction of a Community framework for sustainable mobility* (White Paper). Brussels. Commission of the European Communities (CEC).

Anon. (1993) *North-Eastern Europe Transportation Study, Preliminary Phase.* Helsinki. Ministry of Transport and Communications of Finland, Viatek Group and Technical Research Center of Finland (VTT) on behalf of Economic Commission for Europe (ECE).

Banister, D. (1989) The final gridlock, *Built Environment* **15**(3/4), 163–165.

Barrett, S. (1993) Air transport market, in D. Banister and J. Berechman (eds.) *Transport in a Unified Europe; Policies and Challenges*, North-Holland, Amsterdam, pp. 91–123.

Blum, U. and F. Leibbrand (1993) Transport problems at the juncture between east and west, in D. Banister and J. Berechman (eds.) *Transport in a Unified Europe*, North-Holland, Amsterdam, pp. 315–333.

Button K. (1993) East–west European transport: an overview, in D. Banister and J. Berechman (eds.) *Transport in a Unified Europe*, North-Holland, Amsterdam, pp. 291–313.

Gaponenko, N. (1993) Russia: The way from chaos towards technological progress, *The 13th World Conference of World Future Studies Federation*, Turku, pp. 23–27, August.

Goodwin, P. B., S. Hallett, F. Kenny and G. Stokes (1991) *Transport: The New Realism*, Report to the Rees Jeffreys Road Fund, Transport Studies Unit, University of Oxford, 168 pp.

Hall, D. R. (1993) Impacts of economic and political transition on the transport geography of Central and Eastern Europe, *Transport Geography*, **1**(1), 20–35.

Himanen, V. (1993) Competition between Finnair and SAS. *NKTFK-konferens om konkurrens inom linjetrafik med tonvikt på faktisk och potentiell konkurrens inom järnvägstransport respektive flygtransport*, 1–2 December, Helsinki.

Himanen, V., K. Mäkelä and W. Segercrantz (1993) *Suomen liikenteellinen asema uudessa Euroopassa; Suomen ja Venäjän vuorovaikutus. Espoo. Valtion teknillinen tutkimuslaitos, Tie-, geo- ja liikennetekniikan laboratorio, Tutkimusraportti*, p. 188.

Holt, J. (1993) Transport strategies for the Russian federation, *Studies of Economies in Transformation*, paper number 9, The World Bank, Washington D.C.

Kristiansen, J. (1993) Regional transport infrastructure policies, in D. Banister and J. Berechman (eds.) *Transport in a Unified Europe*, North-Holland, Amsterdam, pp. 221–247.

Levander, K. (1992) The potential for fast ships in European freight transport, *First European Research Roundtable Conference on Shortsea Shipping*, pp. 26–27, November, Delft.

Martellaro, J. A. (1992) Transportation in the USSR, *International Journal of Transport Economics*, **XIX**(1), 3–21.

Nijkamp, P., S. Reichman and M. Wegener (1990) *Euromobile: Transport, Communications and Mobility in Europe*, Avebury, Aldershot.

OECD (1992) *Reforming the Economies of Central and Eastern Europe*, Paris.

Richter, S. and L. Toth (1993) *After the Agreement on Free Trade among the Visegrad Group Countries: Perspectives for Intra-Regional Trade in East-Central Europe*. Wiener Institut für Internationale Wirtschaftsvergleiche.

4 Dissolution of Borders and European Logistic Networks: Spatial Implications and New Trajectories for Service Performers, Border Regions and Logistic Networks

REMIGIO RATTI

Institute of Economic Research, Bellinzona, Switzerland

4.1 INTRODUCTION

The ongoing restructuring process in the field of transport is determined by different interacting factors: new infrastructures and changes in the production system; the application of information technologies to the transport sector; the technological progress in the transportation field; the modification of the political-institutional border framework (Ruijgrok and Wandel 1993); the harmonization and standardization of environmental regulation.

This affects the whole transportation chain and all actors of the logistical sector are concerned by its consequences: the traditional 'point to point' approach is gradually being replaced by the logic of the 'integrated hub and spoke' approach with multi-modal and multi-layer transportation networks on the European level. New logistic platforms are required, capable of carrying out a whole set of complex functions in an integrated way (Janssen 1993).

In the past, borders provoked a multitude of problems in the transportation chain. They had, in particular, determined the transportation networks reflecting the national policy goals and favoured the segmentation of the production structures. Hence, the border regions are *involved* and constitute a privileged observatory for the evaluation of the present restructuring process. In this chapter we will analyse theoretically (in terms of the three functional spaces: market space, production space and support space) and empirically the development trajectory of some traditional 'milieus' in the field of logistical services (forwarding agents) in the border areas.

The analysis examines the effects and consequences of the dissolution process of institutional barriers; the new properties of a unified logistics

European Transport and Communications Networks: Policy Evolution and Change. Edited by David Banister, Roberta Capello and Peter Nijkamp. © 1995 John Wiley & Sons Ltd.

system; the implications for logistic networks, especially for small and medium transportation enterprises and forwarding agents in a border region.

4.2   THE ROLE OF NATIONAL BORDERS IN THE STRUCTURING OF TRANSPORT IN EUROPE

Europe is—excluding the extensive Russian plains—a most variegated territory; her physical geography and political physiognomy is very fragmented in the west as well as in the centre. In this situation the border regions and their problems constitute a familiar reality, which have affected, in particular, the infrastructures of the transport and communication networks.

Before analysing the different effects, we have to clarify the significance and role of the border (Figure 4.1).

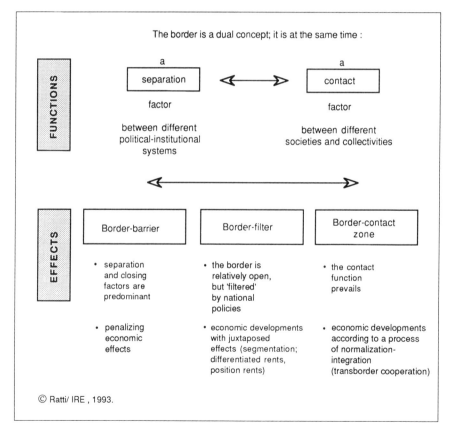

Figure 4.1. Border typology: functions and effects of the border (Ratti and Reichman 1993, with permission)

Throughout European history, three types of border regions with their different roles can be distinguished.

Although the construction of the European Union (Community) has favoured the abolition of the borders, or better, the transition to the border-contact zones, situations characterized by the border-filter (between Switzerland and the European Union and with the Eastern European countries) still persist. Important steps to a full harmonization in special fields of custom controls still remain to be made (United Nations 1993). At the same time new borders are built up in the ex-Yugoslavia, ex-Czechoslovakia and in the area of the Community of Independent States (CIS).

Every case of the following border typology influences the transport structure in a different way (see Figure 4.1):

1. In the case of the 'border-barrier', real economic effects of the 'shipment rupture' type can be observed. This can even lead to a recomposition of the shipment and the emission of new shipment letters, and hence a very important segmentation of national delivery structure (forwarding structure).
2. In the case of the 'border-filter' the persistence of legal, fiscal and customs differences requires special service functions, creating a multitude of intermediate activities. The economic significance is very similar to the 'law of refraction', described by Tord Palander (Moran 1965), in a way that the border introduces elements of distortions of the transport flows, introducing additional nodes in the logistic network.
3. Finally, the case of the 'border-contact' zone as well as the new situation in the European Union must not be underestimated or taken for granted. The border doesn't disappear completely, because there remains the heritage of the old infrastructural situation and important institutional and cultural differences will last (Figure 4.2). The border regions as logistical nodes don't disappear necessarily, because very often, as we will see further on, in these localizations a 'milieu' has emerged, forming a space of particular competences, which will not simply disappear with the customs barrier. However, the contact with the main markets—areas where the major freight flows start or end—must be intensified.

4.3 LIBERALIZATION, LOGISTIC REVOLUTION AND THE FUTURE OF BORDER AREAS OPERATORS

4.3.1 SHIFTING FRAMEWORK FOR THE BORDER AREA OPERATORS

In spite of several current difficulties, the general scenario can still be characterized by a drastic reduction of barriers between countries, peoples and

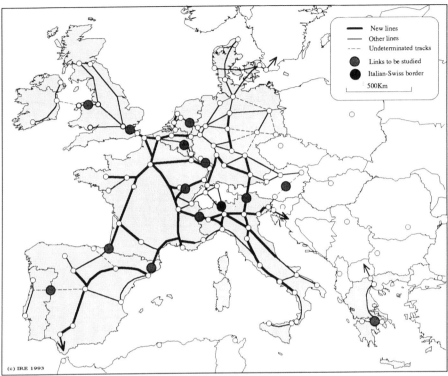

Source: Master scheme for the European railway network (1990)

Figure 4.2. European map with principal railway lines and borders (produced by IRE; reproduced with permission; source: Schéma directeur du réseau européen TGV 1990)

companies. Examples include the common market, the integration and transformation of eastern Europe, the trade and service liberalization within GATT, the greater geographical extension of technical standards, the integration of worldwide financial markets, the faster geographical diffusion of new technologies and the greater international mobility of specialists and managers.

Besides this shifting framework specific mega-trends towards faster transportation systems and the introduction of EDP technologies (OECD 1992) at every stage of the logistic chains could radically change the situation for the forward agents on the borderline (see Figure 4.3). For obvious reasons all factors influencing the development trajectories of the forwarding agents in the border area must be considered.

The main question concerns the capacity of the forwarding agents to integrate these new tendencies and to respond in a competitive way. This of

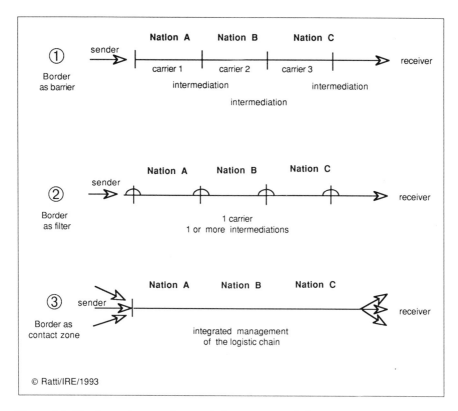

Figure 4.3. Border region typology and the transport chain

course requires major innovations in the conception and production of the whole transport chain. According to the following methodological approach, we will try to understand whether a traditional milieu in the transportation industry in a border region will cope with this new situation, identifying the different trajectories (Figure 4.4).

4.3.2 METHODOLOGICAL PROPOSAL

However, the answer to this question is somehow more difficult than we could expect from a market approach. In order to study the future scenario, we will analyse the evolution of the three functional spaces of the forwarding agents and other connected services, emphasizing the embedded attitude of the border operators (Grabher 1993).

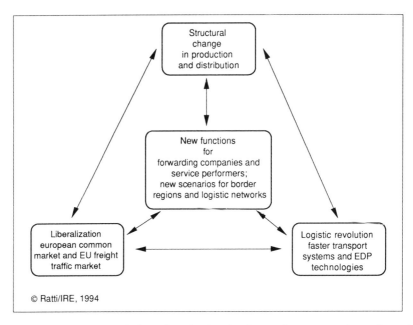

Figure 4.4. Major factors influencing the border forwarding companies and services producers

4.3.2.1   A 'functional space' approach

Among the ways to deal with changes of the economic structures we have chosen an interpretative frame, previously developed within the GREMI studies (Ratti 1991; Bramanti and Senn 1995), which makes use of three 'functional spaces' of a firm's economic activity:

1. The production space of the firm is determined by the technological and organizational characteristics of the economic segment, the locational properties and the innovation field.
2. The market space reflects the relationship of the enterprise to the market, i.e. the intensity, the frequency, the flexibility and the degree of confidence in its service to clients.
3. The supporting space describes three types of relationship outside the firm and the market, and concerns those connections to the production factors, to partners, customers and suppliers and finally to local institutions.

The focus here is on the relevant changes—possible ruptures or exogenous forces—affecting milieus and their internal/external relationships.

4.3.3   RECOMPOSITION AND DECOMPOSITION OF A TRADITIONAL MILIEU

The main indicators of changes concerning the functional spaces of forwarding agents and border localizations regularly discussed in the literature and partly confirmed by empirical studies can be summarized as follows:

(a) *Production space*
- (i) widespread introduction of information services (handling of goods and management of clients);
- (ii) innovation in transport technologies (combined transport);
- (iii) high differentiation in the organizational structure of logistics firms;
- (iv) necessity of new infrastructure (decay of traditional customs warehouse).

(b) *Market space*
- (i) radical change in the services connected to different types of customs transactions (elimination and organizational innovation);
- (ii) increasing demand due to the tendency of outsourcing by the industrial producers;
- (iii) structural change of demand (just-in-time, diminishing share of mass-products, shift to road freight transport);
- (iv) decay of old networks (as intermediate services) and high competence and financial barriers to enter new logistical markets;
- (v) new requirements (flexibility, reliability).

(c) *Supporting space*
- (i) fragmentation of old vested interests and loss of identity;
- (ii) important diminution of support by national institutional agencies;
- (iii) importance of new and privileged relations with specific clientele or operators;
- (iv) new needs of 'logistic culture'.

4.4   THE FUTURE OF BORDER AREAS OPERATORS

The observations of the ongoing dynamics in the activities of the economic actors in border regions in the realm of transport services are those very typical for periods of crisis and deep restructuring processes (Storper and Walker 1989). In this situation there can be a bifurcation in the direction of either a complete collapse, with a partial relocation of activities in a new emerging area or region (main market area) due to a complete restructuring of the market, or product innovation establishing a new competitive edge.

According to our methodology, an agent localized in the frontier region has to rearrange the configuration of his specific functional spaces of production, market and support. It is worth while to start with the analysis of

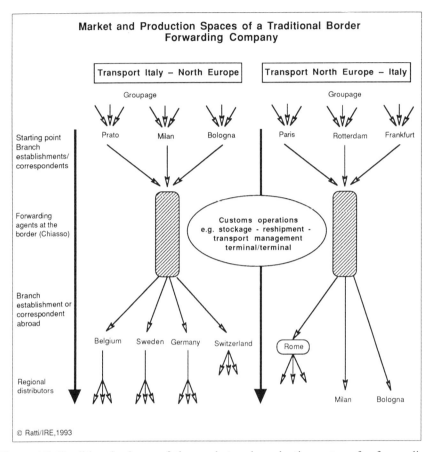

Figure 4.5. Traditional scheme of the market and production space of a forwarding agent in the border area (Example Chiasso (CH-I)). Reproduced with permission

the traditional model (Figure 4.5) marked by the fiscal and customary functions in order to analyse the different types of functions in the new scenario of European transport. Due to the objective empirical contingencies of every localization, possible strategic options can be derived on the level of a single operator as well as on the level of frontier regions and logistic networks.

4.4.1 THE TRADITIONAL MODEL

The evolution of international trade and transport in Europe is sensitive to national and transport policies. In addition, there remained within the transportation system many points of shipment rupture for reasons of intermodal changes (harbour, fluvial navigation, railway and road) or technical reasons (national railway systems partially incompatible, different structure g and electric voltages). To these obstacles can be added other ba deriving from the national fiscal, commercial and customs systems as well as missing network links (Nijkamp and Vleugel 1994).

The result of these points of shipment rupture and barriers had been a development of the activity of intermediation and services constituting a network node, which in part has resisted the European economic integration process of the last 30 years. The forwarding agent, thanks to the border, constituted a network of branch offices or correspondents, which permitted him to be inserted in a large market with differentiated functions. But even if there were such important involvements in the market, they hardly dominated the whole logistical chain. One of the most significant traditional examples is given by the intermediation points of north–south/south–north transalpine transport and the Iberian peninsula and also in those to Eastern Europe and overseas.

These services established a veritable bridge in favour of international trade, which had led to the creation of a culture of intermediation, which today still represents valuable capital stock (Ratti and Alberton 1993). However, in the present scenario of the single European market, the problems remain linked to the external frontiers of the European Union, meanwhile on the inside (in particular in the countries joining the EU) a decisive structural reconversion is taking place. The results of this process are anything but certain (see contributions by Giaoutzi and Kaman and by Garcia and Parellada in Ratti and Reichmann 1993).

4.4.2 THE NEW MODEL FOR BORDER AREAS AGENTS

In general, in the transition phase from the 'border-filter' situation to the 'border-contact zone' the following facts can be confirmed:

1. a radical questioning of the functions of simple intermediation services (cf. Figure 4.5), with the possibility of a clear-cut elimination of these activities;
2. a rupture in the development path with the possibility of a bifurcation in two branches:
   (a) regional operators, whose function will be reduced to regional distribution depending on road vectors or 'subcontracting' relations with respect to logistic operators (cf. Figure 4.6);
   (b) logistic operators able to manage and control the whole set of the function of an integrated logistical chain. In the best case they continue to operate from their original 'transit platform', localized in the border area (cf. Figure 4.2).

In the new scenario of a unified European market—where, however, the

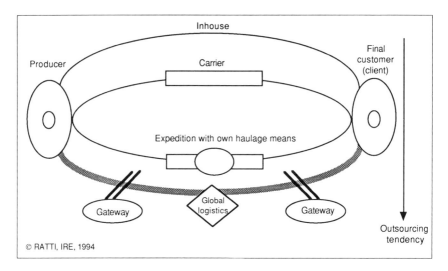

Figure 4.6. Typology of logistic chains

national and regional differences remain still considerable—we can define four ideal types of the logistic chains (cf. Figure 4.6):

1. transport organized and carried out by producer (inhouse transport).
2. transport organized by producer or final client, outsourcing of the function of the physical transport;
3. expedition and transport organized by a traditional forwarding agent;
4. outsourcing of the integrated logistic chain to forwarding agents specialized as a general service producer (logistic agents).

This typology is influenced by the following three fundamental tendencies.

1. In the economic restructuring process, the increased concentration of the core business leads many producers to the *outsourcing* of different functions, among them the organization, transport and distribution of the flow of raw material, semi-worked and final goods.
2. Meanwhile they rely more and more on *just-in-time* manufacturing principles, characterized in the year 2000 by up to 37 per cent of the whole production sector as against 21 per cent in 1990 and only 13 per cent in 1987.
3. Finally the application of the new information technologies to the transportation market allowing logistic integration.

In the first case we find the tendency to a 'contract hire transport', in the second and third the development is linked to the specialization and facilities reserved for special clients. In fact, dedicated contract transport and express transport are growing rapidly. On the other hand, all the facilities deployed for a wide client base (general haulage) or shared between a limited number of regular clients (shared contract transport) are expected to decline (Ruijgrok and Wandel 1993).

Hence, the future of the border operators depends on a strategy of regionalization and the importance of a market in the range of 50–100 km. Thanks to their know-how the forwarders localized in the border area can, in less favourable conditions, operate the transport (second case) or develop for their clients the function of regional distribution from the terminal to the final destination or the shipment from the starting point.

Given the demographic and productive situation in Europe, this type of reconversion is all but impossible. The main competition is provided by nearby metropolitan areas (Milan in the case of Chiasso, Barcelona in the case of La Jonquera), where production space (terminal) and support spaces (telematic services) might be superior.

But the most interesting case is the one of the integrated logistic chain. In fact, today a transport operation is more and more linked to other operations concerning the flow of information, the treatment of commercial data, the control and warehousing and the search for optimal solutions for packing from the environmental point of view (reverse logistics).

A new concept of logistics is emerging. It is called Product Channel Logistics and can be defined as the cross-functional formation, organization and control of all logistical activities needed to bring a product to its customer. Its aim is to maximize overall logistical efficiency in pipelines or networks, rather than individually managing these activities (Janssen 1993).

The complexity and richness of the global logistics is illustrated by Figure 4.7, representing a map of logistical relations of a multinational company present in Europe and overseas, administrated in an integrated way by one or more firm providing services.

Figure 4.7 shows the high degree of interpenetration between the service performer and the producer/factory and the sales organization on both ends of the logistical chain. The processing generally starts with the order from the point of sales to the producer or intermediary organization. From there the transport is organized at a regional distribution centre (RDC) for one or several countries. In many cases, the service provider offers supplementary value added services varying from the control, weighting and numbering of goods on entry to the warehouse to the equipping with bar-code labels (e.g. for a refill system). Services like fiscal and customs clearance are carried out by the same service provider. In our specific case, the packing into shipping units (e.g. nine pairs of shoes with shrink packing to avoid empty material (rolltrainers/shipping cartons) reducing considerably the costs of the waste management as a further service is produced by the same company. These logistic chains require a high level of technical and organizational capacity.

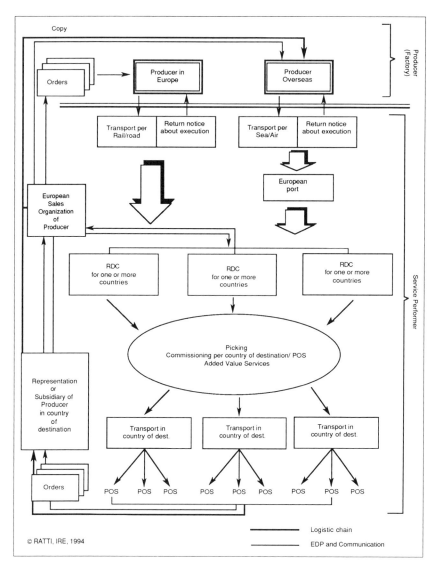

Figure 4.7. Scheme of the production and market space of Integrated Logistics. Scheme of the production and market space of a forwarding operator at customs reconverted to the integrated logistic system

In particular, the logistic operator tries to offer to the customer far-reaching services with a considerable added value improving his or her own position by reducing the risks of a highly competitive market situation.

This kind of approach to the global logistics is not available to all operators and the entrance barriers are particularly high. According to the results of one of our empirical inquiries, more than one year is needed to create an integrated logistic system. It requires relations of complete confidence and discretion with the client as well as a harmonization of the information system at the level of soft- and hardware. Consequently, on the one hand important investments are necessary, while on the other, the typical know-how of the forwarding tradition must be combined with the new possibilities offered by information technologies.

At present, only a few operators of great logistical culture (not necessarily large companies) offer access to this market of high 'confidence' and added value. This type of culture can be found, in general, in localizations exercising the function of a 'gateway' (airports, harbours), without excluding the 'old' localizations at the frontier.

The study of the Swiss case has demonstrated how a few operators at the frontier (Basilea and Chiasso) have overcome the negative tendencies of the new scenario thanks to a valorization of the existing capital (production and market know-how), confidential relations with some clients and a full application of the new information technologies to the logistics field.

The overall results of the regional development remain to be seen. In the final section of our chapter we will start with a short summary of the trajectories of decomposition–recomposition of the economic structures linked to the transport in a border region. Then we will form some hypotheses about the possible roles of the different border regions in the European transport network and the remaining problems.

4.5 THE TRAJECTORIES OF DECOMPOSITION–RECOMPOSITION OF THE ECONOMIC STRUCTURES LINKED TO THE TRANSPORT INDUSTRY IN A BORDER REGION

In this section we want to find out the spatial consequences of the restructuring (redeployment) of the transport market for operators—forwarding agents and service performers, located for historical reasons at the border.

The question involves not only single operators but also the intermediate role of many regions, following different typologies related to specific empirical evidence. In any case, it would be really too easy to think that these operators and regions will simply be pushed out of the market. It rather depends on the organizational factors (Bramanti and Senn 1995) of the economic structure specific to the transport sector and of strategic instruments to be exploited with reference to broader spatial economic planning. The different socio-economic and spatial organizational pathways of the operators in the forwarding sector are shown in Figure 4.8.

The traditional model could be that of a single agent, which could be

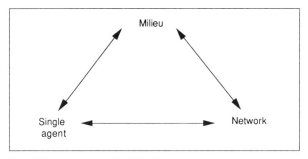

Figure 4.8. Triangle with different pathways of an economic organization of the operators in the forwarding sector

twofold: a large vertical integrated company or a small autonomous firm, acting in an isolated way in the market place. This dominant model had been replaced during the economic restructuring process by a network organization, as well as disintegrated at the level of large firms into many smaller entities (profit centres) with strongly centralized strategic functions, and integrated on the level of small enterprises into a multi-company local network (Capello and Gillespie 1993).

Finally, the third model is that of a 'territorialized milieu'. Under the notion of 'milieu' we understand—in complete correspondence with the Groupe de recherche européen sur les milieux innovateurs (GREMI)—a specific set of territorialized relations capable of giving a unity to the production system by the creation of a collective and dynamically open learning process (Camagni 1991).

The concept of the 'territorialized milieu' is very important for a sector like transport, and even more important considering the complex functions connected to international trade and borders.

In our realm, the existence of a certain milieu can be considered the rule rather than the exception. But the new scenario profoundly modifies these territorialized economic organizations as explained in Figure 4.9.

In the ongoing period of dissolution of some effects of the frontier, the regional impact in the logistics network will depend on the organizational dynamics and the operational strategies at the level of the functional space (production, market and support space) in the barrier regions. Every single case will be determined in a field of forces and tendencies from which different development strategies will ensue. In the following table we exemplify tendencies and current results obtained in the realm of the logistics of international transport between north and south Europe across the Alps.

In general, the localization at the frontier line used to constitute a 'milieu', the result of the external shock of the abolition of the barriers signifying its dissolution. The trajectory of a milieu depends on its innovative capacity,

	Organization Types of the Economic Tissue 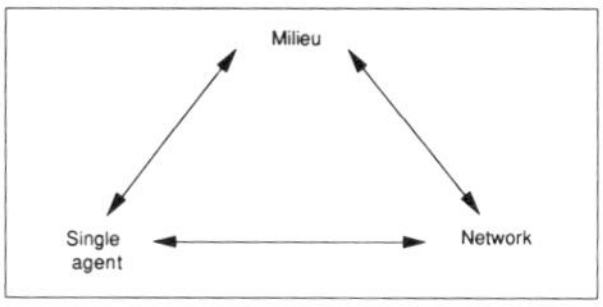	Trends
Nature of space	Milieu / Single agent — Network	
Production Space	Restructuring of the transport chain and new logistic services, determining some central localizations (hub and spoke) at the expense of the decay of traditional localizations.	
	New actors: Emerging of mixed enterprises (combined transport) following a philosophy of network.	
	The principal hubs and transit platforms require a concentration of investments.	
Market Space	A process of liberalization and deregulation and new modalities of the division of labour (just in time) increase the transport demand.	
	Tendency to a specialization of markets (tailor-made transport) but in a logic of integrating several logistic functions (product channel logistics).	
	The market of the regional distribution becomes trivial, the connection of the platforms into a network, whereas the true strategic functions of the logistic services require a construction in terms of a milieu (Rotterdam, Frankfurt, Lyon, Bale).	
Support Space	Dissolution of the support space linked to the customs functions. Necessity of new strategies in the operational field.	
	Crises in the traditional structures. Creating new alliances of the type of new networks or of the type of a milieu in the case of transit platforms (Bale). Increasing the share of facilities reserved to specific clients (dedicated contract transport and distribution).	
© RATTI, IRE, 1993	Process of retreating commitment by the public institutions, with the exception of cases of strategic spatial reorganization and negotiations (platforms).	

Figure 4.9. Trajectories of decomposition–recomposition after observed tendencies of the organization structures at the level of the three functional spaces. Reproduced with permission

in this case only seems possible at the highest level of the hierarchical of continental logistics (Rotterdam, Frankfurt, London) or at the level of large metropolitan distribution areas. Only in a few cases does the 'milieu-gateway' appear around the cargo airport platforms like Amsterdam, Frankfurt or Milan and in particular nodes at or near barriers like Verona, Lyon and Bale.

Otherwise, also due to the minor role of national and transport policy, the trajectory of the operators and the barrier regions is clearly marked by a bifurcation. On the one hand the new scenario in the product, market and support space confronts the operators with the challenge of a truly open market characterized by collapsing prices, the role of the service sub-contractors and the specialization in niche markets. On the other hand we find those operators (not necessarily medium large firms) who take a step in the direction of global logistics and who are successful in the domination of the whole logistic chain, which is becoming more and more complex, multi-modal and varied depending on the commodity.

In this second strategy in particular, the traditional know-how of the experienced operators still has a very important value, especially when combined with the new information technologies. This fact was confirmed by our empirical research on a part of the internationally operating Swiss forwarding agents (Bale and Chiasso), who transform the traditional know-how linked to intermediation in north–south transalpine transport into integrated logistic services.

4.6 CONCLUDING REMARKS FROM THE POINT OF VIEW OF THE EUROPEAN INFRASTRUCTURE AND LOGISTICS NETWORKS

- As pointed out in our theoretical definition of border functions and effects, borders can no longer be seen in the current scenario of western Europe as barriers to development and to communications but rather as windows of opportunity. This is particularly due to their historical characteristics in terms of nodes of exchange of information and knowledge.
- Today, information and know-how are key factors in the determination of new logistic chains. This information represents the main capital of a number of forwarding agents. In the phase of strategical behaviour of the forwarding agents the role of these factors has been clearly under-estimated.
- As illustrated in the example of our contribution, it is certainly worth analysing the transformation of the role of forwarding agents, in particular in the border region, which constitutes a privileged point of view in the following directions:

- functions with respect to the 'interoperability' of new system networks: for example, the role of the forwarding agents in the creation of a support space, the precondition for access to the market (bar-coding);
- functions with respect to the 'interconnectivity' of a new European network still dominated by the national heritage. In addition, the information and the technology of communication build a key element in the determination of new integrated logistic chains;
- functions with respect to the existing 'intermodality' of an infrastructural system. The forwarding agents play not only a role of intermediation in the transport market but truly constitute elements of regulation, able to correct the existing distortion (tariffs, rupture of shipment) and to increase the added value of the logistic–distributional functions.
- Last but not least, we have to remember that the study of the new behaviour of forwarding agents is important in order to be able to test the hypothesis about the trip distribution and the modal split. They also have to be considered as privileged actors in the realm of the strategies of environmental sustainability.

The fact that they cooperate and support the railways for transportation appears crucial not only because there exists a traditional and privileged contact between the forwarding agents in the border area with the railway companies but also because the integrated logistics, as we could observe in preliminary case studies, allow the railways to reclaim shares of the transport market.

In this sense the traditional forwarding agent changes from simple intermediator to an operator of a logistic chain, able to reduce the environmental impact of road transport.

REFERENCES

Bramanti, A. and L. Senn (1995) *The future in north-western Lombardy: understanding structural changes, milieu connections and governance structures*, Gremi IV, Ascona (to be published).

Bühler, C. (1993) *Die neuen Strategien in der Distribution und deren Einflüsse auf bestimmte Regionen*, University of Fribourg (unpublished).

Camagni, R. (1991) In GREMI (eds) *Innovation Networks*, Belhaven Press, London.

Capello, R. and A. Gillespie (1993) Transport, communication and spatial organisation: conceptual framework and future trends, in P. Nijkamp (ed.) *Europe on the Move*, Avebury, Aldershot.

Grabher, G. (ed.) (1993) *The Embedded Firm. On the Socioeconomics of Industrial Networks*, Routledge, London.

Janssen, B.J.P. (1993) Product channel logistics and logistic platforms, in P. Nijkamp (ed.) *Europe on the Move*, Avebury, Aldershot.

Kamann, F. (1993) Bottlenecks, barriers and networks of actors, in R. Ratti and S.

Reichman (eds) *Theory and Practice of Transborder Cooperation*, Helbing & Lichtenhahn, Basle.

Moran, P. (1965) *L'analyse spatiale en science économique*, éditions Cajus, Paris.

Nijkamp, P. and J. Vleugel (eds) (1994) *Missing Transport Networks in Europe*, Avebury, Aldershot.

OECD, (1992) *Advanced Logistics and Road Freight Transport*, OECD Publications Service, Paris.

Prognos (1984) *Güterverkehrsmarkt Europa*, Gemeinschaftsuntersuchung, Basel.

Ratti, R. (1991) Small and medium-size enterprises, local synergies and spatial cycles of innovation, in R. Camagni, *Innovation Networks*, Belhaven Press, London.

Ratti, R. (1993) How can existing barriers and border effects be overcome? A theoretical approach, in R. Cappellin and P. Batey (eds) *Regional Networks, Border Regions and European Integration*, Pion, London.

Ratti, R. and S. Alberton (1993) *Crises, mutations des espaces fonctionnels des entreprises et mutations des réseaux/milieux innovateurs*, Colloque Gremi IV, Ascona (first draft).

Ratti, R. and S. Reichman (eds) (1993) *Theory and Practice of Transborder Cooperation*, Helbing & Lichtenhahn, Basle.

Ruijgrok, C. (1990) Telematics in goods logistics process, in H. Soekkha (ed.) *Telematics—Transportation and Spatial Development*, VSP, Utrecht.

Ruijgrok, C. and S. Wandel (1993) Spatial and structural change in logistics, in P. Nijkamp (ed.) *Europe on the Move*, Avebury, Aldershot.

Storper, M. and R. Walker (1989) *The Capitalist Imperative. Territory, Technology and Industrial Growth*, Blackwell, Oxford.

Thaler, S. (1990) *Betriebswirtschaftliche Konsequenzen des EG-Binnenmarktes und der EG-Güterverkehrsliberalisierung für europäische Speditionsunternehmungen*, P. Haupt, Bern.

Travella, R. (1993) *Strategien der nationalen und internationalen Spedition im Hinblick auf die europäische Einigungsbestrebungen*, dargestellt am Beispiel von Chiasso, University of St Gall (unpublished).

United Nations (1993) *Harmonization of Frontier Controls of Goods* European Commission for Europe. Acts and Proceedings of the International Symposium on Harmonization of Frontier Controls of Goods, Ljubljana, October, 1991. Milan.

5 New Diffusion Mechanisms in Telecommunications Networks: Core and Periphery Responses in Europe

ROBERTA CAPELLO
Politecnico di Milan, Italy

PETER NIJKAMP
Free University of Amsterdam, The Netherlands

5.1 INTRODUCTION

In recent years much attention has been devoted to the diffusion and adoption processes of new information and communication technologies (ICTs). The interest in these technologies originates from the importance they have assumed in the process of defining competitive advantages among firms and comparative advantages among regions. It is in fact a common idea among economists and policy makers that the diffusion of these technologies is of strategic importance for the economic development of less favoured regions. For this reason in 1987 the EC launched a five-year programme, called the STAR Programme, for the implementation and diffusion of these technologies in Objective 1 regions of the Community.

The degree of success of these interventions is related to the diffusion process that these technologies follow in the first phases of their implementation. In this perspective, it is extremely interesting to define which mechanisms best sustain the diffusion of these technologies among users. The aim of the present paper is to offer empirical evidence of the diffusion (and adoption) processes of information and communication technologies. A new concept has recently been introduced by industrial economists as the main explanation of diffusion processes of interrelated technologies: the concept of *network externalities*. The basic idea at the centre of all studies dealing with network externalities is that the rate of growth in the demand for interrelated technologies is dependent on the number of subscribers or clients already using that specific technology.

The first author wishes to thank the Italian National Research Council (project no. 94.00560.CT11) for its financial support. Though the paper is the result of common work, R. Capello has written Sections 5.2, 5.3, 5.5, whilst the remaining Sections have been jointly written.

European Transport and Communications Networks: Policy Evolution and Change. Edited by David Banister, Roberta Capello and Peter Nijkamp. © 1995 John Wiley & Sons Ltd.

From this simple, yet fundamental, observation a series of theoretical and empirical analyses have followed, trying to conceptualize and empirically prove the significance of network externalities. Up to now, the concept has been widely studied and applied, especially with a view to measuring the impact that it has on the utility function of each subscriber or client. In essence, the basic concept stems from the strong interdependence which exists in the utility function of each potential subscriber to join the network, making the decision highly dependent on the behaviour of others.

The aim of the present chapter is twofold. Firstly, it provides a review of the existing literature on network externalities. Many studies have been developed on this concept, and some confusion exists about this term. Secondly, we present a behavioural analysis of the importance of network externalities on diffusion processes in the case of two different economic environments in which these technologies are developed. The STAR Programme provided an interesting area of analysis for less developed regions. However, for our aim of measuring network externality effects on the willingness to adopt in different economic environments, an analysis developed in a rather contrasting economic milieu was necessary. As a developed region we chose the *north of Italy*, representing a rather contrasting case with respect to the south. In fact, the diffusion process of new technologies is more advanced; firms are stimulated by a more developed economic environment, and, last but not least, these technologies, although public technologies, are commercial technologies. The contrasting characteristics give, undoubtedly, much emphasis to a comparative study, since they allow the measurement of network externalities as mechanisms for technological adoption in different stages of the diffusion process.

Some interesting results stem from the analysis. What is especially interesting is the high level of similar replies obtained within the two macro-areas, mirroring a common behaviour by firms located in the same economic environment, as we will see below.

5.2   A TAXONOMY OF NETWORK EXTERNALITIES

The telecommunications industry seems to offer the most appropriate context for studying network externalities and all the economic consequences they provoke. This is because the telecommunications sector is in fact an industry where the concept of externalities, and in particular of network externalities, appears under different guises, which influence both the efficiency of the entire telecommunications system, and, moreover, its dynamics.

The concept of network externality is clearly explained when applied to the telephone network. However, a rigorous analysis needs to go far beyond this basic definition, in order to define precisely what is meant by network externalities. Far too broad a definition is given nowadays of this concept, and therefore the need is felt to organize the existing literature dealing with network externalities in a systematic way. Moreover, as we will see below, the typology we create on network externalities shows how this concept is sometimes in reality similar to other more traditional economic concepts. To overcome confusion, a typology of network externalities and their interpretation in the literature is presented in this section.

In the literature on network externality, it is shown how network externalities apply not only to the explanation of demand dynamics, but also to the interpretation of supply mechanisms, driven by interdependent mechanisms. The telecommunications system is in fact characterized by some strategic features, namely:

1. *Interdependence of consumers' utility*, since the decision of a person to join the network is dependent on the behaviour of other clients; when specifically dealing with the adoption of a new technology, *interdependence between potential adopters and existing users* exists, since through dynamic learning processes the latter may create for the former a reduction in search costs and market prices for complementary inputs, maintenance and skills which stems from their greater experience in using the technology already adopted.

2. *Interdependence between potential users and suppliers.* On the one hand, the know-how and the experience accumulated by suppliers act as a driving force in the adoption process. In fact, the adopting firms are assisted in the search for know-how and complementary inputs (i.e. organizational strategies) because of precise 'guidelines' provided by the suppliers. On the other hand, the higher the number of adopters, the broader the know-how of the supply will be. In other words, the relationship between supply and demand generates cross-learning processes via the bridging interaction between demand needs and supply knowledge.

3. *Interdependence between producers of complementary technical components and products in the telecommunication 'filière'.* The interrelation of submarkets may provoke externalities, since the profit function of a producer is influenced by the economic transactions of other producers whose behaviour affects the market prices of intermediary inputs.

4. *Interdependence between users' productivity*, since the advantages obtained by a firm in terms of its productivity are dependent on the number of already networked firms. In fact, the advantages obtained through the use and exploitation of these technologies are a function of the number of firms already using them.

While the first two of these strategic features (1,2) affect the *utility function of a final individual user*, the last two (3,4) act on the *productivity of firms*, the telecommunication service acting as an input factor in the production function. Moreover, these features are related to both telecommunica-

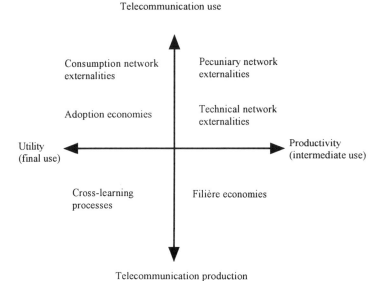

Figure 5.1. A typology of network externalities (*Source*: Capello 1994)

tion manufacturing firms and service providers (the *telecommunication production sphere*) and also to the adopting firms using these technologies as final or intermediate products (the *telecommunication use sphere*).

Figure 5.1 presents a typology of network externalities on the basis of the above-mentioned telecommunication market features. In the top-left quadrant of Figure 5.1, network externalities are related to the telecommunication adopters and are the typical *consumption network externalities* acting on the utility function of an individual final user (i.e. the economic features explained in point (1). Here the interdependence among utility functions of users of telecommunication networks is at the basis of the traditional network externality concept presented above. Telecommunication demand is more and more explained through interrelated decision-making processes of adopters, a situation which in turn influences the growth rate of telecommunication demand.

A well-known example of consumption network externalities is the so-called hardware–software paradigm (Katz and Shapiro 1985, 1986; Stoneman 1990), regarding the strong interdependent preferences dominating the choice of a consumer when buying a certain kind of hardware. In the words of Katz and Shapiro (1985, p. 424):

> ... an agent purchasing a personal computer will be concerned with the number of other agents purchasing similar hardware, because the amount and

variety of software which will be supplied for use with a given computer is likely to be an increasing function of the number of hardware units that have been sold.

Also, concerning this example the benefit that a consumer derives from the use of a good is an increasing function of the number and behaviour of other consumers, in this case the fact that they buy compatible items.

On the users' side, another kind of network externality is present, known in the literature as *adoption economies* (Antonelli 1992), when dealing with the adoption of new technologies (i.e. the economic features explained in point (1) above). In the diffusion processes of new telecommunication technologies a crucial role is also played by collective learning processes, as is common within all types of complex technologies. These processes seem to hide a sort of network externality mechanism, because of non-paid-for advantages that potential users of the technology gain from the experience of long-established adopters. For potential adopters, non-paid-for advantages may emerge from lower search costs of complementary inputs, or from specific know-how on how to use and maintain the technology, stemming from the consolidated experience on the use of these technologies accumulated by previous adopters.

However, these features, recently interpreted as an externality mechanism (Antonelli 1992; David 1992), may in reality be explained only in terms of the traditional concept of dynamic learning processes, which are similar in their effects, but different in nature from the traditional concept of network externalities. Learning processes stem in fact from the concept of dynamic economies of scale (Spence 1981), while network externalities stem from the non-paid-for benefits obtained by interdependent mechanisms. The difference between the two concepts may be more easily explained by recalling the traditional features of externalities mechanisms, i.e. *interdependence* and *non-compensation*. In the case of learning processes the interdependence among users *is* present, and explains part of the diffusion mechanism. The second feature, i.e. non-compensation, is less evident and is what distinguishes learning processes and adoption economies from network externalities. In fact, one may easily argue that even if late adopters may gain from lower search costs for specific know-how on the use of these technologies stemming from consolidated experience of previous adopters, it might well be that:

- these advantages are paid for by late adopters in terms of loss of productivity during the period of non-adoption;
- these advantages may actually be the result of a clear strategy of the first adopter, who could foresee in this behaviour a source of profit, thus eliminating the unintended feature of an externality mechanism, i.e. non-compensation; and, moreover,

- while with network externalities the non-compensation is valid for both
  the late and the previous adopters, in the case of learning processes the
  advantages are only in favour of the latter.

The same can be said for the case of telecommunications product firms
acting on the utility function of telecommunications users through *cross-
learning processes* (bottom-left quadrant in Figure 5.1), linked to the interdependence between potential users and suppliers, described in point (2)
above. Again, users benefit from these learning processes, through dynamic
economies of scale, which are different in nature from the concept of
network externalities, for the same reasons explained before.

Network externalities in the telecommunications sector do not only
affect the final user, by impacting on his utility function. In the telecommunications industry the intermediate user (or supplier) also acts
under certain particular conditions (bottom-right quadrant). As far as the
telecommunication technologies production is concerned, telecommunication networks are built upon an array of interrelated technical components such as terminals, transmission facilities and switching equipment,
as well as intermediate outputs in the extremely complex telecommunication 'filière'. The interdependence (expressed in point (3) above) exists
both in vertical relationships (intermediate inputs for telecommunications
outputs) as well as in horizontal final products markets (advanced terminals whose development stimulates value added services such as minitel
and electronic mail). In both horizontal and vertical interrelationships the
behaviour of each economic agent on the market (reduction of prices,
new market niches) positively affects the profits of the other interconnected producers, generating what can be interpreted as network
externalities. However, these kinds of advantages are typical *filière economies*, stemming from vertical integration in a sector. In other words, these
advantages may be associated with traditional 'economies of scale' generated in a vertically or horizontally strong market relationship (bottom-
right quadrant in Figure 5.1). Another extremely appropriate example of
these kinds of 'filière economies' is presented by the hardware/software
industry. Computers (hardware) and programs (software) have to be used
together, and the greater the sales of the hardware are, the higher the
profits for software producers will be, via the technical interconnectivity
of the two markets.

Finally, an interesting situation concerns interdependence among the productivity of different intermediate users (see point (4) above). In this case, it
is possible to speak of network externalities, this time related to the use of
the service as an input factor for other products, thus having an impact on
the productivity level of firms (top-right quadrant in Figure 5.1). In this
framework, both the concept of *pecuniary (network) externalities* (Scitovsky
1954) and *technical (network) externalities* (Meade 1954) may be useful.

Pecuniary externalities arise whenever the profits of one producer are affected by the actions of other producers. In other words, pecuniary externalities act on input factors decreasing their costs and thus having positive effects on the output. This category differs from the 'technical external economies', defined by Meade (1954) as those advantages obtained by a firm for its output through the non-paid-for exploitation of the output and input factors belonging to other firms. The latter category sees external economies as a peculiarity of the production function, i.e. these external economies act on input factors' productivity. Through the increase in the input productivity these external economies positively influence the corporate output.

For telecommunications network users, the use of the network generates an increase in input productivity (or profit advantages), only partially covered by the costs of joining the network. The non-paid-for advantages obtained by a subscriber joining a network have positive effects on the economic performance of the new subscriber. This holds true also for the already existing subscribers, who obtain non-paid-for advantages on their production functions if an additional member uses the network. *If network externalities represent one of the (economic) reasons for entering the network, a better economic performance of firms is the (economic) effect they produce on the productivity side* (Capello 1994; Capello and Nijkamp 1995).

From the above observations we conclude that in the telecommunications sector the classical concept of network externality is related only to telecommunications (final or intermediate) users (in Figure 5.1 only the upper half). In recent years, the definition given to network externalities has expanded to embrace network externalities in the production sphere (manufacturing firms and service providers), thus broadening the precise meaning to cover yet more traditional economic concepts.

Whilst consumption network externalities in the context of the use of telecommunications (top-left quadrant in Figure 5.1), as well as adoption economies (learning processes) (bottom-left quadrant) and 'filière' economies (bottom-right quadrant) have been widely identified and analysed in the literature, as yet no work has been done on the measurement of the effects of network externalities on the productivity side. The advantages of users joining a network are reflected in the performance of these subscribers via the reduction of input factor costs or the increase in their productivity. These kinds of network externalities and the effects they generate have not yet been investigated. This is the area the authors are working on. This chapter, on the other hand, focuses attention on the top-left quadrant of Figure 5.1, i.e. on consumption network externalities; in particular, a behavioural analysis has been developed, at both a descriptive and an interpretative level, in order to identify the role network externalities play on the decision to adopt.

5.3   TELECOMMUNICATIONS AND REGIONAL DEVELOPMENT: THE STAR PROGRAMME

An increasing awareness of the importance of telecommunications infrastructures and services for encouraging economic development characterized the 1980s, justified by an increasing body of economic literature underlining the importance of telecommunication technologies for economic growth at both a theoretical and empirical level. This increasing awareness has also stimulated normative interventions in that direction, by interpreting infrastructure in general, and telecommunications technologies in particular, as the driving forces for economic growth.

The Special Telecommunication Action for Regional Development (STAR) Programme is a European Community Programme launched in 1987, reflecting the general positive attitude towards telecommunication infrastructures as the driving forces for regional development. In fact, born under these auspices, the STAR Programme has been developed with the aim of encouraging economic development in the less favoured regions of the Community by means of easier and quicker access to advanced telecommunication technologies. The programme was launched in all 'Objective 1' regions of the Community, namely parts of Italy, of Spain, of Great Britain, of Ireland, of France and the whole of Greece and Portugal.

The backward situation in the telecommunications infrastructure and services adoption in which less-favoured regions were lying was the second reason stimulating a direct intervention of the Community. In 1987, when the STAR Programme was launched, the situation of the telecommunication infrastructures in those regions was rather poor. The telephone network density was extremely low in comparison with core regions, and the situation was even worse for advanced telecommunication networks and services.

The intervention of the Community aimed at reducing the disparity in terms of network accessibility between the advanced and less-favoured regions, by installing basic infrastructures (fibre optic networks, satellite networks, ISDN, advanced networks) for advanced telecommunication services and by encouraging supply and demand for such services. Thus, the STAR Programme is a typical supply driven innovation process, with the aim:

1. *Developing a technological supply* in areas where the market rules would have never stimulated the public operator to contribute sufficient financial effort. In fact, the lack of an explicit demand for advanced telecommunication services in these regions does not justify the financial effort required to produce these advanced technologies. This phase has been developed with the full involvement of the public operators for different reasons. First of all, the institutional monopolistic position of most of the European public operators (apart from the British case) would

have otherwise prevented a full implementation of the Programme. Secondly, the 'additionality' principle governing the European Structural Funds would have not otherwise been fulfilled. The Community is in fact expecting, as a principle for its local policies and interventions, the same financial effort from the National bodies in areas where otherwise this would never take place.

2. *Promoting advanced and innovative telecommunication services*, with the help of pilot services in specific micro-areas.
3. *Supporting and stimulating a demand for these services*, in order to create a real market at the end of the five-year programme. For this reason demonstration and promotion centres have been developed in order to facilitate the linkage between technological potentialities of the supply and real needs of the demand. In other words, these centres play a role of intermediaries in the market, with the aim of stimulating a sustainable innovation process. Whilst the promotion of advanced services is based on a sample of firms, this phase is directed to all potential business users.

In order to achieve these aims, the Programme has been organized into two large projects aimed at:

1. developing advanced telecommunication networks (fibre-optic networks, integrated service digital network (ISDN), digital telephone networks) for the development of advanced services (the so-called 4.1 measures);
2. developing promotion and demonstration centres (the so-called 4.2 measures).

The structure of the Programme is presented in Figure 5.2. The Programme was run by DG XVI (regional policy) with a small peripheral role played by DG XIII (technology). DG XVI spent 50 per cent of funds for both the 4.1 measures (infrastructure provision) and 4.2 measures (service promotion). A National STAR Steering Committee has been created for monitoring both 4.1 and 4.2 actions. In the case of service provision, the National Committee was in charge of deciding the kinds of projects to develop under 4.2 measures, which had to be additionally financed with national public funds (for the other 50 per cent). In the case of 4.1 measures, the national public operator was of course playing a strategic role, being obliged to finance the other 50 per cent of the 4.1 measures. The large role played by national bodies in all countries has led to a different configuration of the STAR Programme itself in the different countries. A simple example of this difference is provided by the Italian case in contrast with the Spanish one. The logic underpinning the STAR interventions in the two countries is rather a contrasting one. In the case of Italy, national bodies decided on a 'top-down' intervention, where the decisions of which network or service were made at a national level, and then applied in the different areas. In the case of Spain,

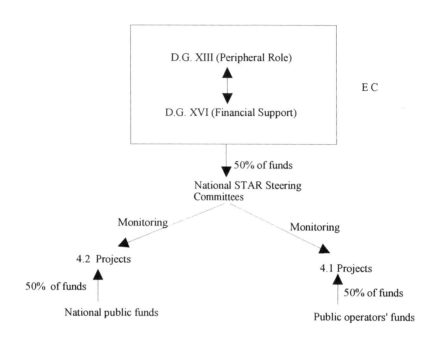

Figure 5.2. Structure of the STAR Programme (*Source*: Capello 1994)

the policy chosen for the promotion of the STAR technologies has been seen to follow the opposite approach, i.e. a 'bottom-up' approach, where the decisions about which networks and services had to be offered were made at the local level by the local responsible bodies. The result in terms of networks and services offered in Italy and Spain has been rather different. In Italy, the top-down approach has brought the development of similar new networks and services in all regions of the Mezzogiorno, with no particular interest for local needs or local requirements expressed by local users, but with the possibility of connecting all southern Italy with well-advanced and modern infrastructures. In Spain, on the contrary, the 'bottom-up' policy has led to the creation of networks and services customized on local requirements and needs, but with no possibility for long-distance communications.

The allocation of resources to different projects reflects the different adoption philosophies and policies, thus following different supply 'philosophies' in the different countries. In this chapter, on the contrary, we concentrate our analysis at the demand side, and present a behavioural analysis of firms in their decision to adopt and use these new technologies. The analysis is run in both the north and the south of Italy, in order to capture regional variations in this analysis. Moreover, the same exercise is run for the service and the industry sector, in order to measure industrial variations in the replies.

5.4 NETWORK EXTERNALITIES AS THE MAIN REASONS FOR ADOPTION: DESCRIPTIVE RESULTS

This section presents descriptive results, which are important for the definition of the empirical validity of the 'economic and spatial symbiosis' framework. In particular, this incorporates a *behavioural analysis* pinpointing the most important decisions to adopt telecommunication technologies. In this way, we are able to test whether our research proposition interpreting network externalities as one of the major reasons for joining the network is empirically valid.

The present section provides contingency table analyses of questions related to the adoption decision process of new networks and services; in this connection the above-mentioned behavioural analysis of corporate users forms the major focus of this work. The contingency tables show the percentage of firms replying positively to each specific choice, dividing the sample between the south and the north. Correlation analyses based on chi-square (χ^2) test statistics and *P*-values are run between each reply and the regional dimension. In this way we are able to show which replies are correlated with the regional dimension. The same analysis is run at the industry level. In Section 5.4.1 we present the descriptive results for the south and for the north. Section 5.4.2 presents the results at the industry level.

5.4.1 REGIONAL VARIATIONS

The results on the level of adoption show higher adoption rates in the north than in the south, especially in the case of infrastructure. Among infrastructures, packet-switching networks and LANs prevail, while in the services data banks and telefax are mostly adopted.

Concerning the results of our *sample in the south*, adoptions are overwhelmingly concentrated in 1991 and involve services much more frequently than infrastructures (Table 5.1). Among infrastructures, packet-switching networks and local area networks (LANs) are the only ones with a significant number of adoptions. This result may be explained by the fact that packet-switching networks can only be used for data transmission because of their technical characteristics, and data transmission is the second reason for communication among firms, after voice. Among services data banks prevail, followed by electronic mail in both service and manufacturing sectors. This means that both interactive services (such as E-mail) and non-interactive services (such as data banks) have been well received. Ranking the importance between these two most popular services, data banks are ranked more highly than E-mail. This is mirrored by actual use: despite highly individual patterns, on average data banks are used on a weekly basis, but E-mail occasionally.

The results from the sample in the *north of Italy* are quite different, as expected. Adoptions are much more distributed over the last 5 years. Among infrastructures, LANs prevail, followed by packet-switching networks (Table 5.1). Their higher penetration rate can easily be attributed to two main reasons: (a) these are typically data transmission networks, and transmission is a necessary aspect of these technologies; (b) moreover, the technical elements associated with these networks were introduced in the market some years before the technical elements supporting other kinds of networks, i.e. ISDN or Broadband Networks. This fact has two consequences: (i) the technology itself is more reliable; (ii) the adoption process has had more time to take place.

Among services, data banks also prevail in the north (54.3 per cent), followed by the videotex service (34.3 per cent) (offering among other things a database system), and by E-mail (28.6 per cent). Video conference is not

Table 5.1. Adopted telecommunication networks and services by macro-areas

	South	North	Entire sample	χ^{2}*	P-value[†]
Networks					
Optical Fibre Network	2.9	5.7	4.3	0.35	0.55
ISDN	2.9	5.7	4.3	0.35	0.55
Packet-Switching Network	11.4	31.5	21.4	**0.15**	**0.04**
LAN	11.4	40.0	25.7	**7.48**	**0.00**
Broadband Network	8.6	5.7	7.1	0.21	0.64
Services					
Videotex	22.8	34.3	28.6	1.1	0.3
Videoconference	2.8	2.8	2.8	0.0	1.0
Electronic Mail	51.4	28.6	40.0	**3.8**	**0.05**
Electronic Data Interchange	17.1	17.1	17.1	0.0	1.0
Databanks	71.4	54.3	22.8	2.2	0.1
Telefax	37.2	100.0	68.6	**32.1**	**0.0**

These values represent the percentage of positive replies obtained for each question in the two areas.
* The χ^2 value represents the degree of dependency between the percentage of positive replies and the regional variables.
The results indicating a significant dependency are reported in bold characters.
[†] The P-value represents the probability to accept the null hypothesis when this is true. The P-value is acceptable up to a value of 0.05 (Loehlin 1987; SAS/STAT 1989).

yet diffused, and, to a lesser extent, this is also true for the EDI service (17.2 per cent), which has a lower percentage of adopters than expected (Table 5.1).

Interesting results stem from the *main reasons for adoption*, especially if a comparison between the two macro-areas is carried out. The most important reasons for adoption in our sample are the 'importance of the technologies for the business' and 'low costs of implementation and use'. However, once the analysis is carried out at a regional level, variations in the replies emerge.

In the *south of Italy* interesting results originate from the analysis, since they support our first research proposition, that the number of already existing subscribers is the main reason for adoption. Reasons put forward for adoption stress very strongly the 'existence of promotion and demonstration centres', but almost as highly are the reasons concerning the 'importance of the technology for the corporate business' and 'low usage and implementation costs' (Table 5.2). These replies show the difficulties of adoption processes, since:

1. Help from the supply side in demonstrating and promoting innovation is of crucial importance. Greater possible connectivity deriving from greater availability is not a sufficient element to give rise to adoption processes.

Table 5.2. Main reasons for adoption by macro-areas

	South	North	Entire sample	χ^2	P-value
High percentage of already networked subscribers in your region	5.7	20.0	12.8	3.18	0.074
High percentage of already networked subscribers in other regions	11.4	17.1	14.3	0.467	0.493
Importance of the network or service for your business	37.1	65.7	51.4	**0.719**	**0.017**
Your suppliers were networked or were using the same service	5.7	11.4	8.6	0.729	0.393
Other firms in the same sector were connected	8.6	14.3	11.4	0.565	0.452
No other advanced communication network or service available	11.4	5.7	8.6	0.729	0.393
Low costs of implementation	42.9	22.9	32.8	3.17	0.075
Low costs of use	45.7	17.1	31.4	**3.17**	**0.01**
Increasing awareness of these technologies through demonstration centres	40.0	5.7	22.8	**11.66**	**0.001**
Image effect	20.0	14.3	17.1	0.402	0.526
Others	31.4	14.3	22.8	2.92	0.088

See notes for Table 5.1

2. Moreover, another main reason for adoption in the first phases of a diffusion process is the 'importance of innovation for the business'. In other words, if a linkage is shown between the technology and the business areas, there is a higher rate of success for an innovation.
3. Also stimuli on the financial side become strategic, once the risks of failure of the adoption have to be borne by the adopters. Risks are of course higher during the first phases of adoption.

Another remarkable result is the total lack of any references to the 'number of connected users (in the same or other regions), suppliers, buyers and competitors'. This result stresses once more the empirical plausibility of our testable hypotheses. In the first phases of the adoption process, such as in the case of the STAR Programme in the decision to adopt, other reasons prevail rather than the 'number of already existing subscribers'. In this area, in fact, there is still too small a number of subscribers, which cannot act as an attracting factor for potential subscribers.

The major reason put forward for adoption by firms located in the *north of Italy* is the 'importance of these technologies for business' (more than 65.7 per cent of replies), followed by 'low costs of implementation' (22.9 per cent) and 'use costs' (17.1 per cent) and by the 'high percentage of already existing subscribers in the region and in other regions' (37 per cent) (Table 5.2). These replies confirm our expectations, since in the case of an area in an advanced stage of diffusion:

1. The number of already existing subscribers is a crucial motive for previous adoption, once these technologies are of any interest to the business and their costs are reasonable.
2. In comparison with the south, where the percentage of responses declaring this motive is around 17 per cent, the number of subscribers plays a more crucial role in the decision process to adopt in the north. In fact, given the more advanced stage of the diffusion process in the north, one can explain this result by reiterating the fact that the higher the number of subscribers (after the critical mass has been achieved), the more interesting the adoption and thus the 'bandwagon effect' occurs.
3. In advanced stages of diffusion, the help from the supply side, through the establishment of demonstration and promotion centres, does not play any role in the diffusion process (only 5.7 per cent of replies).
4. In any stage of diffusion, price incentives (i.e. free of charge technologies in the south) play a significant role as mechanisms which help the adoption process.

The importance of the number of adopters in the first phases of diffusion is also confirmed by another result obtained, i.e. the *reasons put forward by firms for non-adoption*. The results from the entire sample show that the most important reasons put forward by firms for non-adoption are the 'importance of the network or service for corporate business' and 'low percentage of connected subscribers, suppliers and customers'.

Regional variations are once more evident. In the south two such reasons prevail in particular: the 'low percentage of connected local users' and the 'irrelevance for corporate business'. Other significant adoption bottlenecks include, in decreasing order of importance, the 'low percentage of non-local connected users and of connected suppliers and customers', the 'use and adoption costs (including time costs before efficient connection is achieved)', and the 'inability to understand and/or evaluate these technologies' (Table 5.3).

The 'bandwagon effect' in the advanced stages of diffusion processes is confirmed by the results of the north. In the case of the 'reasons for non-adoption' the difference between the north and the south is even greater than the case of the 'reasons for adoption'. While in the south the main reason for non-adoption was the 'low percentage of regional subscribers' (71.4 per cent), in the north the main reason for non-adoption has been identified as the 'non-importance of these technologies for business' (48.6 per cent), followed by the 'high costs of use' (37.1 per cent) (see Table 5.3). The reasons for non-adoption in the north have very little to do with very low levels of subscribers; they are linked rather to business interests and financial constraints.

Table 5.4 presents the *main conditions for future adoptions* for both networks and services. Even from these results it is clear that a 'higher number of subscribers' does not play an important role in stimulating adoption *in the north* (11.4 per cent in general; 8.6 per cent in the case of customers and suppliers), where the interest in future conditions for

Table 5.3. Main reasons for NON-adoption by macro-areas

	South	North	Entire sample	χ^2	P-value
Low percentage of connected subscribers	71.4	14.3	42.8	**23.33**	**0**
Low percentage of connected subscribers in other regions	45.7	8.6	27.1	**12.21**	**0**
None of your suppliers or customers use them	57.1	31.4	44.3	**4.69**	**0.03**
You do not think they may be useful for your business	51.4	48.6	50.0	0.057	0.81
You do not see their importance in your business	28.6	14.3	21.4	2.12	0.14
High costs of use	22.9	37.1	30.0	1.70	0.19
Other	28.6	14.3	21.4	2.12	0.14

See notes for Table 5.1

Table 5.4. Main conditions for future adoption by macro-areas

	South	North	Entire sample	χ^2	P-value
Networks					
A higher number of people connected	28.6	11.4	20.0	3.14	0.07
A higher number of suppliers and customers connected	25.7	8.6	17.1	**3.62**	**0.06**
A better geographical distribution of the network	11.4	5.7	8.6	0.73	0.39
A reduction in the price of use	11.4	11.4	11.4	0.0	1.0
A reduction in the price of access	8.6	8.6	8.6	0.0	1.0
More recent availability of the network	8.6	11.4	10.0	0.16	0.69
Technical progress in the network	25.7	20.0	22.8	0.32	0.57
Increase in the efficiency of the network	20.0	14.3	17.1	0.40	0.53
Good results from the previous adoption	14.2	5.7	10.0	1.43	0.23
Services					
A higher number of people connected	40.0	14.3	27.1	**5.85**	**0.02**
A higher number of suppliers and customers connected	31.4	20.0	25.7	1.19	0.27
A better geographical distribution of the service	17.1	2.9	10.0	**3.97**	**0.05**
A reduction in the price of use	5.7	20.0	12.9	3.2	0.07
A reduction in the price of access	11.4	17.1	14.3	0.47	0.49
More recent availability of the service	11.4	5.7	8.6	0.73	0.39
Technical progress in the service	20.0	17.1	18.6	0.09	0.76
Increase in the efficiency of the service	25.7	17.1	21.4	0.76	0.38
Good results from the previous adoption	20.0	2.9	11.4	**5.08**	**0.02**

See notes for Table 5.1

adoption is more linked to the 'technical progress of networks' (20 per cent), to an 'increase in their efficiency' (14.3 per cent) and a 'reduction in the price of use for services' (11.4 per cent). On the contrary, in the south the main condition for future adoption is related to a 'higher number of subscribers', for networks (28.6 per cent) and services (40 per cent) (see Table 5.4 above). For services, the results are different than in the case of networks. While the 'reduction in the price of use' still remains rather crucial in the north (20 per cent), this has the same weight as a 'higher number of suppliers and customers connected' (20 per cent). 'Technical progress' and 'increase in the efficiency of the service' still both play a crucial role (17.1 per cent). *In the south*, instead a 'higher number of people connected' is stated to be the most important reason, especially in the case of advanced interactive services, such as videotex and E-mail. For data banks too, a 'higher number of people connected' is the main expected condition for future adoption, explained by the fact that a higher number of people connected would assure a higher revenue to the service managers, and thus stimulate a better quality of service (i.e. a larger variety of available information). For E-mail and electronic data interchange (EDI) services the most important reasons stimulating future adoption are also 'technical progress' and 'increase in network efficiency'.

The most interesting result is the strong regional variation in the *reasons for the dissatisfaction with the quality of existing technologies*. As expected, the 'degree of satisfaction with the technologies' among subscribers in the north is higher than in the south (68.6 per cent against 57.1 per cent) (Table 5.5). Moreover, the main reason put forward for a high degree of dissatisfaction is the 'too low number of subscribers' (46 per cent) and the 'too high costs of implementation' (30 per cent).

It is interesting to see that these two reasons are also very much dependent on the regional dimension. While in the north the 'high costs of their implementation' are the most important reasons (54.5 per cent) this does not apply to the south (13.3 per cent) (see Table 5.6). Dissatisfaction does not stem from the 'low number of subscribers' (only 9.1 per cent of positive replies for this reason), as is the case in the south (73.3 per cent).

The sample in the north shows a general greater interest in these technol-

Table 5.5. Degree of satisfaction with the quality of communication by macro-areas

	South	North	Entire sample	χ^2	P-value
Yes	57.1	68.6	62.8	0.98	0.322
No	42.9	31.4	37.1		

Table 5.6. Major persistent user problems, in the case of dissatisfaction by macro-areas

	South	North	Entire sample	χ^2	P-value
Too high costs of implementation	13.3	54.5	30.7	**5.06**	**0.024**
They are no longer free charge services	20.0	0.0	11.5	2.48	0.11
The use of these technologies requires profound organizational changes	0.0	9.1	3.8	1.42	0.23
The number of subscribers using these services and networks is still too low	73.3	9.1	46.1	**10.53**	**0.001**
Suppliers are not using these technologies	33.3	9.1	23.1	2.10	0.15
Competitors are not using these technologies	0.0	26.7	15.4	3.47	0.063
Other	54.5	20.0	34.6	3.35	0.067

See Notes for Table 5.1

ogies, witnessed by a high percentage of replies recording the 'increased intensity of business relationships after the adoption of telecommunication technologies' (62.9 per cent of replies in the north against 20 per cent in the south) (see Table 5.7). For the 37 per cent of 'decreased business relationships after the introduction of these technologies', in the north the reasons justifying this reply are not linked to network externality effects. As Table 5.8 shows, none of the choices embodying a network externality effect (such as 'no suppliers or customers using these technologies') are put forward as the main reasons for constant business relationships. Again, these results witness the fact that at low penetration levels, such as in the south, the 'low number of subscribers' becomes a constraint for the increase in the intensity of use of these technologies. In the case of more advanced adoption processes, the reason explaining the 'low level of intensity of use is the non-importance for business', is represented by 14 per cent of replies.

Table 5.7. Intensity of relationships after telecommunication technology adoption by macro-areas

	South	North	Entire sample	χ^2	P-value
Increased intensity of business relationships	20.0	62.9	41.4	**13.25**	**0.0**
Constant intensity of business relationships	0	0	0		
Decreased intensity of business relationships	80.0	37.1	58.6		

Table 5.8. Main reasons in case of decreased intensity by macro-areas

	South	North	Entire sample	χ^2	P-value
None	20.0	11.4	15.7	0.97	0.32
Costs of initial investment	2.8	0.0	1.4	1.01	0.32
Lack of reliability on the technologies used	2.8	0.0	1.4	1.01	0.34
Level of telecommunication charges	0.0	0.0	0.0		
Lack of staff skills in their use	2.8	14.3	8.6	2.92	0.08
Services are not relevant to business	14.3	17.1	15.7	0.11	0.74
Your customers do not use them	0.0	31.4	15.7	**13.05**	**0.0**
Your suppliers do not use them	2.8	25.7	14.3	**7.45**	**0.006**
Too few subscribers in general	40.0	0.0	20.0	**17.5**	**0.0**
Restricted geographical diffusion of networks and services	8.6	2.8	5.71	1.06	0.30
Other reasons	8.6	5.7	7.14	0.21	0.64

See notes for Table 5.1

The results presented are quite satisfactory, since at least at a descriptive level, our expectations are fulfilled and our testable hypotheses verified with *strong regional variations*, as expected. The importance of the regional dimension in our replies is witnessed by the statistically significant level of the χ^2 and of the *P*-value between the 'network externality' variables and the regional dimension. It is surprising that a significant correlation exists for the regional dimension only for the network externality variables, as is represented by the contingency tables, where the statistically significant dependency is indicated by bold print (Tables 5.2 to 5.8).

These results reinforce our expectations based on the general beliefs in the theoretical literature on consumption network externalities that in the first phases of adoption of new networks and services (such as in the south), the main reason for non-adoption is the 'low percentage of subscribers already linked to the network'. It makes sense, therefore, to find that the main reasons for adoption in the very first phases of the adoption process is *not* the number of already existing subscribers. Nevertheless, one of the major problems still existing in the use of new telecommunication technologies is the 'low number of subscribers'. It should be noted, however, that all these results do not hold in the case of northern Italy, where more advanced diffusion processes lead to opposite results, as expected.

5.4.2 INDUSTRIAL VARIATIONS

A completely different framework emerges when a descriptive analysis at the industry level is presented. A first consideration needs to be made here. The strong regional variations which were revealed by the regional analysis (presented in the previous section) are not at all so evident in the case of the industry analysis. This is witnessed by the very limited number of statistically dependent cases in the sectoral distribution of replies, as we will see below. Thus, the differences between industries do not seem to play a significant role in the explanation of the reasons for adoption.

The sample, divided between the industrial sector (both traditional and advanced sectors) and service sector, shows only some small discrepancies in the replies concerning the *different reasons for adoption*, which can all be explained by the varying nature of the business structure of these economic activities. Among networks, packet-switching networks and LANs remain the most adopted infrastructures for data transmission, while ISDN, broadband networks and fibre optic networks show a very limited number of subscribers, due to their recent introduction in the market. Regarding services, data banks and telefax mainly prevail, followed by E-mail and then by the videotex. The very low video conference adoption is entirely concentrated in the industry sector, while this technology is non-existent in the case of the service sector (Table 5.9).

The *main reasons for adoption* show industry differences which can be

Table 5.9. Adopted telecommunication networks and services by industry and service industry

	Industry	Service industry	Entire sample	χ^2	P-value
Networks					
Fibre optic network	2.7	6.1	4.3	0.48	0.49
ISDN	2.7	6.1	4.3	0.48	0.49
Packet-Switching Network	18.9	24.2	21.4	0.29	0.59
LAN	24.3	27.3	25.7	0.08	0.79
Broadband Network	5.4	9.1	7.1	0.36	0.55
Services					
Videotex	24.3	33.3	28.6	0.69	0.40
Videoconference	5.4	0.0	2.8	1.84	0.17
Electronic Mail	48.6	30.3	40.0	2.45	0.12
Electronic Data Interchange	21.6	12.1	17.1	1.11	0.29
Databanks	64.6	60.6	62.8	0.14	0.71
Telefax	62.2	75.7	68.6	1.49	0.22

See notes for Table 5.1

easily attributed to sectoral characteristics (see Table 5.10). In the case of the *service sector*, the 'importance of these technologies for business purposes' is to a large extent the most crucial reason for adoption (more than 66 per cent of replies). This result demonstrates the importance of these technologies for the service sector. The number of subscribers does not represent an important reason.

In the case of the *industry sector*, the 'low costs of implementation and use' are the main reason for adoption, followed by the 'support provided by demonstration centres', i.e. by an active marketing policy. The 'high number of subscribers in the network' as a reason to adopt has greater importance in the industry sector, than in the service sector. This fact may easily be explained by two factors: (a) more external relationships are established by firms belonging to the industry sector (with suppliers and customers). Service activities, however, are much more characterized by internal flows of information and thus the number of subscribers using the network is less important; (b) in most cases the service sector bases its internal flows of information on private networks whose adoption process, by definition, is not supported at all by network externality effects.

Regarding the *main conditions for future adoption*, the 'number of subscribers' represents the most important reason for future adoption in the case of both networks and services. In the case of networks, 'technical progress' also plays an important role for future adoption. The industry sector is slightly more inclined to a 'higher number of people or suppliers and customers connected' than is the service sector. The structural features

Table 5.10. Main reasons for adoption by industry and by service industry

	Industry	Service industry	Entire sample	χ^2	P-value
High percentage of already networked subscribers in your region	16.2	9.1	12.8	0.79	0.37
High percentage of already networked subscribers in other regions	18.9	9.1	14.3	1.37	0.24
Importance of the network or service for your business	37.8	66.6	51.4	**5.80**	**0.02**
Your suppliers were networked or were using the same service	10.8	6.1	8.6	0.50	0.48
Other firms in the same sector were connected	13.5	9.1	11.4	0.34	0.56
No other advanced communication network or service available	8.1	9.1	8.6	0.02	0.88
Low costs of implementation	40.5	24.2	32.9	2.10	0.15
Low costs of use	37.8	24.2	31.4	1.49	0.22
Increasing awareness of these technologies through demonstration centres	18.9	27.3	22.8	0.69	0.41
Image effect	16.2	18.2	17.1	0.05	0.83
Others	29.7	15.1	22.8	2.10	0.15

See notes for Table 5.1

of the business in the two macro-sectors can explain this result. In fact, the industry sector has considerably more contacts with other external agents, especially subscribers, whilst information in the service sector is more internal information. On the other hand for the service sector, the main conditions for future adoption is linked to the 'technical characteristics of the network' and to the 'increase in the efficiency of the network', as well as to the 'number of firms connected'. For advanced services, a higher number of subscribers represents the main reason for future adoption for both the industry and the service sector. This result again shows very limited *industry variations* in the replies (Table 5.11).

The *intensity of relationships after telecommunication technology adoption* shows a strong sectoral dependency. The industry sector shows a very high decrease of business relationships (72.9 per cent of replies), in contrast to only 42.4 per cent for the service sector (Table 5.12). What is interesting is that the most important reason explaining the very high degree of decreased intensity of business relationship is represented by 'network effects', i.e. by suppliers, customers and other firms in general not using them (Table 5.13). In the case of the service sector, the most important reason expressed is the irrelevant role of the service for the business.

Table 5.11. Main conditions for future adoption by industry and by service industry

	Industry	Service industry	Entire sample	χ^2	P-value
Networks					
A higher number of people connected	12.8	15.1	20.0	0.917	0.388
A higher number of suppliers and customers connected	24.3	9.1	17.1	2.85	0.091
A better geographical distribution of the network	10.8	6.1	8.6	0.502	0.479
A reduction in the price of use	13.5	9.1	11.4	0.337	0.562
A reduction in the price of access	10.8	6.1	8.6	0.502	0.479
More recent availability of the network	13.5	6.1	10.0	1.077	0.299
Technical progress in the network	27.0	18.2	22.8	0.774	0.379
Increase in the efficiency of the network	18.9	15.1	17.1	0.174	0.676
Good results from the previous adoption	13.5	6.1	10.0	1.077	0.299
Services					
A higher number of people connected	29.7	24.2	27.1	0.27	0.61
A higher number of suppliers and customers connected	27.0	24.2	25.7	0.07	0.79
A better geographical distribution of the service	13.5	6.1	10.0	1.08	0.30
A reduction in the price of use	10.8	15.1	12.8	0.29	0.59
A reduction in the price of access	13.5	15.1	14.3	0.04	0.84
More recent availability of the service	10.8	6.1	8.6	0.50	0.48
Technical progress in the service	18.9	18.2	18.6	0.00	0.94
Increase in the efficiency of the service	21.6	21.1	21.4	0.00	0.97
Good results from the previous adoption	16.2	6.1	11.4	1.78	0.18

See notes for Table 5.1

Table 5.12. Intensity of relationships after telecommunication technology adoption by industries and by service industry

	Industry	Service industry	Entire sample	χ^2	P-value
Increased intensity of business relationships	27.0	57.6	41.4	**6.71**	**0.01**
Constant intensity of business relationships	0.0	0.0	0.0		
Decreased intensity of business relationships	73.0	42.4	58.6		

Table 5.13. Main reasons in case of decreased intensity by industry and by service industry

	Industry	Service industry	Entire sample	χ^2	P-value
None	16.2	15.1	15.7	0.01	0.90
Costs of initial investment	2.7	0.0	1.4	0.90	0.34
Lack of reliability on the technologies used	2.7	0.0	1.4	0.90	0.34
Level of telecommunication charges	0.0	0.0	0.0		
Lack of staff skills in their use	8.1	9.1	8.6	0.02	0.88
Services are not relevant to business	13.5	18.2	15.7	0.29	0.59
Your customers do not use them	29.7	0.0	15.7	**11.6**	**0.00**
Your suppliers do not use them	18.9	9.1	14.3	1.37	0.24
Too few subscribers in general	32.4	6.1	20.0	**7.58**	**0.00**
Restricted geographical diffusion of networks and services	8.1	3.0	5.7	0.83	0.36
Other reasons	8.6	5.6	7.1	1.59	0.21

See notes for Table 5.1

The behavioural analysis at the industrial level has shown very limited industry variation in the responses obtained. What differences there are can be related to the business structure, characterized by the following.

1. The nature of information flows; the service sector is in general characterized much more by *internal flows of information*, while for the industry sector the strategic flows are *external flows of information*, with suppliers and customers.
2. Internal flows are much more supported by *internal networks*, such as LANs (which in our sample are also the most adopted networks (27.27 per cent of replies) compared to all other possibilities). These networks, by definition, are not based on network externality effects (Capello 1994).

Our aim is now to see whether our results may also be explained through an interpretative analysis. First of all we present the methodology used for our interpretative analysis (Section 5.5.1); Section 5.5.2 contains the results of the estimated logit model for the industrial and regional levels of analysis. Section 5.6 presents some concluding remarks.

5.5 AN INTERPRETATIVE ANALYSIS

5.5.1 THE DISCRETE CHOICE MODELLING APPROACH

The interpretative analysis of our first research proposition is based on a standard discrete choice modelling approach, with economic random utility theory as the underlying theoretical rationale and revealed preferences as the empirical orientation. Discrete choice models such as multinominal logit, nested multinominal logit, and multinominal probit models are now well-established model approaches that are applied in a wide range of fields.

The importance of these models for our analysis stems from the fact that in most cases the decision of a firm to join and to use a network is of a discrete nature; in our case, too, the behavioural analysis is based on revealed preferences and the database obtained may be applied only for discrete models.

The logit models in our empirical analysis will be based on the complete database. In the next section the willingness to join a network will be analysed, highlighting industry and regional variations in the decision to join the network. The results of the analysis go further than the exercise of testing our first research proposition. They also have policy implications, especially in the case of southern Italy, where the results will be able to prove whether the financial effort made by the EC to promote the use of these technologies in less-developed regions has, in fact, generated a willingness for future adoptions by local firms in the south.

A number of *explanatory variables* characterizing the reasons to join a network have been selected, namely (Table 5.14):

1. The *price incentives* for a firm. Here a distinction is made between firms having replied that low implementation and use costs have been a crucial variable in the decision to adopt (PRICE = 1), and firms which in previous adoptions have not recognized low financial costs as a basic reason for adoptions (PRICE = 0).
2. The *role of the supply* in supporting the adoption of new networks and services, through, for example, demonstration and promotion centres. The question here is whether firms have recognized the efforts made by the supply side as helpful in their decision-making process of previous

Table 5.14. Definition of variables

	1	0
Price effect	If low implementation and use costs have played an important role in previous adoptions	Otherwise
Role of the supply	If the existence of promotion and demonstration centres has played an important role in previous adoptions	Otherwise
Bandwagon effect	If the existence of a high number of users has played an important role in previous adoptions	Otherwise

adoptions (ROLE = 1) or whether they have never taken the supply efforts into consideration (ROLE = 0).
3. The *bandwagon effect* in the decision-making process to adopt. In this respect, a distinction is made between firms which have recognized the number of adopters as a crucial variable in the decision-making process for previous adoptions (NET = 1) and firms which have assigned no role at all to the number of adopters in decision-making processes for previous adoptions (NET = 0).

The choice of these explanatory variables has been based on a selection of a great number of potential explanatory variables, such as the size of firms, the sector firms belong to, innovation capacity of firms, their flexibility with respect to changes, the importance of the technology for the business, learning processes, etc. Among all plausible categorical variables, we have chosen those having the highest degree of dependency with the dependent variable. The results of the dependency analysis are shown in Table 5.15. Among all possible categorical variables, it is interesting to underline that the *size of firms*, and the *sector firms belong to*, are independent from the willingness to adopt. This means that contact patterns do not differ between small- and medium-sized firms, as well as the reasons for adoption not varying among sectors. This second result is in line with what has been found in the descriptive analysis at the sectoral level presented above.

The above-mentioned variables (points (1)–(3) above) represent the expected explanatory variable of the willingness to adopt. A measure of the willingness to adopt is given in our database by the revealed interest to adopt advanced (interactive) services in the near future. A 0–1 variable has thus been built on the distinction between firms which have revealed a preference for future adoption of advanced interactive services (i.e. electronic mail) (WILL = 1) and those which have not (WILL = 0).

A way of estimating logit models without incurring the risk of infinite solutions is to make particular assumptions on parameters. In our case, the

Table 5.15. Degree of dependency between the dependent variable and the categorical variables

Willingness to adopt*	$\chi^{2\dagger}$	P-values
PRICE1	5.66	0.017
ROLE1	3.81	0.066
NET1	3.02	0.082

* Dependent variable
$^\dagger \chi^2$ shows a certain degree of dependency between the categorical variables and the dependent variable

models have been estimated by assuming that the sum total of the parameters over the various categories of all main and interaction effects is equal to zero.

The research strategy for estimating the above-mentioned logit model was based on an initial estimation of the model including all main effects, except for the variables 'regional dimension' and 'sectoral dimension'. Besides the estimated parameter values for the main effects, also the P-values and the χ^2 of the likelihood ratio will be presented in our results. Next the 'regional dimension' and the 'sectoral dimension' variables were introduced separately into the model. In this way, the extent to which these variables lead to a better fit of the model has been estimated, and the estimated parameter values are also presented in Section 5.5.2 below.

5.5.2 ESTIMATED LOGIT MODEL FOR THE REGIONAL AND INDUSTRY LEVEL OF ANALYSIS

In this section we consider the estimated logit model with respect to the willingness to adopt in the near future. Table 5.16 presents the results of the estimated logit model with respect to the regional dimension, while in Table 15.17 the sectoral characteristics have been introduced. In this way, the need to test the statistical importance of the sectoral and regional dimension has been fulfilled.

As stated in the previous section, in the tables presented in this section, it should be recalled that our logit models have been estimated under the hypothesis that the sum total of the parameters over the various categories of all main effects is equal to zero (Agresti 1990). Taking the example of the price effect variable, it can be derived from Table 5.16 that the estimated parameter for those firms replying that in previous adoptions price incentives have been strategic for their adoption decision-making process (i.e. PRICE = 1) equals 0.49. Given the above-mentioned restrictions, the esti-

Table 5.16. Estimated logit model with respect to the willingness to adopt at the regional level of analysis

Variable	Estimated parameter	χ^2	P-value
PRICE1	0.4913	3.29	0.069
ROLE1	0.3879	1.28	0.257
NET1	0.1037	0.07	0.793
REG1	−0.3226	1.25	0.264

$\chi^2 = 4.18$
P-value $= 0.6526$

Table 5.17. Estimated logit model with respect to the willingness to adopt at the industrial level of analysis

Variable	Estimated parameter	χ^2	P-value
PRICE1	0.74	6.4	0.011
ROLE1	0.57	2.36	0.124
NET1	0.03	0.01	0.939
SET	−0.86	4.91	0.027
ROLES * SET	−0.53	2.08	0.149

$\chi^2 = 7.23$
P-value = 0.405

mated parameter for firms which deny giving any role to financial support in their decision-making processes (i.e. PRICE = 0) then becomes –0.49.

In discussing the results of the estimated logit model, we first concentrate on the main explanatory variables, leaving till the end any comments on the regional and sectoral dimensions.

From a *statistical point of view*, the results are rather satisfactory for the *estimated logit model at the regional level*. This model has a 0.65 probability value to explain the willingness to adopt. Moreover, the good fit of the model from a statistical point of view is also represented by the low number of significant categorical variables and by the lack of significant interaction effects among these categorical variables.

The *economic interpretation* of the model is interesting. A first conclusion drawn from Table 5.16 is that firms having chosen 'low implementation and use costs' as an important reason for adoption are also the most dynamic firms in terms of future adoption, as is shown by the positive sign of the estimated parameter of the PRICE variable. Consequently, *financial incentives represent a very important stimulus* for future adoption. This result has very important policy implications, since it shows that in the first diffusion stages, the price variable plays an important role in the decision to adopt. Moreover, this result assumes even more importance when linked to the STAR Programme. The STAR Programme provided these technologies free of charge and our analysis reveals how strategic this choice has been in stimulating a local demand for these technologies. However, this result also represents a strategic lesson for future transitions to a commercial phase; a gradual move towards market prices is required in order to maintain and increase the adoption level achieved in the first instance with a financial incentive policy.

As far as the supply role is concerned, it is evident from our results that the existence of promotion and demonstration centres has positive effects on the willingness to adopt. This is indicated by the significant positive estimated parameter value with respect to the variable ROLE. This result stresses once again an established idea in the literature about the necessity for a *bridging mechanism* between demand and supply for the successful adoption processes of these technologies. The profound adjustments in terms of technological and organizational changes required in order to adopt and exploit these new technologies are overcome only if technical and organizational support is provided by the supply side.

Another quite interesting result is presented by the existence of a *bandwagon effect* in the willingness to adopt. Contrary to what we expected, the number of already existing subscribers does not seem to stimulate future adoption, as is witnessed by the estimated parameter of the variable NET. The parameter of this variable, in fact, has a *P*-value of 0.79.

As far as the regional dimension is concerned, we introduced into our model a variable reflecting the location of firms (REG), assuming a value 1 when located in the north and 0 when located in the south. This variable assumes a negative value, thus underlining the fact that firms in the north of Italy have less willingness to adopt than the ones located in the south. This result is not surprising at all, since the north of Italy presents higher adoption rates than the south, for those networks and services commercially available (see Table 5.1 above). Low adoption rates in the north are typical for those networks and services which are either in an experimental phase (such as ISDN or videoconference) or are still very limited in their geographical extension (such as fibre optic networks).

From the same estimated parameter (with opposite sign) we deduce a clear interest in future adoption for firms located in the south. This is an extremely positive result when we analyse it in the framework of the STAR Programme. One of the aims put forward by the EC was to stimulate an interest for these technologies among firms in the south, and to show their importance for the business activities of these firms and for the future of these firms. This aim of the Programme seems to have been achieved, as the sample in the south has demonstrated the positive attitude of southern firms towards future adoption. However, before being sure of the positive results of the STAR Programme, it is also necessary to test whether the other extremely important aim of the Programme, i.e. an economic revitalization of backward regions, has been achieved.

At *a sectoral level of analysis*, the estimated logit model is less satisfactory, although rather interesting from an economic point of view. The statistically less satisfactory results in comparison with the regional case are witnessed by a *P*-value equal to 0.405 and by the presence of a statistically significant interaction effect. In any case, the sectoral results confirm what was previously proved at a regional level. The financial incentives, as well as the supply support, explain quite clearly the willingness to adopt, while the 'bandwagon effect' loses much of its explanatory effect.

As far as the *sectoral component* is concerned, it appears quite clearly that firms belonging to the service sectors are more in favour of future adoption, as is witnessed by a statistically significant negative estimated parameter for the industry sector. Moreover, the statistically significant interaction effects ROLES * SET demonstrates that especially those service firms supported by the supply are oriented towards future adoption.

The interpretative results obtained are satisfactory, although they do not seem to support our idea that *consumption network externalities play a role in the diffusion process of these technologies*. However, the results show that it is interesting to develop the analysis at a territorial level. In fact, the *regional dimension explains part of the innovative behaviour* of firms and it is not just an additional variable in an already complex interpretative framework.

The present behavioural analysis has some policy implications, especially in terms of successful innovation policies for developing a local innovative demand. This is only the first part of the innovative process and of an innovation policy. To be successful, an innovation policy is expected to generate positive results on the production side, stimulating productivity and economic growth.

5.6   CONCLUSIONS

In this chapter we have focused on a behavioural analysis of the main reasons for adoption of advanced telecommunication technologies. In particular, we have presented both a descriptive analysis, through contingency tables analysis, and an interpretative analysis, through the estimation of logit models.

The analyses have led to slightly different results. The descriptive analysis strengthens the role of the number of already existing subscribers as one of the main reasons for adoption, while the interpretative analysis shows 'price mechanisms' and 'help from supply side' as the major explanatory variables.

Both the descriptive and interpretative analyses have shown a *regional variation* in the results. Concerning the descriptive analysis, this shows a strong regional difference in consumption network externalities. In backward regions, where a critical mass has not yet been achieved, the reasons for adoption are not the number of already existing subscribers. On the contrary, in advanced regions, the number of adopters represents one of the most important reasons for adoption. As expected, the results are the opposite when dealing with the main conditions for future adoption. In backward regions, the most important reasons for future adoption lie in an increase in the number of subscribers connected, while for advanced regions other reasons, such as low implementation and use costs, are relevant for future adoption. Moreover, an interesting result obtained from the contingency tables analysis is the statistical dependency of 'consumption network externalities variables' on the regional dimension. This result is even more important when one remarks that consumption network externality variables are *the only ones* showing a statistical dependency on the regional dimension. This statistical dependency has not been found at a sectoral level.

The interpretative analysis does not confirm the importance of consumption network externalities as a crucial explanatory variable of the willingness to adopt. This analysis, run on the basis of an estimation of multinomial logit models, points out two critical variables explaining the willingness for future adoption, namely 'price incentives' and 'support from the supply side'; the 'number of already existing subscribers' does not seem to be an important explanatory variable of the willingness to adopt.

Even in the interpretative exercise, the regional dimension is important and underlines the fact that there is a positive linkage between firms located in the south and the willingness to adopt. The interpretative analysis also emphasizes some important policy implications, since the strategic elements of future adoption are highlighted.

The results at the industrial level have been less satisfactory. The expected differences in the replies between the industrial and the service firms have not emerged in reality. These results have therefore acted as a disincentive to carry out the analysis at the industry level.

As far as the STAR programme is concerned, our analysis clearly demonstrates its successful results in stimulating the willingness to adopt in the near future. However, to evaluate its full degree of success, it is necessary to test its effects on the performance of (adopting) firms and regions. In other words, whether *production network externalities are exploited by firms and regions* must be tested empirically. This aspect undoubtedly deserves attention of future research activities.

REFERENCES

Agresti, A. (1990) *Categorical Data Analysis*, Wiley Interscience, New York.

Allen, D. (1988) New telecommunications services: network externalities and critical mass, *Telecommunications Policy*, September, 257–271.

Allen, D. (1989) Competition, cooperation and critical mass in the evolution of networks, paper presented at the *8th ITS Conference*, Venice, 18–24 March.

Antonelli, C. (1989) The diffusion of information technology and the demand for telecommunication services, *Telecommunications Policy*, September, 255–264.

Antonelli, C. (1990) Induced adoption and externalities in the regional diffusion of information technology, *Regional Studies*, **24**(1), 31–40.

Antonelli, C. (1992) (ed.) *The Economics of Information Networks*, Elsevier, Amsterdam.

Ben-Akiva, M. and S. R. Lerman (1985) *Discrete Choice Analysis: Theory and Application to Travel Demand*, MIT Press, Cambridge, MA.

Bental, B. and M. Spiegel (1990) Consumption externalities in telecommunication services, in M. de Fontenay and D. Sibley (1990) (eds), *Telecommunications Demand Modelling*, Elsevier, Amsterdam.

Bishop, Y. M. M., S. E. Fienberg and P. W. Holland (1977) *Discrete Multivariate Analysis: Theory and Practice*, MIT Press, Massachusetts.

Cabral, L. and A. Leite (1992) Network consumption externalities: the case of Portuguese telex service, in Antonelli (1992), pp. 129–140.

Camagni, R. (1992) *Economia Urbana*, Nuova Italia Scientifica, Rome.

Camagni, R. and R. Capello (1991) Le Caratteristiche delle Nuove Tecnologie di Comunicazione e loro Interazione con la Domanda, in R. Camagni (ed.), *Computer Network: Mercati e Prospettive delle Tecnologie di Telecomunicazione*, Etas Libri, Milan, pp. 3–42.

Capello, R. (1994) *Spatial Economic Analysis of Telecommunications Network Externalities*, Avebury, Aldershot.

Capello, R. and P. Nijkamp (1995) Corporate and regional performance of ICTs development: the role of network externalities, in A. Rallet and A. Torre, *Industrial Organisation and Spatial Economics*, Economica, Paris, forthcoming.

Curien, N. and M. Gensollen (1987) A functional analysis of the network: a prerequisite for deregulating the telecommunications industry, *Annales des telecommunications*, **42**(11–12), 629–641.

David, P. (1985) Clio and the economics of Qwerty, *AEA Papers and Proceedings*, **75**(2), 332–337.

David, P. (1992) Information network economics: externalities, innovation and evolution, in Antonelli (1992), pp. 103–106.

Domencich, T. A. and D. McFadden (1975) *Urban Travel Demand: a Behavioural Analysis*, North-Holland, Amsterdam.

Farrell, J. and G. Saloner (1985) Standardisation, compatibility, and innovation, *Rand Journal of Economics*, **16**(1), 70–83.

Farrell, J. and G. Saloner (1986) Installed base and compatibility: innovation, product preannouncements and predation, *The American Economic Review*, **76**(5), 940–955.

Fischer, M. and P. Nijkamp (1985) Developments in explanatory discrete spatial data and choice analysis, *Progress in Human Geography*, **9**, 515–551.

Fischer, M., R. Maggi and C. Rammer (1992) Stated preference models of contact decision behaviour in Academia, *Papers in Regional Science, The Journal of RSAI*, **71**(4), 359–371.

Griguolo, S. and A. Reggiani (1985) Modelli di scelta tra alternative discrete: alcune note introduttive, *Archivio di Studi Urbani e Regionali*, **22**, 47–86.

Hayashi, K. (1992) From network externalities to interconnection: the changing nature of networks and economy, in Antonelli (1992), pp. 195–216.

Katz, M. and C. Shapiro (1985) Network externalities, competition and compatibility, *The American Economic Review*, **75**(3), 424–440.

Katz, M. and C. Shapiro (1986) Technology adoption in the presence of network externalities, *Journal of Political Economy*, 822–841.

Leonardi, G. (1985) Equivalenza Asintotica fra la Teoria delle Utilità Casuali e la Massimizzazione dell'Entropia, in A. Reggiani (ed.) *Territorio e Trasporti: Modelli Matematici per l'Analisi e la Pianificazione*, Franco Angeli, Milano, pp. 29–66.

Loehlin, J. C. (1987) *Latent Variable Models*, Hillsdale, New York.

Markus, M. (1989) Critical mass contingencies for telecommunications consumers, paper presented at the *8th ITS Conference*, Venice, March 18–24.

McFadden, D. (1983) Econometric Analysis of Qualitative Response Models, in Z. Griliches and D. Intriligator (eds) *Handbook of Econometrics*, North-Holland, Amsterdam, pp. 1396–1450.

Meade, J. E. (1954) External economies and diseconomies in a competitive situation, *Economic Journal*, **62**, 143–151.

Nijkamp, P., H. Leitner and N. Wrigley (eds) (1985) *Measuring the Unmeasurable*, Martinus Nijhoff, Dordrecht.

Rohlfs, J. (1974) A theory of interdependent demand for a communication service, *The Bell Journal of Economics and Management Science*, **5**, 16–37.

SAS/STAT (1989) *User's Guide*, Version 6, 4th edition, Vol. 1, Cary, SAS Institute Inc.

Scitovsky, T. (1954) Two concepts of external economies, *Journal of Political Economy*, **62**, 143–151.

Spence, M. (1981) The learning curve and competition, *Bell Journal of Economics*, **12**(1), 49–70.

Stoneman, P. (1990) *The Intertemporal Demand for Consumer Technologies requiring Joint Hardware and Software Inputs*, Working Paper No. 355, University of Warwick.

Part II

APPLIED NETWORK ANALYSIS

6 Analysis of the Evolution of Transport Networks: the Case of the Dutch Railway System

PIET RIETVELD
JOOST VAN NIEROP
Free University of Amsterdam, The Netherlands

6.1 INTRODUCTION[1]

The analysis of the location of *facilities* has attracted much attention from scientists in the field of operational research, regional science, physical planning, geography and spatial economics (see for example Current and Ratick 1992; Rushton 1993). The analysis of the location of *networks* has received much less attention. This is surprising since the average lifetime of transport networks such as railway lines, canals and highways is usually at least as long as the average life time of facilities. In addition, facility location modelling is based on the existence of transport networks so that network location is a driving force behind facility location. On the other hand, one has to take into account that decisions on network location do not take place independently from decisions about the location of facilities.

In the current chapter we provide a concise review of some contributions in the field of network location modelling. These contributions usually have a normative orientation: given a certain objective function and constraints in terms of, for example, available budget, minimum service level, potential demand for transport, etc., one determines the optimal shape of the network. Actual network development will not always follow the optimal patterns generated by the network location models. There are several reasons for this. Models give an analytical representation of decision problems which may not entirely coincide with the perception of the decision-makers at the time the network location decisions were made. For example, expectations about future developments will play an important part in network decisions; modelling these expectations is not easy. In addition, one must be aware that network location decisions are determined not only

[1] The authors thank Bert van der Knaap for making available population data of Dutch municipalities since 1840.

European Transport and Communications Networks: Policy Evolution and Change. Edited by David Banister, Roberta Capello and Peter Nijkamp. © 1995 John Wiley & Sons Ltd.

by transport related variables, but that other variables also play a part as follows:

1. Political considerations may favour the development of certain links in a network because of electoral reasons.
2. Infrastructure investments may take place in the context of regional policies with the explicit aim of boosting lagging regional economies.
3. Network infrastructure may be improved in particular regions for strategic–military reasons.
4. Status considerations sometimes play a part in the construction of networks as can be illustrated by the case of the German Autobahnen before World War II (cf. Hall 1990).
5. Political and language borders may play an important part in network location decisions, so the number of border crossing connections is much smaller than one would expect on the basis of pure transport related variables (cf. Wolfe 1962; Haggett et al 1977; Bruinsma and Rietveld 1993).

The aim of this chapter is to investigate the development of the Dutch railway system in the nineteenth century and to see to what extent this development can be replicated by means of network location models. In addition, attention will be paid to the role of non-transport-related factors as indicated above.

6.2 NETWORK DESIGN; COST MINIMIZATION ALGORITHMS

The following definition of a network will be used in this paper:

A *network* N is a pair $N = (V,A)$, where V is a set of nodes i with weight m_i, $(i = 1, 2, \ldots)$ and A is a set ordered pair of nodes a_{ij} called arcs with length d_{ij}.

The practical implementation to railway networks is obvious: each city is represented by a node with a weight equal to the number of inhabitants of the city and each link between two cities is represented by an arc with a (symmetric) length equal to the distance between these cities. There are of course many drawbacks in this representation since it neglects many social, economic and geographic factors, but even in this simple case, the problem of building an optimal network appears to be quite intractable.

The first step in building an optimal network is to answer the question of what objective function has to be maximized or minimized, and what are the constraints to which this optimization is subjected. When a cost minimizing approach is followed, we have several choices: if the objective is to minimize *user costs* of a network, the resulting *maximum connected network* problem can be defined by

$$\text{(MCN)} \qquad \min \Sigma_i \Sigma_j p_{ij} \qquad (6.1)$$

where p_{ij} is the length of the shortest path between node i and j. p_{ij} can recursively be defined by

$$p_{ij} = \min_{k,\, a_{kj} \in A} (p_{ik} + d_{kj});$$

if nodes i and j are not connected, $p_{ij} = \infty$.

The solution to this unconstrained problem is trivial: a direct link between each pair of nodes, resulting in a complete network. Note that we assume user costs to be proportional to the distance travelled, and that the demand is equal between all nodes. If construction of a link is cheap in relation to the user costs, such a complete network can be desirable. In most practical applications, the MCN is subjected to a budget constraint:

$$\min \Sigma_i \Sigma_j p_{ij}$$
$$\text{subject to}$$
$$\Sigma_i \Sigma_j \lambda_{ij} d_{ij} \leqslant B;$$
$$\lambda_{ij} = \lambda_{ji}, \forall i, \forall j;$$
$$p_{ij} \leqslant \infty, \forall i, \forall j;$$

where p_{ij} is the length of the shortest path between node i and j;
$\quad\ \lambda_{ij}$ is a binary variable, assuming values 0 and 1;
$\quad\ d_{ij}$ is the physical distance between nodes i and j and
$\quad\ B$ is the budget available.

When it is the objective to minimize the construction costs of a network connecting all the nodes, we have a *minimum spanning tree* problem:

$$\text{(MST)} \qquad \min \Sigma_i \Sigma_j \lambda_{ij} d_{ij} \qquad (6.2)$$
$$\text{subject to}$$
$$\lambda_{ij} = \lambda_{ji}, \forall i, \forall j;$$
$$p_{ij} < \infty, \forall i, \forall j;$$

where λ_{ij} is a binary variable assuming values 0 and 1;
$\quad\ p_{ij}$ is the length of the shortest path between node i and j.

In this formulation production costs are assumed to be proportional to the length of the trajectories built. The resulting network is a tree. A tree is a subnetwork of nodes and arcs, such that for every pair of nodes there is exactly one path that connects them. A spanning tree is a tree that contains all the nodes of a network. Spanning trees are used in the design of communication networks in which each node must be able to communicate with every other node. If the communication links are expensive, then it is desirable to have just one path between each pair of nodes so that the resulting network is a spanning tree. The Minimum Spanning Tree problem can be solved easily with a greedy algorithm (Kruskal, 1956). The idea of the greedy algorithm for solving the minimum spanning tree problem is quite straightforward:

Given a network $N = (V, A)$, start building a subnetwork $N' = (V', A')$ with any arbitrary node $i \in V$ and add the nearest node $j \in V$ to N', as well as the arc a_{ij}. In the next step, consider all the arcs $a_{kl} \in A$, which connect a node $k \in V'$ with a node $l \in V \backslash V'$. Let a_{kl} be the arc with minimum length, then add the node l and the arc a_{kl} to the subnetwork N'. Repeat this step until $V' = V$.

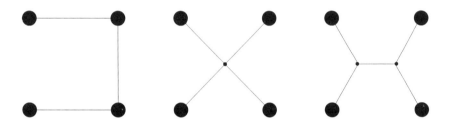

Figure 6.1. The effect of auxiliary nodes on a network

If it is allowed to add new auxiliary nodes to the network, an important improvement can be achieved with regard to the original minimum spanning tree problem. The minimum spanning tree on the four nodes in Figure 6.1 has length 3. When one auxiliary node is added, the total length decreases to $2\sqrt{2} \approx 2.82$. A second auxiliary node results in a total length $1 + \sqrt{3} \approx 2.73$. Unfortunately, it is very hard to determine the optimal position and optimal number of the additional nodes. This problem is called the *Steiner problem* or the *floating-point location problem* and is well solved for the case of three nodes only. In that case, the optimal location of a node added to the network is that place where the angle of the links between the new node and each pair of the old nodes is $120°$. It is known that the maximum number of additional nodes needed to construct an optimal network on a network with $n = |V|$ nodes is $(n - 2)$ (Miehle 1958). In most cases this number is smaller since additional nodes can coincide with the original nodes. The structure of

120° angles still holds in more complex networks, but the main problem is which nodes should be added. In an n-node network there are $\Sigma_{m=1}^{(n-2)}$ $2^{-m}\binom{n}{n+2}(n + m - 2)!/m!$ arrangements possible. This can be solved by a brute-force enumerative algorithm, but the time this requires increases exponentially with the size of the problem. The Steiner problem is recognized to be NP-hard, that is: there is no efficient algorithm known for solving the Steiner problem, and it is very unlikely that a good solution method that provides an optimal solution in a reasonable amount of time ever will be found. For an extensive treatment of NP-hardness and the complexity of (general) optimization problems see, for instance, Garey and Johnson (1979) or Papadimitriou and Steiglitz (1982). Some heuristic algorithms for the Steiner problem are developed by Morgan (1967).

The former two approaches minimize either user costs or construction costs of a network. Usually, however, and particularly in transportation networks, we want to take both into consideration. Minimizing a weighted sum of construction costs and discounted user costs is one possibility. But several other formulations are possible. For example:

1. minimize construction costs subjected to a maximum detour factor S for each pair of nodes:

$$\min \Sigma_{ij} \lambda_{ij} d_{ij} \qquad (6.3)$$
$$\text{subject to}$$
$$p_{ij} \leqslant S d_{ij}, \forall i, \forall j;$$
$$\lambda_{ij} = \lambda_{ji}, \forall i, \forall j;$$

where d_{ij} is the physical distance between i and j:
   λ_{ij} is a binary variable, assuming values 0 and 1;
   p_{ij} is the length of the shortest path between i and j and
   S is the maximum detour factor for each pair of nodes ($S \geqslant 1$).

It is not difficult to see that when the detour factor is finite, the resulting network is connected.

2. minimize user costs subjected to a budget constraint for the construction of links:

$$\min \Sigma_i \Sigma_j p_{ij} \qquad (6.4)$$
$$\text{subject to}$$
$$\Sigma_{ij} \lambda_{ij} d_{ij} \leqslant B;$$
$$\lambda_{ij} = \lambda_{ji}, \forall i, \forall j;$$
$$p_{ij} \leqslant \infty, \forall i, \forall j;$$

where p_{ij} is the length of the shortest path between node i and j;
   λ_{ij} is a binary variable, assuming values 0 and 1;

d_{ij} is the physical distance between i and j and
B is the budget available.

In the latter problem the sum of the lengths of the shortest paths is minimized under the condition of a limited budget. However, this problem cannot be solved by a brute-force enumerative algorithm, since in a network on $n = |V|$ nodes, there are $2^{n(n-1)/2}$, that is exponentially many, configurations possible. Although some work is done in order to improve exact algorithms, we still depend on heuristic approaches. One such heuristic algorithm has been developed by Scott (1969) and MacKinnon and Hodgson (1970) for a slightly different problem formulation: the *forward approximation algorithm*:

1. Determine which links of the possible network $N = (V, A)$ are very unlikely to occur in an optimal network and remove them.
2. Let $N' = (V', A')$ be a (the) minimum spanning tree on $N = (V, A)$.
3. Add that arc a_{ij} to the network that results in the largest improvement of the objective function, subject to the constraint not being violated.
4. Find a 1-optimal network by systematically removing each arc $a_{ij} \in V'$ and substituting it with an unused arc $a_{kl} \in V \backslash V'$ if this would result in an improvement of the objective function and the constraint is not violated.
   Repeat (3) and (4) until no improvement can be made.

Step (1), although it improves the efficiency of the algorithm, is quite arbitrary since it is hard to give a general criterion which decides whether an arc is implausible in an optimal network or not. It is obvious that the resulting network is not necessarily optimal. The results are assured to be closer to the optimal solution when a more sophisticated interchange step (4) is used in order to build a 2-optimal or a 3-optimal network. The problem is that this step is the most time-consuming step of the algorithm and interchange procedures which take into consideration substitution of three or more arcs at once are unfortunately far too slow.

Another approach suggested by Scott (1971) works the other way around: it starts with a complete network and eliminates successive nodes until a feasible solution is obtained.

6.3   NETWORK DESIGN: INTRODUCTION OF THE DEMAND SIDE

The problems as defined so far are not very realistic, because they assume each node to be of the same size. When a link is to be made to transportation networks, it is obvious that a link between two big cities is much more desirable than a link between two small cities, when the distances are more or less equal. Another point that must be taken into consideration is that the demand of transportation between two cities strongly depends on the length of the shortest path between them. A more demand-orientated formulation such as maximizing the total use of the network is preferred above a formulation in the terms of minimizing costs. The demand of transportation measured as the number of trips between two cities i and j can be approximated by the following gravity-type model:

$$D_{ij} = c\, m_i m_j / p_{ij}^{\alpha} \qquad (6.5)$$

where c is a constant value and α is a distance decay parameter.

A drawback of this demand function is that every link added to the network results in a rise of the total demand of transport, while it is reasonable to assume that the total demand of transport in a city is limited. The unconstrained gravity model does not include the possibility of substitution between destinations. To allow for such a possibility a constrained gravity model can be used. In that case the variables A_i and B_j are added to the demand function as balancing factors, resulting in the following double-constrained model:

$$D_{ij} = c\, A_i B_j m_i m_j / p_{ij}^{\alpha}$$

where c is a constant;
$$A_i = (\Sigma_{j \neq i} B_j m_j / p_{ij}^{\alpha})^{-1}; \qquad (6.6)$$
$$B_j = (\Sigma_{i \neq j} A_i m_i / p_{ij}^{\alpha})^{-1} \text{ and}$$
α is a distance decay parameter.

The total use of the network can be measured by the number of trip kilometres:

$$\Sigma_i \Sigma_j c\, A_i B_j m_i m_j p_{ij} / p_{ij}^{\alpha} \qquad (6.7)$$

In the case of a railway company with uniform tariffs this is proportional to the total receipts. When we assume in addition that operational costs are proportional to the number of trip kilometres, the above criterion is proportional to total receipts net of operational costs. The resulting network optimization problem is

$$\max \Sigma_i \Sigma_j A_i B_j m_i m_j p_{ij}^{1-\alpha} \qquad (6.8)$$
$$\text{subject to}$$
$$\Sigma_i \Sigma_j \lambda_{ij} d_{ij} \leqslant B$$

where $\lambda_{ij} = \lambda_{ji}, \forall i, \forall j$;
   p_{ij} is the length of the shortest path between node i and j;

λ_{ij} is a binary variable, assuming values 0 and 1;

B is the budget available.

This problem is more complex than the problems mentioned above and since there is no efficient algorithm known, we shall focus on heuristic methods. The forward approximation algorithm as proposed in Section 6.2 can, in an adapted form, be used again and now consists of a repetition of steps (3) and (4).

In the present context, the algorithm can be formulated as follows:

input: a network $N = (V,\emptyset)$ and a set A',
           where V is a set of nodes i with weight m_i, the set A of arcs a_{ij} between pairs of nodes of V is empty and A' is a set of all possible arcs a_{ij} between pairs of nodes in V.
output: a 1-optimal network N', on the nodes N subjected to the budget constraint $\Sigma_i \Sigma_j \lambda_{ij} d_{ij} \leqslant B$. Thus, there is no link $a_{ij} \in A$, and a link $a_{kl} \in A'$ such that substitution of the two arcs both improves the objective function and satisfies the budget constraint.

1. $\lambda_{ij} := 0, \quad \forall a_{ij} \in A'$
2. $p_{ij} := \infty, \quad \forall a_{ij} \in A'$
3. for all $a_{ij} \in A$, determine what would be the value of $\Sigma_i \Sigma_j A_i B_j m_i m_j p_{ij}^{(1-\alpha)}$ after adding a_{ij} to the network, and add that arc to the network for which the value of this objective is maximal, provided it does not violate the budget constraint
4. for all $a_{ij} \in A$, examine whether there exists an arc $a_{kl} \in A'$ such that the network $N'' = (V,\{a_{kl}\} \cup A \backslash \{a_{ij}\})$ satisfies the budget restriction and improves the objective function
5. if such a pair a_{ij} and a_{kl} exists then
   $A' := \{a_{ij}\} \cup A' \backslash \{a_{kl}\}$ and
   $A := \{a_{kl}\} \cup A \backslash \{a_{ij}\}$ and repeat (4)
   if there does not exist such a pair, go to (6)
6. if the network N' is a maximum connected network or if no arc a_{ij} can be added without violating the budget constraint then STOP.
   The resulting network N' is the approximation of the optimal network, otherwise go to (3).

The computation time of this algorithm is dependent to a considerable extent on the efficiency of the algorithm used to find the shortest path between each pair of nodes. Floyd (1962) has developed an efficient algorithm that finds the shortest path between each pair of nodes in a network simultaneously. In the network optimization algorithms, in each iteration the shortest paths are recomputed several times after adding or deleting one edge to the existing network. It is not necessary to repeat the computation completely. Algorithms that use information of the already existing network are developed by Loubal (1967) and Murchland (1970). For an overview of shortest path algorithms see Steenbrink (1974).

Other approaches of the unconstrained version of this problem are made by Boyce et al (1973), with a branch-and-bound algorithm and by Hodgson (1974, 1975) who uses dynamic programming.

6.4   THE RATE OF RETURN ON INVESTMENT IN NETWORK INFRASTRUCTURE

Consider the following objective function:

$$\max s = (\Sigma_i \Sigma_j A_i B_j m_i m_j p_{ij}^{(1-\alpha)})/(\Sigma_i \Sigma_j \lambda_{ij} d_{ij}) \tag{6.9}$$

The numerator is proportional to the number of passenger-kilometres travelled and is assumed to represent receipts of the railway company net of variable costs. The denominator is the length of the whole network and is assumed to be proportional to the total investment sum. Thus the objective function is proportional to the return on investment. Decision-making behaviour on the basis of the rate of return on investment can be assumed to take place in the following sequential way.

At the start of the process, that pair of nodes can be determined which leads to the highest values of s. In a second step, one can determine the next pair of nodes which leads to the highest value of s, given the arc that already exists. Railway construction comes to a halt when the return on adding a link in the network is lower than the rate of return outside the railway sector.

The resulting network is not required to be a connected network. So it is possible that a choice is made for linking two major cities, while there already exists a path between them, instead of connecting a small village with the rest of the network.

Consider the problem of choosing the first link to be built in the network. Based on the measure of the rate of return on investment one can rank all the potential links in order of attractiveness. Suppose that the link with the best performance is built. Then, given this fact one can again generate a rank order of links in terms of which is the most attractive one to build as the second link. Then, because of the interdependencies within networks, it is not guaranteed that the second ranking corresponds with the first ranking. The realization of a certain link may have an impact on the rates of return of other possible links.

Network interdependencies have also other implications. For example, if a certain link a_{ij} is the most attractive link to be built as a first link, it is not certain that it will be part of the optimal network if one considers a network consisting of two links. It may appear that the optimal two-link network does not contain the first-mentioned link. If decision-making takes place in a sequential way one may easily arrive at such suboptimalities leading to path dependence. In an algorithmic sense, this problem can be overcome by using a 1-optimal (or i-optimal; see preceding sections) approach so that one takes into account that an interchange of links in a network may lead to a higher performance. Another consequence of network interdependencies is that rate of return on investment on one link is influenced by whether or not another link is built. This is especially important when more than one railway company is involved in constructing and operating links. This occurs, for example, in international networks, and in the early stages of network development many countries have witnessed the existence of several railway companies on the same territory. Situations may occur where a link a_{ij} does not yield the requested market rate of return to firm 1 whilst the same holds true of link a_{kl} for firm 2 when only one of the links is built. But because of network interdependencies it is not impossible that if both links were built the rate of return of the combination a_{ij} and a_{kl} would be above the requested rate of return. If this holds true for both firms, there is no basic problem which prevents the construction and operation of both links. However, a problem emerges if for one firm the rate of return remains below the requested level if both links are built whereas it increases above that level for the second firm. In that case the former firm needs a compensation from the latter one in order to induce it to build and operate the link concerned.

6.5 APPLICATION: THE DEVELOPMENT OF THE DUTCH RAILWAY SYSTEM

The evolution of the Dutch railway system started in 1839 with the opening of a line between Amsterdam and Haarlem, two important cities in the western part of the Netherlands. Within 8 years this line was extended to the cities of Leiden, The Hague and Rotterdam. During this period a line was opened connecting Amsterdam with Utrecht and Utrecht with Arnhem near the German border. Thus in the year 1847, a considerable number of major cities of the Netherlands already had a railway connection. After 1847 the development slowed down for about 15 years. Some developments took place in peripheral parts of the country which can be considered as extensions of the German and Belgian railway systems which were more developed at that time (Figures 6.2 and 6.3). After 1863 railway development was speeded up. As the maps show, in 1865 the railway system still consisted of a number of disconnected parts, but in 1875 the Dutch railway system consisted of an almost fully connected network.

In the application of the algorithm we have assumed that the costs of

Figure 6.2. The Dutch railway system in 1855 (*Source*: Jonckers Nieboer 1938)

construction of the system are equal everywhere with one major exception. Crossing of long rivers such as the rivers Rhine, Waal, IJ, IJssel and Meuse are supposed to give rise to an additional cost equivalent to constructing 20 km of an ordinary connection. The value of the parameter α has been chosen as 2.

The 35 largest Dutch municipalities are taken into consideration, together with Antwerp, Brussels, Aachen and Cologne. By also considering these big foreign cities, some of the developments in the peripheral parts of the country can be explained. Cologne is chosen as the centre of the Ruhr area.

Figure 6.3. The Dutch railway system in 1865

Population totals have been used for 1860. The borders of the municipalities have been based on the definition in 1970. By doing so we include the population of surrounding villages which merged with larger centres between 1860 and 1970 (see Van der Knaap 1980).

The first 20 links generated by the railway-constrained model version presented in Section 6.3 can be found in Table 6.1. Maps of the simulated networks are given in Figures 6.4–6.7. The networks given are the 1-optimal networks after 5, 10, 15 and 20 links. No link interchange has occurred. When we consider the performance of the model, we notice that the model

Table 6.1. Simulated links in the Dutch railway network

	Simulated link	f_n		Simulated link	f_n
1	Rotterdam–Schiedam	0	11	Leiden–Gouda	0.45
2	Middelburg–Vlissingen	0	12	Rotterdam–Dordrecht	0.42
3	Amsterdam–Haarlem	0.33	13	Deventer–Zutphen	0.39
4	Delft–Schiedam	0.25	14	Utrecht–Amersfoort	0.43
5	The Hague–Delft	0.4	15	The Hague–Schiedam	0.4
6	Amsterdam–Zaandam	0.66	16	Utrecht–Leiden	0.44
7	The Hague–Leiden	0.57	17	Amsterdam–Rotterdam	0.41
8	Amsterdam–Leiden	0.5	18	Veendam–Hoogezand-Sap.	0.39
9	Amsterdam–Utrecht	0.55	19	Groningen–Hoogezand-Sap.	0.42
10	Heerenveen–Sneek	0.5	20	Dordrecht–Breda	0.4

tends to build short lines before long lines between major cities. For example, the second link generated is Middelburg–Vlissingen (6 km). These municipalities are respectively 17th and 28th in the ranking of the municipalities according to the number of their inhabitants. The link between two major cities such as Amsterdam (no. 1) and Leiden (no. 5) (46 km) is the 9th generated. There are some isolated edges, but in the historical development of the network, there were also isolated clusters. The development of the network is concentrated in the western part of the country. This conforms to the historical development. The path between Amsterdam and Rotterdam is built in more or less the same way as the historical development. The model builds the path from Rotterdam to Amsterdam, whereas the historical path is built from Amsterdam to Rotterdam. The model generates a link between Amsterdam and Leiden, but it was impossible to build that link before 1852. In that year the Haarlemmermeer was reclaimed. The Haarlem–Leiden link, however, was finished in 1842, so Amsterdam and Rotterdam were already connected. The model can be improved by taking this into consideration. It is interesting whether the improved model will first add the link Haarlem–Amsterdam or Leiden–Utrecht. This would be an alternative connection between Amsterdam and Rotterdam.

Besides the double-constrained model, a simulation with the unconstrained model is also made. As expected, the unconstrained model has a strong tendency to build links between large cities: of the first 10 links, 5 links were between Amsterdam and other big cities. An interesting point of the unconstrained model is that it generates links from Arnhem and Maastricht to the German border.

It is not easy to judge whether the development of the railway system as simulated by the model developed in Section 6.3 is similar to the actual development of the railway system. In order to measure the correspondence between the simulated and the actual network one can proceed as follows.

Figure 6.4. The Dutch railway system simulated after 5 links

Let B_1 denote the first arc realized in the system. It is followed by the arcs $B_2, \ldots, B_N$. Similarly, C_1 is the first arc generated by the algorithm. This arc is followed by the arcs $C_2, \ldots, C_N$. Clearly, when $B_1 = C_1$, $B_2 = C_2$, etc. there is a complete correspondence between the actual and the simulated development of the system. However, this would be an exceptional case. As a measure of the correspondence between $\{B_1, \ldots, B_N\}$ and $\{C_1, \ldots, C_N\}$ we propose to use the share f_n of the number of nodes in $\{C_1, \ldots, C_N\}$ which correspond with a node in $\{B_1, \ldots, B_N\}$. This can be computed for any value of $n = 1, \ldots, N$. Thus, if the first arc was predicted incorrectly $f_1 = 0$. Further, if $B_1 = C_2$ and $B_2 = C_1$, then $f_2 = 1$, etc. The f_n function of

Figure 6.5. The Dutch railway system simulated after 10 links

the double-constrained model of course alternates strongly for low values of n, after 20 links it is 0.4, so about 40 per cent of the links are rightly predicted. The values of f_n can be found in Table 6.1.

6.6 CONCLUDING REMARKS

The simulated results indicate that most of the early railway developments take place in the western part of the Netherlands. This is indeed what

Figure 6.6. The Dutch railway system simulated after 15 links

occurred in reality. However, for a considerable number of links the simulated links do not correspond to actually realized links. About 40 per cent of the first 20 links were predicted correctly. It is probable that a further calibration of the model will lead to better results.

The present analysis can be extended in several directions. The transport demand model can be refined by introducing other transport modes (walking, coach). The number of nodes can be extended by taking into account more than the 40 largest nodes used so far. Another interesting extension would be to pay explicit attention to the different railway com-

Figure 6.7. The Dutch railway system simulated after 20 links

panies in the nineteenth century. To what extent does the occurrence of multiple actors lead to the development of railway networks which are different from those generated when there is one railway company?

Finally, the model can be further developed by including feedbacks of the network development on urban growth. It is quite probable that the place of the cities in the evolving network is an important determinant of population growth in these cities. The integrated analysis of urban population growth and network development is a challenging theme for future research.

REFERENCES

Boyce, D. E., A. Fahri, R. Weischedel (1973) Optimal network problem: a branch-and-bound algorithm, *Environment and Planning*, 519–533.

Bruinsma, F. and P. Rietveld (1993) Urban agglomerations in European infrastructure networks, *Urban Studies*, **30**, 919–934.

Current, J. and S. Ratick (1972) Facility location modeling, *Papers in Regional Science*, **71**, 193–197.

Floyd, R. W. (1962) Algorithm 97, shortest path, *Communications of the Association of Computing Machinery*, **5**(6), 345.

Garey, M. R. and D. S. Johnson (1979) *Computers and Intractability: A Guide to the Theorem of NP-completeness*, Freeman, San Francisco, CA.

Haggett, P., A. D. Cliff and A. Frey (1977) *Location Analysis in Human Geography*, Arnold, London.

Hall, P. (1990) Keynote address on orbital motorways, in D. Bayliss (ed.), *Orbital Motorways*, Thomas Telford, London, pp. 1–31.

Hodgson, M. J. (1974/1975) Stating interurban highway construction by dynamic programming, *Annals of Regional Science*, **8**, 123–136, 1974. Reply on comments: *Annals of Regional Science*, **9**, 105–108, 1975.

Jonckers Nieboer, J. H. (1938) *Geschiedenis der Nederlandsche Spoorwegen*, Nijgh and Van Ditmar, Rotterdam.

Kruskal, J. B. (1956) On the shortest subtree of a graph and the travelling salesman problem, *Proceedings of the American Mathematical Society*, **7**, 48–50.

Loubal, P. S. (1967) A network evaluation procedure, *Highway Research Record*, 205.

MacKinnon, R. D. and M. J. Hodgson (1970) Optimal transportation networks: a case study of highway systems, *Environment and Planning*, **2**, 267–284.

Mandl, C. (1979) *Applied Network Optimization*, Academic Press, London.

Miehle, W. (1958) Link-length minimization in networks, *Operations Research*, 232–243.

Morgan, R. (1967) in R. J. Chorley, F. Honywill and P. Haggett (eds) *Models in Geography*, Methuen, London.

Murchland, J. D. (1970) *A Fixed Method for all Shortest Distances in a Directed Graph and for the Inverse Problem*, Ph. D. thesis, University of Karlsruhe.

Nemhauser, G. L. and L. A. Wolsey (1988) *Integer and Combinatorial Optimization*, Wiley, New York.

Papadimitriou, C. H. and K. Steiglitz (1982) *Combinatorial Optimization; Algorithms and Complexity*, Prentice-Hall, Englewood Cliffs.

Rushton, G. (1993) Lessons from the debate on location analysis in rural economic development, *International Regional Science Review*, **15**, 317–324.

Scott, A. J. (1969) The optimal network problem, *Transportation Research*, **3**, 201–210.

Scott, A. J. (1971) *Combinatorial Programming, Spatial Analysis and Planning*, Methuen, London.

Steenbrink, P. A. (1974) *Optimization of Transport Networks*, Wiley, Chichester.

Van der Knaap, G. A. (1980) *Population Growth and Urban Systems Development; a Case Study*, Martinus Nijhof, Boston.

APPENDIX 1: DEFINITIONS OF PROPERTIES OF GRAPHS

In order to formalize the concept of a network, some tools of graph theory are used. Graph theory has appeared as a tool to solve a large class of mathematical problems. The language of graphs and networks and the results of graph theory play an important part in this paper. Therefore, some terminology of graph theory is presented in this Appendix. An extensive treatment of graph theory with respect to optimization is given in, for example, Nemhauser and Wolsey (1988).

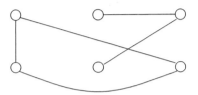

$$A = \begin{bmatrix} 1 & 0 & 0 & 1 & 0 & 0 \\ 0 & 1 & 1 & 0 & 0 & 0 \\ 1 & 0 & 0 & 0 & 0 & 1 \\ 0 & 0 & 0 & 1 & 0 & 1 \\ 0 & 0 & 1 & 0 & 1 & 0 \end{bmatrix}$$

graphical representation of G incidence matrix of G

Figure 6.8. Two representations of the graph G

A *graph* $G = (V,E)$ consists of a finite nonempty set $V = \{1,\ldots,n\}$ of *vertices* and a set $E = \{e_1,\ldots,e_m\}$ of unordered pairs of different vertices, called *edges*. Thus, $e_k = \{i,j\}, i \neq j \in V$.

An edge e_k is incident to a vertex i if $i \in e_k$. Notation: $e_k Ii$. The graph $G = (V,E)$ can be represented by its incidence matrix

$$A = (a_{ij})$$

where

$$a_{ij} = \begin{cases} 1 & \text{if } e_j Ii \\ 0 & \text{otherwise.} \end{cases}$$

The graph $G = (V,E)$ also can be represented in a plane by points and lines: The points of vertices are arbitrary placed in the plane. A line connects points i and j if $\{i,j\} \in E$. Thus, the matrix A and the picture in Figure 6.8 are two representations of the same graph. The number of edges incident to a vertex i is called the degree of i. Notation: $d(i)$. Two vertices i and j are adjacent if $\{i,j\} \in E$. Notation: iAj.

A graph $G = (V,E)$ is called complete if $E = \{\{i,j\} \mid \forall i \neq j\}$. Notation: $G = K_{|V|}$, where $|V|$ is the cardinality of V, i.e. the number of vertices in V. If $V' \subseteq V$ and $E' \subseteq E(V')$, then $G' = (V',E')$ is said to be a subgraph of G. $G' = (V',E(V'))$ is the subgraph of $G = (V,E)$ induced by $V' \subseteq V$. Notation: $G' = <V'>$.

A sequence $0,\ldots,k$, $k \geq 1$, of vertices is called a 0–k *walk* if $\{i-1,i\} \in E$, $\forall i \in Z_k$, $Z_k = \{1,\ldots,k\}$. A walk can also be represented by its edge sequence. The length of a walk is the number of edges in it. A *path* is a walk without vertex repetition. There is a unique partition of the vertices of the graph $G = (V,E)$ into subsets $V_1,\ldots,V_p$, such that vertices i and j are in the same subset V_k if and only if G contains a i–j

path. Let V_k be a subset of the partition. The subgraph $G' = <V_k>$ is called a *component* of G. Two vertices i and j are *connected* when they belong to the same component. The graph $G = (V,E)$ is said to be connected if it has only one component.

The notion of a graph can be extended by allowing multiple edges, i.e. a pair $\{i,j\}$ occurring several times in E, loops, i.e. a pair $\{i,i\}$, or by defining a weight m_i on each vertex i and a length d_{ij} on each edge e_{ij}. It can be restricted by assuming the edges to be ordered pairs of vertices. In the latter case the graph is called a directed graph or digraph. By convention, the vertices in a directed graph are called nodes and the edges are called arcs.

In graph theory a network $N = (V,A)$ is a directed graph together with a source $s \in V$ and a terminal $t \in V$ and a capacity or length b on each arc.

APPENDIX 2: 35 DUTCH MUNICIPALITIES IN 1860

 1 Amsterdam
 2 Rotterdam
 3 The Hague
 4 Utrecht
 5 Leiden
 6 Groningen
 7 Maastricht
 8 Leeuwarden
 9 Haarlem
10 Arnhem
11 Dordrecht
12 's-Hertogenbosch
13 Zwolle
14 Breda
15 Delft
16 Nijmegen
17 Middelburg
18 Schiedam
19 Deventer
20 Gouda
21 Tilburg
22 Den Helder
23 Kampen
24 Zutphen
25 Heerenveen
26 Enschede
27 Amersfoort
28 Vlissingen
29 Veendam
30 Sneek
31 Zaandam
32 Drachten
33 Apeldoorn
34 Eindhoven
35 Hoogezand-Sappemeer

7 Freight Traffic Through the Alps: Peculiarities and Impacts of Abnormal Routing

FABIO ROSSERA

Institute of Economic Research, Bellinzona, Switzerland

7.1 INTRODUCTION

In the years from 1965 to 1990 freight traffic through the Alps more than trebled. The figures have evolved from 18 to 63 million tonnes (Ratti, 1993). This increase has proceeded in parallel with a substantial restructuring of the modal split. During the sixteen-year-long period from 1964 to 1980 four highway tunnels crossing the Alps were opened to traffic. Since then the decrease in the level of competitiveness of the railway system, a general evolution visible at all levels of transport, has spread to this segment as well.

However, this evolution has not taken effect completely, as far as Switzerland is concerned. This is a consequence of the regulations adopted on road haulage in this country. In fact, after the opening of the Gotthard tunnel, Switzerland held to her previous limitation of lorry tonnage to 28 tonnes and to the restrictions concerning circulation during the night hours, on Sundays, and workdays. The consequences of this political attitude may be inferred directly from the statistics of the transport evolution between the different countries. Based on data published by the Swiss Ministry of Transportation one can see that road freight through the Swiss Alps does not reach 10 per cent of that of neighbouring France and Austria.

The problem of freight crossing the Alps has become an important topic in negotiations at the EU level. Until now the European authority has taken into serious consideration the arguments produced by the Swiss government, about environmental protection on the one hand—a matter presenting particular requirements in the Alpine context—and relating to the peculiarities of its political system on the other hand. At the present moment the attitudes of the different parties are far from being clearly established.

For decision-making organs an interesting question concerns the evaluation of the consequences that these restrictive measures may have had and still have on the volume of goods transported. This kind of assessment may be also considered as a necessary point of departure for enlarging the

European Transport and Communications Networks: Policy Evolution and Change. Edited by David Banister, Roberta Capello and Peter Nijkamp. © 1995 John Wiley & Sons Ltd.

analysis to the impact generated in the broader economic and environmental context.

Carrying out this kind of estimate has long been hindered by the lack of reliable statistical documentation. However, recently things have changed so that this gap is being gradually filled.

The purpose of the present chapter is to focus on the main aspects of this problem and to produce a first appraisal of the order of magnitude of the restructuring involved.

In Section 7.2 we will summarize the principal points in the evolution of the transport of goods through the Alps that has occurred recently and give a short presentation of the database we will use for our calculations. In Section 7.3 we will present the methodology we will use and explain our choice. The results derived from our calculations will be presented in Section 7.4. In a final short section we will try to synthesize the main conclusions of this study.

7.2   EVOLUTION 1979–1989

During the 1980s the Swiss Ministry of Transportation gathered an important mass of statistical data giving a detailed description about flows of goods crossing the Alps. Although taking into account that this database is not updated yearly but after a period of five years, this may represent an important source for carrying out detailed investigations about the characteristics of these kinds of flows. Moreover, one can have at one's disposal useful information about their origins and destinations, the modes and routes chosen, and the kinds of goods transported.

For the present analysis we will resort to this set of data (GVF 1994). We will restrict our scope, concentrating only on the points that may be considered as the most interesting for a first characterization of the problems under discussion. All data will refer to the total volume of goods exchanged during a period of a year between both sides of the Alpine chain, without discriminating between the different categories. A distinction will only be drawn as to the direction of the flows. Two shortcomings of these statistics should be taken into account when interpreting results drawn from these data.

The first one concerns the geographical limitation of the arc in the Alp which has been considered. The crossings included, as can be seen from Figure 7.1, extend from those of Mt Cenis and Fréjus in the west side to that of the Brenner in the central-eastern part of the chain. In spite of this limitation one can suppose that the commercial exchanges occurring between countries on the north-western side of the Alps and Italy are reported almost in their entirety. On the other side, a gap remains open as regards contacts between Italy and central Europe, and to a lesser extent with

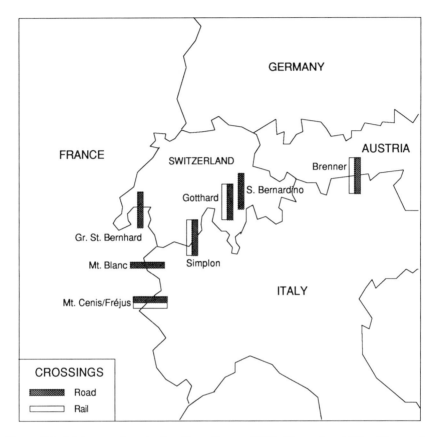

Figure 7.1. Main crossings in the Alps. *Source*: GVF, Sigmaplan 1991 with permission

eastern Europe. More precisely, we are speaking about the goods transiting through the crossings of Tarvisio, Villa Opicina and Ferretti. In conclusion one can say that the data at our disposal are well suited for a study centring on matters which confront Switzerland. Obviously, we need not regard this as a drawback as these problems represent the main focus of interest for our research.

The second shortcoming revealed by our dataset concerns the statistics gathered in relation to Austria. In this case, only flows of transport in transit through this country have been considered. Once again, this need not represent a drawback for our research. We will also remove from it the data originating and terminating in Switzerland and will thereby be considering Switzerland and Austria only as countries of transit which, in fact, largely represents their actual role at an international level.

In synthesis, we can specify that our analysis will concern the transportation of goods between northern and southern parts of the Alps, and will take into account all kinds of land transport. Besides railways and road, the forms of combined transport between rail and road will be considered. (Container transport is included within railway data.) The data we will use refers to three periods of time: 1979, 1984 and 1989, the last date considered in our database. During the whole period considered, the total volume of goods transported increased by about 50 per cent (Table 7.1). This increase was particularly conspicuous during the second subperiod. The total volume lies at a level somewhat under 10 per cent of all international land transport in Europe (Ratti and Rudel 1993, p. 12). We know that during the following years—we have data in aggregated form up to 1992—this growth has been continuing, although at a reduced pace.

From these data it is clear that differences in the growth of the individual modes and routes considered separately may be conspicuous. Transport by railway has been almost stagnating. As universally known, the real segment of growth is represented by road transportation. During this period this has grown up to 70 per cent above its initial level. From the point of view of the rate of growth, combined transport has witnessed a spectacular expansion. However, its importance in comparison with other modes is still limited. Its share of total transport is oscillating around 5 and 6 per cent and has stayed at this level in subsequent years.

Also changing are the shares acquired by individual countries. France's share is growing. This occurred in particular during the first subperiod. This has been paid for by Switzerland which took no part in the important growth of the flows on the road.

Percentage figures in the lower half of the table will enable us to discuss some further details.

First, we can establish how far the Swiss decline is only attributable to the railway component. As far as Switzerland is concerned, one must always highlight how the opening of a highway tunnel in 1980 has shown very limited consequences on the road traffic through this country. Its importance does not compare with the experience of neighbouring countries. Moreover, considering that the length of the crossings should make the Swiss routes the most convenient crossings, one is able to gain a first impression of the effectiveness of the restrictions imposed.

Therefore, as far as road transportation is concerned we can establish the existence of clear sidestepping effects. France seems to have been the main beneficiary of this state. It also represents the only country being considered here without restrictions on haulage.

It also appears clearly from these data how Switzerland represents the only country which up to now has made considerable efforts in order to find in the combined transport an alternative solution for attaining some control over the worrying increase in the volume of road transport. The Swiss com-

Table 7.1. Freight traffic through the Alps

	France	Switzerland	Austria	1979 Total
Railway	5626	10505	2393	18524
Road	8243	302	12522	21067
Combined	446	712	122	1280
Total	14314	11519	15038	40871
	France	Switzerland	Austria	1984 Total
Railway	7192	9152	2727	19071
Road	10501	751	15168	26420
Combined	695	1251	473	2419
Total	18388	11154	18368	47910
	France	Switzerland	Austria	1989 Total
Railway	6001	11669	2917	20587
Road	16226	1551	18679	36456
Combined	659	2339	726	3724
Total	22886	15559	22322	60767

Structure of flows				Percentages
	France	Switzerland	Austria	1979 Total
Railway	14	26	6	45
Road	20	1	31	52
Combined	1	2	0	3
Total	35	28	37	100
	France	Switzerland	Austria	1984 Total
Railway	15	19	6	40
Road	22	2	32	55
Combined	1	3	1	5
Total	38	23	38	100
	France	Switzerland	Austria	1989 Total
Railway	10	19	5	34
Road	27	3	31	60
Combined	1	4	1	6
Total	38	26	37	100

Only flows in transit through the crossings reported in Figure 7.1
Source: GVF 1991

mitment to this objective is remarkable. Shortly, works for fitting the railway lines of the Gotthard and Simplon to the new transport mode will be carried out. By the year 2015, the construction of two new transversal lines through the Alps should be completed. They will represent a substantial increase in the capacity supplied. In spite of all these considerable efforts results are still slow to materialize.

Some further important characteristics shall be mentioned, before ending these short introductory notes.

There exists an important imbalance between flows in opposite directions. In fact, those directed from north to south represent about two-thirds of the total volume. Also the modal split is somewhat different, as can be deduced from Table 7.2. The importance of road transportation is further stressed in the flows toward northern countries. This may partially reflect the peculiarities of the haulage sector in Italy and the attitude of the Italian government.

Considering the distribution among the countries of transit, we observe that Switzerland is—always as a consequence of the same motive—at a disadvantage in the flows in the same direction. Differences do not exist as far as exchanges between Italy and France are concerned. Also in this case the peculiarities of the respective haulage sectors may be invoked for an explanation.

Given the preliminary character of the present study, we choose to restrict our analysis to a limited number of contacts and to the flows directed from north to south. As we will see, this will provide us with sufficient means in order to synthesize in a satisfactory way the conclusions that could be drawn by a more complete investigation.

In Table 7.3 we give a list of the regions we have considered. The possible combinations would allow us to build a matrix with 121 entries. Taking into account the fact that we are interested only in the connections

Table 7.2. Direction of flows

By mode				
	Railway	*Road*	*Combined*	*Total*
North–south	45%	50%	6%	100%
South–north	24%	69%	7%	100%
By nation				
	France	*Switzerland*	*Austria*	*Total*
North–south	38%	29%	33%	100%
South–north	37%	19%	44%	100%

Source: GVF 1991 with permission

Table 7.3. Regions considered by the study

Oberbayern
Karlsruhe – Stuttgart
Hessen
Ruhrgebiet
Alsace – Lorraine
Ile-de-France – Normandie
Belgium – Luxembourg – Netherlands
Piemonte
Lombardia
Veneto – Friuli
Emilia-Romagna – Toscana

---

across the chain of the Alps, the number of combinations comes down to 28. The total of goods in tonnes transported along these connections represented 53 per cent of the total in 1989 (considering only traffic in transit).

7.3 METHODOLOGICAL NOTES

Our problem refers to route choice. In order to evaluate the impact of policy restrictions the obvious choice should be the usual distribution–assignment methodologies. In fact, in transport planning it is usual to solve this task in three steps. First, minimum cost routes are computed. Second, trips between origin and destination nodes are assessed. Lastly, traffic is assigned to the minimum cost routes. The required computations may be performed in a number of separate steps (Harvard–Brookings model, Kresge and Roberts 1971) or in an integrated algorithm (Evans 1976, and many others).

The inconvenience we face lies in the fact that we dispose only of data on flows between origins and destinations but not on individual links. However, some peculiarities of the problem at hand should allow us to look for some simplifying adaptations.

A first point relates to the fact that the variable under consideration— regulatory restrictions—unfolds its effects in a uniform way which may be quantified by simple dichotomous variables. It may also be noticed that the effect produced is of such importance that an approximate evaluation may also be of interest.

As a second point, it has to be remarked how our analysis is restricted to flows crossing the Alps. In such cases, the impact is spread along the whole route, from the origin to the final destination, and not concentrated on individual links. Moreover, the choice among routes is very limited since the number of crossings amounts to three per mode.

Following the reasoning of the planning steps mentioned above, we propose the following revision. We will keep to the first step, although we use the length of trip, a rough approximation for costs in this context, to find out the most suitable routes. The last step, traffic assignment, will be suppressed as such. The information required for its completion will be partially integrated into the gravitational model, whose calibration represents the second step of our procedure. This last model will be adapted with the introduction of a series of categorical measures related to the kinds of variables considered in the assignment problem.

More formally, our model may be presented along the following lines. Its derivation is from an entropy function, so that we will have to solve the following optimization problem:

$$\max_{t_{ij}} - \sum_i \sum_j t_{ij} \ln t_{ij} + \sum_i \sum_j t_{ij} \tag{7.1}$$

with the usual restrictions

$$\sum_j t_{ij} = t_i \tag{7.2}$$

$$\sum_i t_{ij} = t_j \tag{7.3}$$

$$\sum_i \sum_j t_{ij} c_{ij} = C \qquad i = 1,\ldots,I, \quad j = 1,\ldots,J \tag{7.4}$$

to which we add the more specific restrictions

$$\sum_i \sum_j t_{ij} \delta_{ij.k} = t_k \qquad k \in K \tag{7.5}$$

$\delta_{ij.k} = 1$ if the route from i to j belongs to category k, $\delta_{ij.k} = 0$ otherwise. The set K consists of all possible combinations of modes of transport and choice of routes. In our case rail, road, and a combination of the two on the one hand, and France, Austria, Switzerland on the other. We will always have nine categories and nine restrictions. Obviously, the categories may be built in several ways, depending on the kinds of hypotheses we wish to test.

In the present context two schemes seem to be particularly appropriate. A first scheme will represent a saturated factorial design, defined as follows:

$$\kappa_{l,m} \equiv \mu + \alpha_l + \beta_m + \lambda_{l,m} \qquad l = 1,2,3; \ m = 1,2,3 \tag{7.6}$$

μ is a measure of scale, α and β are the main effects, while the parameters λ represent the interaction between categories. In an analysis of variance context (Searle 1971), κ represents the expected value for an individual category (l,m). In our case, it represents one of the components in an analysis of covariance context, as we shall see. This design is appropriate for analysing and measuring the interactions existing between modes and routes.

For estimation purposes, we will suppress redundant parameters. More precisely, we will eliminate, as usual, a parameter for each main effect, and the corresponding parameters at interaction level. We will have

$$\kappa_{l,m} \equiv \mu^* + \alpha_l^* + \beta_m^* + \lambda_{l,m}^* \qquad l = 1,2,3; \; m = 1,2,3$$

and

$$\alpha_l^* = \beta_m^* = \lambda_{l,m}^* = 0 \qquad \text{for} \quad (l = 3) \cup (m = 3) \tag{7.7}$$

Limiting our interest to the interaction effects, the estimated parameters will have the following interpretation:

$$\lambda_{l,m}^* = (\lambda_{l,m} - \lambda_{l,r_m}) - (\lambda_{r_l,m} - \lambda_{r_l,r_m}) \tag{7.8}$$

where the two indices r_l and r_m refer to the reference categories of the two main effects. This represents the usual statistics for testing independence. For impact evaluation we will resort to a second specification, using a hierarchical design:

$$\kappa_{l,m} \equiv \mu + \alpha_l + \alpha_{l,m} \qquad l = 1,2,3; \quad m = 1,2,3 \tag{7.9}$$

In this case we concentrate on a single main effect, and examine successively what are its interactions with the second factor. In our case, the mode of transport is an obvious candidate for the main effect. For estimation we will use:

$$\kappa_{l,m} \equiv \mu^* + \alpha_l^* + \alpha_{l,m}^* \qquad l = 1,2,3; \; m = 1,2,3$$

and

$$\alpha_l^* = \alpha_{l,m}^* = 0 \qquad \text{for} \quad (l = 3) \cup (m = 3) \tag{7.10}$$

In this case the interaction effects will differ from those of the previous model and will be measured by

$$\alpha_{l,m}^* \equiv \alpha_{l,m} - \alpha_{l,r_m} \tag{7.11}$$

For instance, let us suppose we are considering the rail as our main effect (index 1). We could measure with this coefficient the difference between the intensity of the transit through France (category m) and Switzerland as a reference category (r_m).

For carrying out our simulations we will be interested also in the parameters of the model concerning the main effects. But first we change our model somewhat to

$$\kappa_{l,m} \equiv \alpha_l^* + \alpha_{l,m}^* \qquad l = 1,2,3; \; m = 1,2,3$$

and

$$\alpha_{l,m}^* = 0 \qquad \text{for} \quad (l = 3) \cup (m = 3) \tag{7.12}$$

The interpretation of the $\alpha_{l,m}^*$ does not change, that of the main effects becomes

$$\alpha_l^* \equiv \mu + \alpha_l + \alpha_{l,r_m} \tag{7.13}$$

It should be noted that these last coefficients are of scarce relevance *per se*, as they are a composition of different effects. The role they can play will become clear when we discuss the out-of-sample adjustment method.

Let us now come back to our optimization problem. We derive from it the following gravitational model:

$$t_{ij} = a_i t_i b_j t_j \exp\left(-\beta \ln c_{ij} + \sum_k \gamma_k \delta_{ij,k}\right) \tag{7.14}$$

$$i = 1,\ldots,O; \; j = 1,\ldots,D; \; k = 1,\ldots,K$$

This is the specification habitually used for doubly constrained models, with the addition of one of the qualitative designs mentioned before. In this last respect, all previous parameters have now been redefined as γ parameters. In the gravitational specification β and the γ are Lagrange multipliers derived from the solution of problem (7.1)–(7.5). The a and b represent balancing factors. All parameters will be estimated by the criterion of maximum likelihood (Sen 1986).

Our second methodological problem concerns the projections outside the sample used for parameter estimation. Substantially, we look for an answer for this kind of question: what will happen to the traffic distribution among modes and routes if intervention from the outside is suppressed?

First, let us make a simplifying assumption. We will suppose that demand and supply conditions at origins and destinations will stay the same. This seems to be an acceptable simplification, considering that we are mainly interested in route choice.

A second assumption is more questionable. Suppressing restrictions hindering road haulage will change the general supply conditions, and may cause some change in the volume of traffic. We will suppose that this fact would not modify our conclusions substantially, and, as a consequence, consider also the volume of traffic as fixed.

Hence, the problem of evaluating the impact of alternatives is twofold. On the one hand, there is the problem of changing the impact of the qualitative variables. This can be done by appropriate transformations of the estimated parameters, which we will present further on. On the other hand, the new estimates of the flows should be adjusted so as to satisfy the original origins–destinations constraints.

Following Erlander (1980), the latter point may be formalized as follows. To simplify our argument, let us suppose that we are changing the conditions for only one mode, e.g. road haulage. We should modify three parameters: γ_h, referring to road transport, $\gamma_{h.1}$ and $\gamma_{h.2}$ measuring the difference of impact in the case of France and Austria, as compared with Switzerland.

Let t_{ij}^0 represent the expected flow between i and j on the basis of our first estimated model, that used for parameter calibration. The new estimate of our flow between i and j, t_{ij}^1, say, will be

$$t_{ij}^1 = r_i s_j t_{ij}^0 \exp\left\{(\gamma_h^1 - \gamma_h^0)\delta_{ij.h} + (\gamma_{h.1}^1 - \gamma_{h.1}^0)\delta_{ij.h_1} + (\gamma_{h.2}^1 - \gamma_{h.2}^0)\delta_{ij.h_2}\right\} \quad (7.15)$$

In the exponential function we have introduced the parameters modified *ex ante*. The r and s represent new balancing factors that should consent the realization of restrictions (7.2) and (7.3).

We shall now define the criteria in order to calculate the new γ parameters and come to the first problem mentioned. As mentioned before $\gamma_{h.1}$ and $\gamma_{h.2}$ represent deviations and so may be dropped. In this way we have

$$\gamma_{h.1}^1 - \gamma_{h.1}^0 = -\gamma_{h.1}^0 \qquad \text{and} \qquad \gamma_{h.2}^1 - \gamma_{h.2}^0 = -\gamma_{h.2}^0 \quad (7.16)$$

For the main effect, by using our former notation (7.13), we have

$$\gamma_h = \alpha_h^* = \mu + \alpha_h + \alpha_{h.r_m}$$

Now, it seems appropriate to add to this coefficient the coefficients corresponding to $\alpha_{h.m}^*$ defined in (7.11) that we have eliminated in (7.16). Then, supposing M to be the number of transport modes, we could use as new parameters

$$\gamma_h^1 = \frac{1}{M}\sum_m^{M-1}[\alpha_h^* + \alpha_{h.m}^*] = \alpha_h^* + \frac{1}{M}\sum_m^{M-1}\alpha_{h.m}^*$$

$$= \mu + \alpha_h + \frac{1}{M}\sum_m^{M}\alpha_{h.m} \quad (7.17)$$

We repeat this transformation for all modes whose interaction effects we want to neutralize. These parameters should not differ too much from those estimated on a new sample reflecting the newly imposed conditions.

7.4 ESTIMATIONS AND PREDICTIONS

We present here the results for three adjustments. The first two refer to estimation of the gravitational models, the third represents an evaluation of the restrictions brought about by changes in the qualitative variables.

The first specification we use for the gravitation model is to incorporate factorial design. This gives the best opportunity for evaluating the interactions present in our sample. The formal definition of this specification has been given in equations (7.7) and (7.8). We will limit our comment here to the results obtained. We present them in Table 7.4.

In order to simplify our exposition, we have also introduced into our table a reference to the original parameters. The adjustment has been carried out on the flows concerning the eleven regions presented in Section 7.2, and considering only the traffic in the north–south direction.

All coefficients results are particularly significant. The first of them is of limited interest. The following four relate to the main effects. As interaction effects have been included in our model, the interpretation of these parameters differs somewhat from that usually reserved for other models. They not only measure the deviation of the main effect in question from that taken as a reference, but also include some interaction effects.

Take, for instance, α_1^*. It measures the difference between rail and combined transport, our reference category here. But it considers this deviation at the level of main effects in general and at the more limited level of the reference category for the second main effect, in this case Austria.

Considering that combined transport represents the mode less frequently present on these trips, the estimates for the first factor seem quite plausible.

Table 7.4. Parameters of factorial design

Codes:	A. Modes:	1 = Rail	2 = Road	3 = Combined	Reference = 3
	B. Routes:	1 = France	2 = Switz.	3 = Austria	Reference = 3

Parameter	Parameters of original model	Estimates	St. Dev.
μ^*	$\mu + \alpha_3 + \beta_3 + \gamma_{33}$	3.57	0.0059
α_1^*	$(\alpha_1 - \alpha_3) + (\gamma_{13} - \gamma_{33})$	1.05	0.0020
α_2^*	$(\alpha_2 - \alpha_3) + (\gamma_{23} - \gamma_{33})$	2.51	0.0018
β_1^*	$(\beta_1 - \beta_3) + (\gamma_{31} - \gamma_{33})$	−0.09	0.0024
β_2^*	$(\beta_2 - \beta_3) + (\gamma_{32} - \gamma_{33})$	−0.19	0.0020
γ_{11}^*	$(\gamma_{11} - \gamma_{13}) + (\gamma_{31} - \gamma_{33})$	0.4	0.0027
γ_{12}^*	$(\gamma_{12} - \gamma_{13}) + (\gamma_{32} - \gamma_{33})$	0.7	0.0023
γ_{21}^*	$(\gamma_{21} - \gamma_{23}) + (\gamma_{31} - \gamma_{33})$	0.17	0.0025
γ_{22}^*	$(\gamma_{22} - \gamma_{23}) + (\gamma_{32} - \gamma_{33})$	−3.41	0.0025
β	distance	−3.62	0.0008

Those for the route choice are less pronounced. In this case, main and inter-action effects may compensate each other.

The most interesting results are of course those concerning interaction effects. With γ^*_{11} we consider the difference existing between France and Austria, at the level of railway and at that of combined transport, once more our reference category. This parameter shows a slight pre-eminence for France. In the case of road haulage this difference is more restricted (coefficient γ^*_{21}). If we subtract the second from the first parameter we will be able to see how much better France performs than Austria on rail, compared with the road.

Regarding Switzerland the differences are, as expected, much more marked. Already the railway comes off better than in Austria. This is not surprising as it has always represented the mode par excellence for Switzerland (Bertschi 1986). But what is of particular relevance in this context is the order of magnitude measured by the coefficient γ^*_{22}, relating the interaction between road transport and transit through Switzerland, which shows the extent of the impact of the limitations imposed by the Swiss government.

In synthesis, we can conclude from the results of this first model that interaction effects are greatly concentrated on the negative correlation existing between road haulage and transit through a Swiss crosspoint. This shows us the way to follow for our simulations. Moreover, it indicates that changes on the coefficients of the road mode may be envisaged without disturbing the structure of the other modes.

To carry out these simulations, we will start from another specification, that of the hierarchical model presented in (7.12) and (7.13). Results are shown in Table 7.5.

The first three estimates are of scant interest *per se*, as they include different original parameters. They will become useful only for our predictions. Obviously, the interactions evidenced in the first model are confirmed by the present specification, so that no further comment is needed. For all modes, the position of Switzerland is antithetical to those of the two other countries. On the other hand, these two present similar characteristics.

Also in the present case the great contrast between Switzerland and the other two nations lies at the level of road transport. At this level, there is no substantial difference between France and Austria.

The last step in our calculations will be dedicated to the evaluation of alternative configurations in the route assignment. We are interested in assessing the impact of suppressing the restrictions on road haulage on the Swiss side. We suppose that the total volume of goods transported on the road should stay at the same level and that the shares between the modes should in principle remain without changes. On the other hand, flows on the individual routes of transit should undergo important modifications. An important share should flow back to the Swiss axes, taking advantage of their geographical location.

Table 7.5. Parameters of hierarchical design

Codes:	A. Modes:	1 = Rail	2 = Road	3 = Combined	Reference = 3
	B. Routes:	1 = France	2 = Switz.	3 = Austria	Reference = 2

Parameter	Parameters of original model	Estimates	St. Dev.
α_1^*	$\mu + \alpha_1 + \alpha_{12}$	3.57	0.0059
α_2^*	$\mu + \alpha_2 + \alpha_{22}$	1.05	0.0020
α_3^*	$\mu + \alpha_3 + \alpha_{32}$	2.51	0.0018
α_{11}^*	$\alpha_{12} - \alpha_{12}$	−0.09	0.0024
α_{13}^*	$\alpha_{13} - \alpha_{12}$	−0.19	0.0020
α_{21}^*	$\alpha_{21} - \alpha_{22}$	0.4	0.0027
α_{23}^*	$\alpha_{23} - \alpha_{22}$	0.7	0.0023
α_{31}^*	$\alpha_{31} - \alpha_{32}$	0.17	0.0025
α_{33}^*	$\alpha_{33} - \alpha_{32}$	−3.41	0.0025
β	distance	−3.62	0.0008

Table 7.6. Projections results (tonnes)

Situation 1989				
	France	*Austria*	*Switzerland*	*Total*
Rail	1 526 981	960 110	5 939 075	8 426 166
Road	5 130 708	4 438 226	431 800	10 000 734
Combined	359 455	336 815	1 037 053	1 733 323
Total	7 017 144	5 735 151	7 407 928	20 160 223

Equalization on road				
	France	*Austria*	*Switzerland*	*Total*
Rail	1 647 650	1 116 191	5 680 233	8 444 074
Road	1 978 961	2 053 772	5 912 130	9 944 863
Combined	387 861	391 570	991 855	1 771 286
Total	4 014 472	3 561 533	12 584 218	20 160 223

General equalization				
	France	*Austria*	*Switzerland*	*Total*
Rail	1 760 029	1 582 410	4 989 397	8 331 836
Road	1 978 585	1 992 117	5 929 938	9 900 640
Combined	407 220	366 124	1 154 403	1 927 747
Total	4 145 834	3 940 651	12 073 738	20 160 223

In Table 7.6 we first present the decomposition of the flows along modes and routes, as registered for our year of observation, 1989. These figures correspond to those replicated by our saturated model.

The first projection that has been carried out is based on an equalization of conditions on the three transit axes, limited to road haulage. This is in line with the problem we stated from the beginning. The figures in the table show us that the consequences will be of considerable impact. Transit across France and Austria is reduced to less than a half. On the contrary, the volume of goods transported through Switzerland soar by a factor of 14. This order of magnitude is in line with estimations produced elsewhere.

Judging from the figures in the table this realignment shows some subsequent effects on the other categories. The changes show some plausibility, but it must be said that they do not originate from an explicit formulation in the model but they result from the logic of the balancing procedure. Important changes in the flows on the road may unbalance the total of individual nodes. The action of the balancing procedure will look for compensations among flows which are complementary to those primarily touched by our parameter changes. We note that transport through France and Austria by rail and combined transport is somewhat increased, which need not necessarily be in line with what could be empirically assessed.

On the other hand, if we consider the volume transported along the Swiss rail axes, we notice how it is decreasing. Therefore, the balancing procedure must get rid of the surplus generated on these routes by the enormous increase on the roads. This indirect effect seems to be quite plausible, but for a different motive. Some rail routes actually used through Switzerland are chosen because of the fact that a deviation by road would be so much harder. Unregulated freight haulage on the road would capture a considerable share of rail transport, maintaining, however, the transit through Switzerland.

In view of these considerations we opt for a second prediction, the last presented in the table, in which all interaction effects are eliminated. Only peculiarities of an economic and/or geographical kind will unfold their effects: the importance of the individual region of origin/destination and their respective geographical location.

Results do not change much, so we can stay with our previous conclusions, but their derivation comes from an explicit modification of the parameters of all the categories.

7.5   CONCLUSIONS

In the present analysis we have concentrated upon the consequences of an important decision of the Swiss government in the matter of transport policy: that of imposing severe restrictions on road haulage. By this, the structure of flows in transit through different countries has been deeply modified. It seems evident that bringing the Swiss regulation in line with that of the neighbouring countries would imply a deep restructuring, difficult to implement.

For our part, we were interested in the question of the order of magnitude of the implied changes and the resulting new structure of the flows. We have been able to verify how coming back into the old track, where only economic and geographic characteristics may be of consequence, would heavily change the actual shares along the transit routes on road: 13 to 14 times more traffic for Switzerland, less than half as much for the other two countries.

The model used to reach these conclusions represents a simplification of the procedure habitually used for freight predictions. Only the existence of particular conditions in the case we focused upon may confer a plausibility to the adoption of such heavy restrictions. In our case, the two main peculiarities we exploited were the following. The structure of flows is particularly unbalanced on a single mode: that of using the road. On the other hand, the effects of the restrictions on the traffic unfold their effects in an uniform way along entire routes of transit, not on individual links.

REFERENCES

Bertschi, H. P. (1985) *Der alpenquerende Verkehr dargestellt am Ausbau einer neuen Eisenbahntransversale durch die Schweiz*, P. Haupt, Bern.

Erlander, S. (1980) *Optimal Spatial Interaction and the Gravity Model*, Springer Verlag, Berlin.

Evans, S. P. (1976) Derivation and analysis of some models for combining trip distribution and assignment, *Transportation Research*, **10**, 37–57.

GVF (1991) GS EVED, *Dienst für Gesamtverkehrsfragen; Transalpiner Güterverkehr*. Auswirkungen des Gotthard-Strassentunnels auf den Güterverkehr, Bern.

GVF (1994) *Database on freight traffic through the Alps, special delivery*, Bern.

Kresge, D. T., P. O. Roberts (1971) *Techniques of Transport Planning*, Vol. II, The Brookings Institution, Washington D.C.

Ratti, R. and R. Rudel (1993) Tableau de l'évolution des transports dans l'arc alpin, *Revue de Géographie Alpine*, **LXXXI**(4), 11–26.

Searle, S. R. (1971) *Linear Models*, Wiley, New York.

Sen, A. (1986) Maximum likelihood estimation of gravity model parameters, *Journal of Regional Science*, **26**, 461–474.

8 A General Equilibrium Analysis of the Italian Transport System[1]

ROBERTO ROSON

University of Venice, Italy

8.1 INTRODUCTION

MITER is a multiregional/network computable general equilibrium (CGE) model of Italy, developed with the aim of providing a tool for the analysis of interaction processes between the economic and the transport system. Using this model it is possible to highlight how demand for freight transport is generated by trade relationships, how transport costs vary to match demand and supply of transport infrastructures, and how transport costs, via market prices, influence economic competitiveness of regions and industries.

The complex mathematical structure of the model, integrating a CGE model, a dispersed spatial price equilibrium (DSPE) model and a freight network equilibrium (FNE) model, as well as several issues associated with the development of the database and the estimation of the model's parameters, are fully described in Roson (1994). For the sake of simplicity, we shall only illustrate here the logical structure of the model, in order to focus on the results obtained by a simulation exercise.

The simulation was carried out by interfacing the MITER model with a regional econometric model (MORE), independently developed by a team from GRETA Econometrics srl, led by Professor Giorgio Brunello of the University of Venice. This model was used chiefly to obtain exogenous values for final demand (public expenditure, investments, exports), forecast for the year 2000 and used as inputs in the quantity system. The MITER model was therefore run to derive estimated congestion levels in the transport system, equilibrium values for relative prices, regional and industrial production levels consistent with the given exogenous scenario, assuming fixed and invariant characteristics of the transport network at their 1988 values, which is the calibration year of the model.

[1] Technical assistance by Piero Vianelli and Stefano Pomaro, and financial support from the Italian National Research Council (CNR—Progetto Finalizzato Trasporti 2) are gratefully acknowledged.

European Transport and Communications Networks: Policy Evolution and Change. Edited by David Banister, Roberta Capello and Peter Nijkamp. © 1995 John Wiley & Sons Ltd.

Clearly, this assumption of invariance in the stock of transport infrastructures makes our experiment unsuitable for drawing conclusions about
the future evolution of the Italian transport system. Why we introduced
such an assumption is apparent in our willingness to test the MITER
model against other approaches, such as the Leontief-Costa PGT model
(Leontief and Costa 1987), where supply constraints in the transport system
are not explicitly taken into account. In other words, we are seeking to
ascertain whether the determination of a simultaneous equilibrium in the
economic and transport system would significantly affect the model's
results, from a qualitative point of view. Figure 8.1 shows the model's
general framework.

The MORE econometric model provides a forecast of exogenous variables
which is used as an input in the quantity system, appearing in the upper
right corner of the figure. Besides public expenditure, investments and
exports, the model requires a level of real social services used in order to
determine part of household consumption.

The quantity system is based on a multiregional extended input–output
model, estimating trade flows and production levels for 17 industries in 20
regions. Interregional trade flows are computed for every pair of regions and
for every type of demand (intermediate, consumption, etc.). Since constant
returns to scale are assumed and relative prices are—at this stage—given,
the system is a linear one. Regional household consumption and imports are
endogenous variables, determined by regional income.

Aggregating by industry, trade flows may be computed among regions
and to/from foreign countries. Both of these flows may also be expressed in
quantity units of measure and split into transport modes, producing a series
of origin–destination matrices which express the potential demand for transport services generated by the economic system, under the implicit assumption of excess supply in transport infrastructure. The output of this
'unconstrained' model may then be taken as a benchmark for comparison
with the results obtained from the complete model.

In the current version of the model, modal splitting is carried out using
fixed shares (for each origin/destination pair) and, as congestion effects are
considered only for road transport, only road traffic flows are really needed
during iterations. Since all flows are expressed in tonnes, they can be aggregated in a single 20 × 20 o/d matrix, comprehensive of traffic to and from
national borders (import, export, transit). Passenger traffic flows, obtained
by applying specific growth rates to the base traffic volumes, can also be
added once freight flows have been translated into vehicle units.

Allocation of total traffic flows to the interregional network is then performed by the network system, where arc costs are updated so as to
balance supply and demand. The transport network is an aggregated one,
that is, a virtual graph connecting regional centroids, whose parameters
have nonetheless been derived from data referring to real road networks.

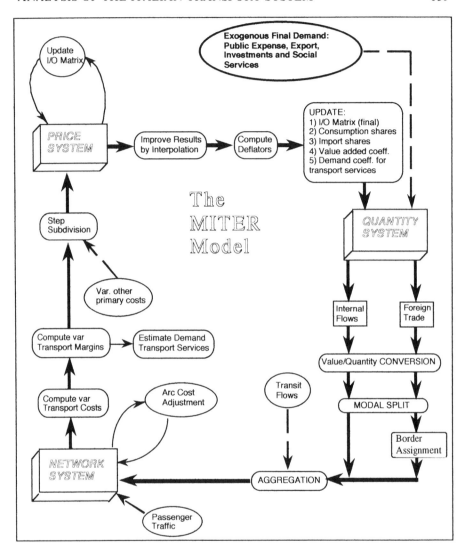

Figure 8.1. General structure of the MITER model

For each o/d pair, it is assumed that there exist one or more alternative paths, which are sequences of arcs connecting origin to destination, and that these paths are known in advance. Route choice is based on decentralized cost minimization (Wardropian user equilibrium), where the cost to be minimized, however, comprises all direct and indirect costs associated with moving goods, such as delays, safety, uncertainty about delivery times, and so on. This 'generalized' cost is not unique, but depends on the char-

acteristics of the transported items, so that optimal choices may differ among agents.

Alternatively, traffic flows may be assigned to the set of available paths according to total transport cost minimization (system optimum). This is a case of efficient centralized planning, which can be achieved by using an appropriate road pricing system.

Transport costs in the various regions can then be computed as averages of path costs, with the relative incidence of road transport and the existence of fixed costs also being taken into consideration. Market prices, comprehensive of transport cost margins, may then be computed.

Equilibrium prices are determined by means of the price non-linear system, which is solved by iteration, subdividing variations in transport margins and interpolating the observed variations in unitary prices. The price system determines equilibrium unitary prices for goods and services produced in each of the 340 regional industries, by equating prices with cost functions. These functions are in turn obtained by cost minimization from a set of nested CRESH-CES production functions. Input factor costs are evaluated at market prices, including transport cost margins. When congestion phenomena are reflected in higher transport and market costs, substitution occurs in both the industrial and regional composition of factors.

Before repeating the cycle, several parameters are updated and some new variables are computed. These include a new multiregional input/output matrix, made consistent with the new set of relative prices, consumer price indices, consumption and import shares, and value added coefficients. The demand for transport services and the input/output coefficients of the transport industries are also modified in order to take variations in costs and trade patterns into account.

At the end of the first cycle, the economic structure has been modified, and a 'rebound effect' is reflected in reduced system productivity and efficiency, which nonetheless also lowers demand for mobility. Further iteration cycles can then be performed to approximate the simultaneous equilibrium state in the economic and transport systems. Iterations may be stopped once endogenous variables have become approximately constant, without exhibiting relevant variations from the previous cycle. Convergence of the iteration process is normally assured by the model structure.

8.2   SYSTEM OPTIMUM VS. USER OPTIMUM

All choices associated with transport activities, like mode, time, route choice, and others, are usually modelled under the assumption of cost-minimizing behaviour. In particular, it is normally assumed that agents choose a minimum cost path when moving something or somebody from a given origin point to a given destination point.

However, if traffic in some network links has already reached the congestion level, then the cost of using these links, and of following all paths sharing these links, depends on traffic flows. In this case, path choices may create negative cost externalities, and decentralized decision-making may bring about inefficiencies in the distribution of traffic flows, implying higher total transport costs.

This phenomenon, which is just a special case of market failure in the presence of externalities, has long been known in transport modelling (Wardrop 1952; Beckmann et al 1956; Gartner 1980). Two different concepts of network equilibrium have therefore been introduced: user optimum, that is decentralized cost minimization (equivalent to a Nash game between shippers), and system optimum, that is overall cost minimization and centralized flow assignment, given a demand for transport services by origin–destination pairs.

Whereas the first type of network equilibrium turns out to be suited to model actual choices in transport markets, and for this reason it has been adopted in MITER to assign freight flows to the interregional network, the second one may be used as a benchmark for the evaluation of road pricing policies, devoted to the internalization of congestion externalities and, consequently, to the improvement of system efficiency.

Necessary conditions associated with either user or system optimum equilibrium may be compared and easily interpreted. User equilibrium implies that chosen paths are all minimum cost paths (consistency between path choices and costs) whereas, on the other hand, overall cost minimization implies that chosen paths are all minimum *marginal* cost paths. In other words, in an optimal flow distribution some agents may be forced to choose a route that does not minimize their own transport costs, but that minimizes the impact on total costs. This target could be achieved by introducing a specific taxation/subsidization system.

The economic impact of inefficiencies in flow distribution is clearly proportional to the amount of congestion already present in the system, because in the case of excess supply of transport infrastructure there would be no difference between system and user optimum. A model like MITER, designed to assess the macroeconomic implications of traffic congestion, can also be used to compare the two situations and to identify some structural weaknesses of a given transport network. This can be done by carrying out two simulation exercises, using alternative assumptions for flow assignment in the network submodel.

In this submodel, the adopted congestion function is a strictly convex, differentiable one (Cascetta 1990):

$$c_a = \bar{c}_a \left(1 + \beta \left(\frac{f_a}{\bar{f}_a} \right)^{\lambda} \right) \tag{8.1}$$

where c_a and f_a are arc cost and flow, and all remaining symbols are parameters.

The two parameters with the bar on top represent, respectively, the basic uncongested cost and the reference congestion flow. β is the percentage increase in cost when the congestion level is reached, whereas λ determines the function convexity.

It is interesting to note that the function's degree of convexity is not a merely technical question, because the flows we are considering cover a fairly long period (a year). When some level of congestion is reached during the year, the agents may still consider changing the time distribution of demand, thereby avoiding peak periods but incurring extra costs associated with suboptimal departure times. Consequently, assumptions on parameter values imply specific hypotheses on the agents' behaviour and on their possibilities of substitution.

Following some examples drawn from literature, parameters for this function were arbitrarily set at $\beta = 4$ and $\lambda = 0.1$, thereby ensuring that costs are 10 per cent higher than their base values when the capacity level is reached, 50 per cent higher when traffic flow exceeds the capacity by 50 per cent, and 160 per cent higher when the actual flow is double the reference capacity flow.

The marginal cost function associated with (8.1) has the same functional form, with a steeper gradient:

$$c_a = \overline{c}_a \left(1 + \beta(\lambda + 1)\left(\frac{f_a}{\overline{f}_a}\right)^{\lambda}\right) \tag{8.2}$$

Both functions are displayed in Figure 8.2, where the reference capacity flow and the uncongested cost have been normalized to one.

It is interesting to note that in the case of congestion, externalities can be considered as a special example of externalities associated with transport activities and, from a qualitative point of view, the results of the simulation exercise remain valid even when other externalities are taken into account. Internalization of environmental externalities, for example, can be simulated by using, instead of the marginal cost function, a social cost function having a similar shape, but shifted upward.

8.3 AN ECONOMIC EVALUATION OF CONGESTION AND SYSTEM INEFFICIENCY IN THE ITALIAN TRANSPORT SYSTEM

Figure 8.3 graphically displays forecasted national growth in GDP and its components, produced by the MORE multiregional econometric model, between 1988 and 2000. In this period, real GDP is expected to rise by

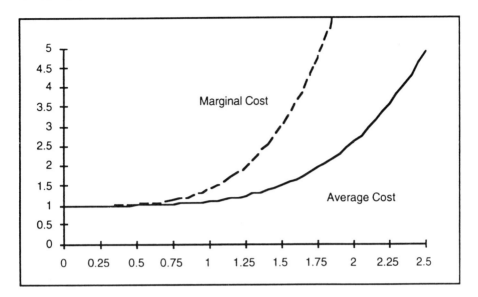

Figure 8.2. Average and marginal congestion functions

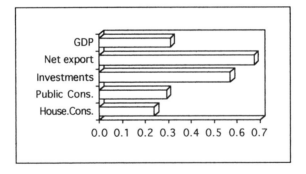

Figure 8.3. Relative growth in the aggregated GDP components

about 30 per cent, averaging rather different growth rates in final demand elements.

The highest growth rates are assumed to occur in import and export flows, testifying to the increasing internationalization of the Italian economy; on the other hand, consumption by households and private institutions, the most important component of GDP, is expected to slow down the growth process observed in the 1980s.

Total increase in GDP amounts to an average yearly growth rate of 2.26 per cent (Figure 8.3). Higher or lower growth rates are assigned to the

Figure 8.4. A map of Italy, showing the twenty regions and their abbreviations

regional economies, depending on the composition of regional industrial structures.

To help pinpoint the geographical position of each region, Figure 8.4 shows a map of Italy, subdivided into administrative regions. This map also includes the abbreviations of the names of regions adopted in this study.

The macroeconomic scenario produced by MORE was used to estimate national and regional economic variables, like GDPs, household consumption, imports, price levels, and so on, as well as to simulate an equilibrium distribution of traffic flows in the interregional road network.

Before going on to illustrate how rising transport costs may affect the economic system, consider the flow distribution obtained by the model, under the assumptions of constant infrastructure stock and of user optimum assignment rule.

Figure 8.5 shows freight flow distribution by the year 2000. Traffic flows

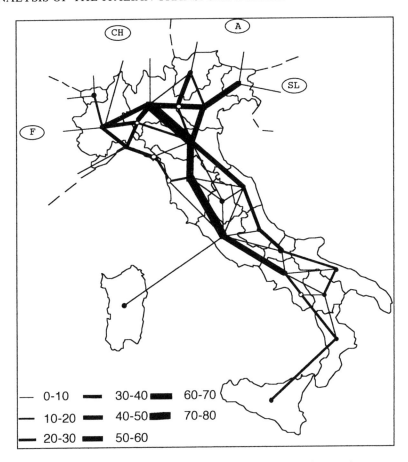

Figure 8.5. Total forecast freight traffic—Year 2000 (millions of tonnes)

appear to be polarized in a few main interregional 'lines', such as an east–west line in the Po valley, a line crossing the Apennines near Bologna and ending in Naples, and a line following the Adriatic coast.

This distribution is rather similar to the current one, but if one looks at differences in flow patterns for 1988 and 2000, it is possible to see that additional freight traffic is more evenly distributed (Figure 8.6). This is the result of rising congestion costs on heavily used links, making alternative routes more convenient.

From the economic point of view, the phenomenon of rising congestion costs may be considered as a reduction in the technological efficiency of the economic system, implying that more primary resources are needed to satisfy a given level of final demand or that less final demand (e.g. consumption, hence welfare) can be sustained with an existing stock of primary resources.

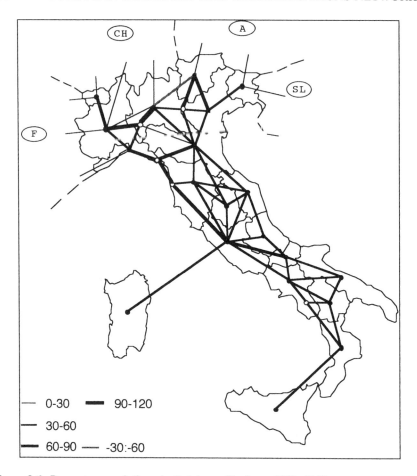

Figure 8.6. Percentage variations in freight traffic flows 1988–2000

In our simulation exercise, all nominal prices of capital and labour inputs are held fixed but, since intermediate inputs become more expensive, market prices for all goods and services rise and real prices of primary factors fall. This induces a substitution between intermediate and primary inputs.

Despite the implicit assumption of perfectly elastic supply of labour and capital, reduction of real per capita income is not offset by the increase in total employment, so that real GDP declines (on average). This is also due, of course, to the substitution taking place between domestic and imported goods.

It should be observed that by ruling out additional constraints in the markets of primary factors and foreign trade, we are actually under-estimating the impact of scarcity of transport infrastructure in the medium to long term. Further reductions in total welfare could be triggered by multiplicative processes generated by supply constraints or currency devaluation.

Industrial production levels exhibit a general tendency towards reduction, which is larger for industries requiring more transport services, in relative terms. This tendency is made less clear by the overlapping of an additional intermediate demand generated by the transport industry, whose production levels increase in a significant way.

Since demand for transport services is rather inelastic, the increase in production of transport services is quite similar to the average increase in the index of total transport costs. Rising volumes of transport services also create positive production variations in some industries that sell inputs to transport firms. These remarks apply both to the user optimum case and to the system optimum case (Figure 8.7). The only difference concerns the relative impact of congestion costs which are, of course, higher for the decentralized route choice hypothesis.

Somewhat more interesting are the results produced by the model at the regional level, because congestion phenomena have a spatial dimension and, since substitution occurs in factor and consumption demand, a few regions actually benefit from rising transport costs. Indeed, positive variations of production volumes are observed in two southern regions (Calabria and Basilicata) where a situation of excess supply of transport infrastructure is likely to endure, and in Lazio, probably due to the rather low economic growth assumed in the exogenous scenario (Figure 8.8).

When the system optimum assignment rule is used, traffic flows are lowered in some links and increased in some other links. The improvement of system efficiency is reflected in a smaller decrease of production levels (−0.51 per cent instead of −0.59 per cent) at the national level. At the regional level, however, results may be different: some regions (VDA, TAA, FVG, ABR, PUG, SIC) exhibit the same production level in the two cases, whereas three regions (LIG, MOL, CAM) are actually made worse off by the more efficient traffic allocation.

Benefits obtained by single regions under system optimum assignment are, anyway, better described in terms of reduction of regional GDP (Figure 8.9). From this point of view, all regions are better off (or, at least, do not show variations in their GDPs), and larger benefits are observed in the two most industrialized northern regions (PIE and LOM).

This effect is due to the substitution occurring between imported and domestically produced goods and services, lowering all regional GDPs.

Table 8.1 summarizes the results obtained by our model for some national macroeconomic variables, relative to the situation of absence of congestion (excess supply in transport infrastructure) and for the two cases of user optimum and system optimum flow assignment.

Savings obtained under system optimal assignment amounts to a signifi-

168

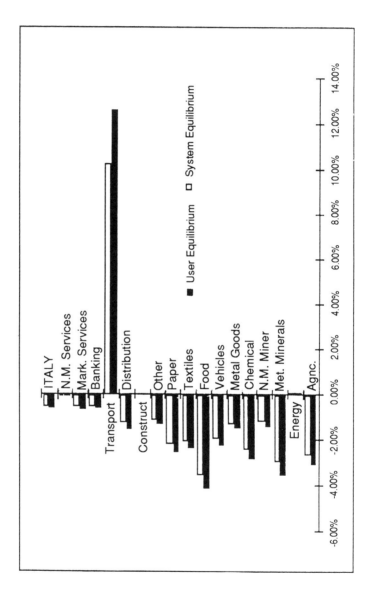

Figure 8.7. Variations of industrial production due to traffic congestion

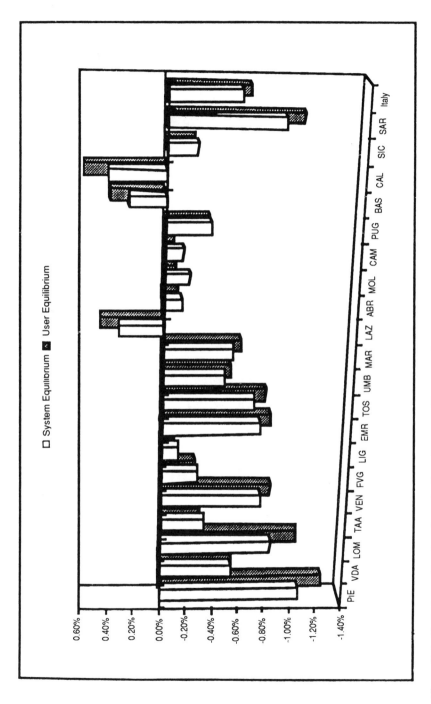

Figure 8.8. Variations of regional production due to traffic congestion

170

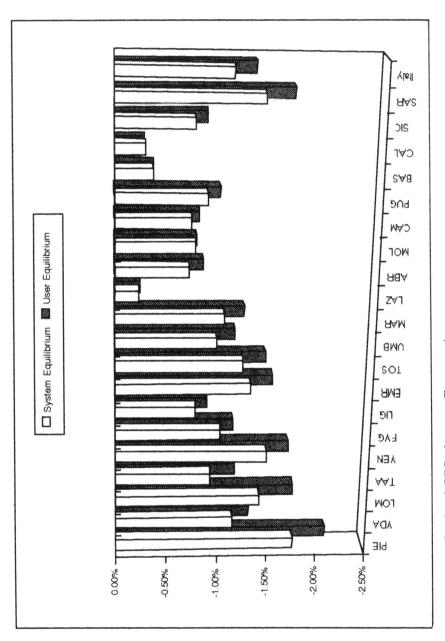

Figure 8.9. Variations of regional GDPs due to traffic congestion

Table 8.1. Main national economic indicators (variations)

	User optimum (%)	System optimum (%)	Diff. (%)
GDP	−1.31	−1.09	0.22
Production	−0.59	−0.51	0.08
Imports	2.02	1.64	−0.38
Intermediate demand	0.23	0.15	−0.08
Household consumption	−1.33	−1.14	0.19

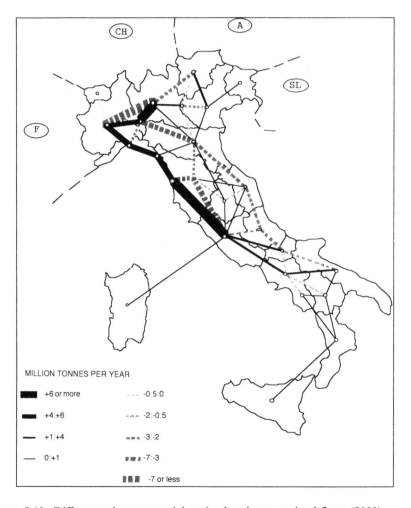

Figure 8.10. Differences between social optimal and user optimal flows (2000)

cant 0.22 per cent of national GDP, which is approximately equal to 4000 billion lire at 1994 values (US $2.5 billion).

Perhaps the most interesting result obtained by our simulation exercises is given by a direct comparison of traffic flows obtained under the two alternative assignment criteria. Figure 8.10 shows a graph depicting differences in freight flows: darker lines mean that system optimal flows exceed user optimal flows, whereas the opposite is indicated by grey lines.

A fiscal policy devoted to the internalization of externalities should therefore introduce subsidies for under-utilized links, proportional to the thickness of the line in the graph, whereas over-utilized links should be taxed. If there are other externalities, e.g. pollution, and if these are themselves dependent on congestion levels, then it could be socially optimal to apply specific taxes for all roads, but with different rates, still proportional to the difference between social and user costs.

It should be noted that most under-utilized interregional links form a continuous line, starting in Lombardy, creating a detour to Piedmont, and continuing southward along the Tyrrenean coast.

From this figure it can be clearly seen that the weak side of the Italian road system is given by the Tyrrenean coastal line, which is much less used than the corresponding line on the Adriatic coast (Figure 8.5). This suggests the adoption of policies devoted to promote this north–south corridor.

Observe, however, that this should not be considered as a recommendation for new investments in infrastructure, because the system optimized flow distribution has been computed by assuming as given the characteristics of the interregional road system.

8.4 CONCLUDING REMARKS

The demand for transport services is derived from demands for goods and services and from the activity patterns of individuals. For this reason, the analysis of a transport system must take into account the complex framework of interdependencies linking industries and regions.

In this chapter, some simulation exercises carried out with an operational general equilibrium model of Italy have been presented and discussed. These provided useful insights on the interaction between economic and transport systems, and on the economic impact of congestion costs. Furthermore, it was possible to make a comparison and an evaluation of effects due to suboptimal traffic flow distribution in the interregional network. Simulation results indicate that the Italian road system is under-utilized along the Tyrrenean coast.

Divergence between system and user optimal flow distributions, as well as potential savings that can be obtained by consistent road pricing policies, are proportional to congestion levels. From 1970 to 1992 demand for freight transport has risen by about 111 per cent in Europe and 214 per cent in Italy, and the market share of road transport has risen even faster. It is therefore easy to predict higher congestion and externality costs, making necessary the introduction of specific policies, devoted to correct market failures in the market for transport services.

REFERENCES

Beckmann, M., C. B. McGuire and C. B. Winsten (1956) *Studies in the Economics of Transportation*, Yale University Press, New Haven.

Cascetta, E. (1990) *Metodi quantitativi per analisi dei sistemi di trasporto*, Cedani, Padua.

Costa, P. and R. Roson (1988) Transport margins, transportation industry and the multiregional economy, some experiments with a model for Italy, *Ricerche Economiche*, 2, 237–287.

Gartner, N. H. (1980) Optimal traffic assignment with elastic demands: a review (Part I: Analysis framework, Part II: Algorithmic approaches), *Transportation Science*, 14(2), 174–208.

Leontief, W. and P. Costa (1987) *Il trasporto merci e l'economia italiana. Scenari di interazione al 2000 e 2015*, Sistemi Operativi, Venice.

Roson, R. (1993a) Spatial computable economic equilibria and freight network models, *International Journal of Transportation Economics*, XX, 51–66.

Roson, R. (1993b) The development of a multiregional network CGE model of Italy, in *Problems of Building and Estimation of Large Econometric Models—Macromodels 93*, Polish Academy of Sciences and University of Lodz.

Roson, R. (1994) *Transport Networks and the Spatial Economy: A General Equilibrium Analysis*, Umeå Economic Studies, University of Umeå.

Takayama, T. and G. G. Judge (1964) Equilibrium among spatially separated markets: a reformulation, *Econometrica*, 32, 510–524.

Wardrop, J. G. (1952) Some theoretical aspects of road traffic research, *Proceedings of the Institute of Civil Engineers*, 1, 325–378.

9 A Typology of Barriers Applied to Business Trip Data

ULRICH BLUM
FRANK LEIBBRAND
Technical University of Dresden, Germany

9.1 INTRODUCTION

The unification process of Europe and the emerging new trade zones in America and the Pacific have led to a reassessment of border effects, especially with respect to transportation infrastructure. The potential of high-speed railway systems (HSR) has caught public and private attention as a means to foster economic and political integration (Blum et al 1992; ERTI 1992). A recent publication edited by Cappellin and Batey (1993) focuses on emerging regional networks in Europe as a means of integration from a theoretical and political view. These articles, however, fail like others to measure the barrier effect of borders, which we think is important if we want to remove obstacles against economic integration.

In our chapter, we will focus on only one type of spatial interaction, namely the flow of persons for business purposes using the car. We will formulate and estimate a trip distribution model for Europe centred on the Federal Republic of Germany, which allows us to measure different types of barrier effects. This will be preceded by the definition of a new typology of spatial interaction and its economic as well as econometric effects. Finally, we will comment on our findings and point out new fields of econometric research in this area.

9.2 INQUIRY INTO SPATIAL INTERACTION

9.2.1 BORDERS, BARRIERS AND THE DEFINITION OF THE MODEL

Let us start with an intuitive definition of *borders* as limits of access or control in a system; in a mathematical sense, they are the *boundary* of a set. In most cases, we associate this term with geographical space. In this sense, borders in different qualities will exist, for instance between Bavaria and

European Transport and Communications Networks: Policy Evolution and Change. Edited by David Banister, Roberta Capello and Peter Nijkamp. © 1995 John Wiley & Sons Ltd.

Hesse, or between Germany and France, or between the European Community and the Commonwealth of Independent States. We thus ask ourselves by what phenomenon borders are described. Borders are often associated with border regions; this term is mostly used in a negative sense, i.e. as a synonym for a backward region because of having less positive opportunities which can be validated through econometric analysis especially in the case of formerly divided Germany (Blum 1984). This relates to the availability of goods and income. However, the exchange of people, goods and ideas may lead regions adjacent to borders to prosperity if they are able to exploit the opportunities of interaction.

We have proposed three basic determinants of regional structure and regional interaction: *metric*, as a measure of space, *resources/procurement* as an indicator of the starting or the existing conditions and *economies of scale/ of scope* as forces driving the system through the pressure of reducing costs in a competitive environment (Blum and Leibbrand 1994); the theories of Christaller (1933) and Lösch (1941) are specific nested cases as are the concepts proposed by Cattan and Grasland (1993). Rietveld (1993) points out that two main determinants of location in space exist: *distance* and *borders*. Borders lead to heterogeneities between places at different sides of the border and to discontinuities in flows. The impact of distance is seen to be similar, but more gradual. We have concluded that, as spatial interaction is not only influenced by one, but by a portfolio of metrics (i.e. distance, cost, tax, language, culture, etc.), misspecifications when building models and in the process of estimation are very likely to occur.

In our typology, a barrier exists only in terms of a 'wrong' or insufficient metric. *Barriers* in our definition are places where we measure a strong deviation from the normal, i.e. the interaction as a function of the 'normal' metric. What we experience as different types of spaces (geographical, preference, product, legal/regulatory, ethnic, linguistic, etc.) is included in our concept of metric which thus unifies different approaches of spatial interaction.

We explain spatial interaction through resources and procurement, through economies of scale and of scope and through metric. The metric closes the system in a tautological way, which is an analogue of the way Becker (1982, p. 6) finalizes his theory of individual behaviour: he explains different behaviour through different prices and costs on the level of the individual, i.e. opportunity costs and shadow prices. Important conclusions can be derived from systems closed in such a tautological way; a validation can be performed by defining an operationalized theory.

We define *borders* as the maximal distance of a good with respect to the underlying metric; it thus relates to the willingness to pay for the supply of a good offered at a certain location. In fact, we notice borders because of the calculation of optimality in different spaces using different metrics than the metric considered by us.

Natural borders are defined by the point where, for a given distribution and, consequently, underlying spatial metric, the marginal costs of controlling space are matched by the marginal willingness to pay for the product (an externality like security or other property rights) offered. The stability of states has been analysed by Blum and Dudley (1991). Why will this state remain stable? If geographic distance does not limit control, a barrier (relative to the geographic metric) must be constituted through other means. Note that this barrier could be constituted in a different metric, for instance a very distinct monetary system or a very different language. The first explains why a country like Luxembourg is able to extract a considerable rent even though the geographical barrier does not exist: militarily, Luxembourg does not exist, geopolitically it is small and its financial space is gigantic. For Ratti (1993) borders should be overcome by the institutionalization of contract areas.

9.2.2   ECONOMIC IMPLICATIONS

If we assume that a portfolio of metrics determine spatial distribution and interaction, it is not self-evident that this portfolio is constant over the whole domain of the spaces covered and analysed. To be more precise with respect to a geographic explanation: it is not *a priori* clear whether the ratio between distance as measured in kilometres and distance as measured in time is the same in Germany and in France, or in a central and in a border region. In our chapter, we use travel time as proxy for the correct metric and later try to estimate the error (i.e. barrier in the metric) which stems from this approach.

The phenomenon of *distorted spatial networks* was already observed by Thünen (1826) when he changed the geographical transport cost metric with respect to the geographical distance metric by introducing a better means of transportation, a river. Later Christaller (1933) and Lösch (1941) analysed this phenomenon explicitly with respect to the distribution of activities in geographical space. Their use of space rests on an Euclidean metric imposed on geography.

Until now we have explained interaction through the metric, resources and procurement under assumptions such as an efficient spatial division of labour. Because of the present limitations in the data available, we now simplify in the direction of a fixed trip time budget.

If the portfolio of metrics is not invariant in space and with respect to relations, this implies that, compared to the normal ('average'), trip patterns will vastly differ. In a region where interaction is partly blocked, for instance by the sea or the former Iron Curtain, trip patterns, i.e. the number of trips per person and its average length, may be very different to the national average. The resulting time budget of spatial mobility may even remain uniform and constant (in Germany, e.g. about 1 h per day; Spiegel 1993). As

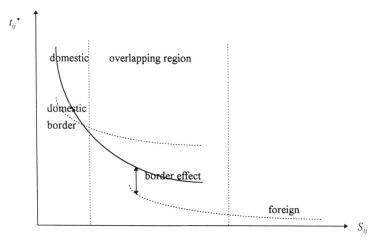

Figure 9.1. Structural Break of the Impedance Function

a consequence, the *impedance function* which relates trips to the average portfolio of metrics will be flatter with smaller decrements near the origin. Furthermore, it is interesting to investigate structural breaks of the impedance function itself. In Figure 9.1, t_{ij}^* are the trips between region i and region j per unit of heterogeneity, i.e.

$$t_{ij}^* = f(s_{ij}), \quad t_{ij}^* : \text{trips between region } i \text{ and } j \text{ corrected for effects of}$$
$$\text{heterogeneity;} \tag{9.1}$$
$$s_{ij} : \text{metric.}$$

If the true relationship between a metric and the number of trips corrected for heterogeneity were known and the share of each metric in the total impedance function were also given and invariant over the metric spaces, then a respective sample of trips would suffice to estimate the total interaction model. However, the following problems have to be handled simultaneously, namely

1. the function of heterogeneity and its (proxy) variables;
2. the mix of metrics and the (proxy) metric used;
3. statistical observation errors.

Even if statistical observation errors are assumed to be normally distributed, the first two reasons for errors imply that it is impossible to trace a deviation from the 'true' model; thus, in the approximate model structural breaks in the impedance function and overlapping impedance functions may occur.

9.2.3 ECONOMETRIC IMPLICATIONS

If a model of spatial interaction is formulated, it becomes very likely that not all spatial variables of heterogeneity and of metric, especially those describing deviations from the 'normal pattern', are properly accounted for. Then, as known from time-series analysis, an error correction is possible through the analysis of spatial autocorrelation. We have employed three types of corrections: (1) a correction through a dummy structure; (2) a 'stretching' of the impedance values; (3) spatial autocorrelation.

A stretching of the impedance function implies that we have increased the distance measure by a constant value to account for problems of a mis-specification of the metric, especially in the sense of the *activity space concept*.

Autocorrelation analysis in our approach thus gives us a possibility to include similiarity effects not accounted for yet (Rietveld 1993). By defining *impact matrices* (Blum 1987; Blum et al 1990), which are hypotheses of spatial interaction, we are also able to circumvent the problem of first observation. Through a *direct order*, i.e. an order directed by the hypotheses of the researcher, it is defined which region should relate to which other region. In our case, we defined a measure of direct vicinity between two adjacent regions.

9.3 ECONOMIC MODEL AND ECONOMETRIC SPECIFICATION

9.3.1 THE DISTRIBUTION MODEL

Our analysis is based on a gravity model:

$$t_{ij} = f\{h(M_i, M_j), m(S_{ij})\} + v_{ij}, \qquad i = 1, 2, \ldots, n; \quad j = 1, 2, \ldots, n. \quad (9.2)$$

where t_{ij} are flows from region i to region j, M_i and M_j variables of masses of regional heterogeneity mapped by a function h, S_{ij} variables of different metrics (in the case of our estimation only one) mapped by function m; the two functions are combined by a mapping f. v_{ij} are residuals.

9.3.2 ECONOMETRIC SPECIFICATION

We used an algorithm that simultaneously allows us to estimate the para-meters, the functional form (through a Box–Cox transformation) and the error structure of a model. The cross-section data is discussed at length in Gaudry and Blum (1988) and Blum and Gaudry (1990). A general definition of autocorrelation was proposed by Blum (1987) and Blum et al (1990) through a residual impact criterion (RIC) given by a matrix P which has a unit value if a certain criterion is met (in our case: a region borders another region) and a null if the opposite is the case. We normalize this matrix by the sum of its rows and columns in such a way that each element is divided by the square root of the product of the columns and the row sum and obtain matrix R for the error term:

$$v = \sum_{k=1}^{K} \rho R_k v + w$$
(9.3)

where v are the residuals from the systematic equation of the model, w are white noise residuals, ρ_k are correlation coefficients of order k and R_k is a normalized non-negative *residual impact matrix* with elements r_{hl}.

If we are uncertain about the different orders of autocorrelation, it is possible to use a formula which lets the data decide on the level of vicinity, such as

$$v = \rho(aRv + a^2 R^2 v + \cdots) + w, \qquad a \in [0,1]$$
(9.4)

where the first term in the parentheses describes first degree spatial auto-correlation. If R is a matrix of simple neighbourhood, we obtain, by multiplying the impact matrix with itself once, all neighbours of neighbours, that is second degree contiguity, etc. The coefficient a describes the length of the regional 'tail' of the autoregressive spatial process, i.e. the intensity of backward and forward linkages. By simple transformation, we obtain the *distributed spatial autocorrelation formula*:

$$v = \rho(1-a)R[I-aR]^{-1}v + w$$
(9.5)

If the value of a tends to zero, we obtain a simple autoregressive process, if a tends to unity, we obtain a stationary matrix in which each line element takes a value identical to its relative share of number of borders if R is an adjacency matrix. Figure 9.2 shows an example.

9.4   RESULTS OF THE ESTIMATION FOR WEST GERMANY, 1985

9.4.1   OPERATIONALIZATION AND DESCRIPTION OF THE DATA SET

We have used transport flows of 1985 between 95 European regions as endogenous variables, 79 of which are in West Germany. These transport

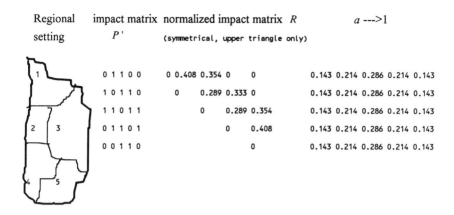

Regional setting	impact matrix P'	normalized impact matrix R (symmetrical, upper triangle only)	$a \dashrightarrow 1$
	0 1 1 0 0	0 0.408 0.354 0 0	0.143 0.214 0.286 0.214 0.143
	1 0 1 1 0	0 0.289 0.333 0	0.143 0.214 0.286 0.214 0.143
	1 1 0 1 1	0 0.289 0.354	0.143 0.214 0.286 0.214 0.143
	0 1 1 0 1	0 0.408	0.143 0.214 0.286 0.214 0.143
	0 0 1 1 0	0	0.143 0.214 0.286 0.214 0.143

Figure 9.2. Example of a specification

regions are delimited for the purpose of the German Highway Masterplan of 1985. We selected road traffic for business purposes, as this matrix is complete in terms of observations. The exogenous variables were taken from public statistics and aggregated if necessary. The following variables were used as proxies of heterogeneity:

1. population at origin i and at destination j, BEV_i and BEV_j;
2. area at origin and destination, $FLAE_{ij}$, as product;
3. per employee incomes at origin and destination, $PKER_{ij}$, as product.

The data is not able to distinguish between outward or return trips, i.e. a trip between zone i to zone j could also be a return trip from zone i to zone j which formerly originated in j. Consequently, we added the by and large symmetrical matrix elements $t_{ij} + t_{ji}$ to an upper triangle matrix and then reduced it to a form which only included flows from German regions to other German regions, and from German regions to foreign regions. We computed dummies where the values are 1 if a barrier exists (e.g. the Alps, or the former Iron Curtain). Dummies were included for

1. the Alps, $ALPZ$;
2. East Germany, $DDRZ$;
3. Eastern Europe (including East Germany), $OSTZ$;
4. the European Community, EGZ;
5. the sea, $MEERZ$.

The variable of impedance, $ZEITD_{ij}$, is the travel time between two regions, as computed from an assignment model of 1985.

9.4.2   RESULTS OF THE ESTIMATION

9.4.2.1   Reporting Format

Table 9.1 presents results of each experiment, identified by a column number. The rows give summary descriptions of the independent variables and their code names; code names that are underlined twice refer to dummy variables.

In the first part of the table, we find, for each variable, the elasticity associated with it. Elasticities of the expected value of the dependent variable are computed at the mean values of the positive observations of each variable. For a dummy variable, the change of the estimated dependent variable due to the presence of that dummy variable is defined as its elasticity (i.e. 0.01 = 1 per cent). Below the elasticities, the t-statistics in parentheses are conditional upon the optimal values of the Box–Cox parameters. The indications LAM under the t-statistic indicate that a Box–Cox transformation has been applied to that variable; the numbers, e.g. $LAM1$, indicate the group of variables which were subjected to a common transformation.

In the second part of the table, we find the optimal values (LAM) of the functional form, as well as Pseudo-R^2, the value of the log-likelihood and all other information.

The results and the tables are obtained with the L-1.1 algorithm in the TRIO program package (Gaudry et al 1993).

9.4.2.2   Model Structure

We have computed three groups of models in order to inquire into the secrets of metrics and heterogeneity. The first set of three models only includes the usual set of variables of a distribution function: proxies for variables of heterogeneity and of metric. In the second set, we added dummies to correct a misoperationalization of the metric, and analysed the impedance function by splitting it into a domestic and a foreign part. In the third set we investigated spatial autocorrelation.

In each of the model groups, we computed a linear model, which was statistically far inferior to the logarithmic model; it is not included in either Table 9.1 or 9.2. We always display the logarithmic models ($\lambda = 0$ and fixed in the respective column) and models in which data is allowed to decide on one common λ. A further breaking down of the functional structure (two independent but flexible λs for the endogenous and all exogenous variables) did not improve the model in terms of log-likelihood (gain of 2 points for change of one degree of freedom). Thus, the best model tends to the logarithmic form.

In the first set of experiments not displayed in a table we tested the structure of the model with respect to domestic, international, and total (i.e.

domestic and international in one model) interaction. We found that domestic impedance was −2.647, foreign impedance was 1.005. Impedance for both domestic and foreign trips was 2.278, all with a value of the corresponding t-statistic above 8.0. In an additional experiment, we increased all trips to foreign destinations by two hours; this 'crude' transformation of the metric improved the results found in terms of statistical quality and is in accordance with results found by Bröcker (1984).

Table 9.1 contains the second set of experiments in which we included dummy variables and split the impedance variable, thus differentiating between domestic and foreign destinations allowing for different elasticities.

Table 9.2 summarizes two experiments on spatial autocorrelation: in the first column, we supposed an impact matrix where each German region is related to its adjacent German region and where each foreign region is related to its adjacent foreign region (the latter thus builds a ring around Germany). The second column supposes a lag of $a = 0.99$ (formula 5).

9.4.2.3 Results from the Systematic Part

In Table 9.1, we discover a certain asymmetry between the elasticities of population at origin (about 0.9) and at destination (about 0.75), which ideally should be identical. We suspect that this result is the property of the data sample in which part of the incoming traffic from abroad may be statistically difficult to monitor.

Area has an elasticity of about −0.2 and a negative impact, as it increases transport distance (for given population and fixed travel time budget; this implies reduced density) and as the share of long-distance traffic inside a region, which we do not cover, increases with the size of the region. This means that the larger the region, the smaller becomes the volume of traffic that can be observed. Income per employee increases traffic volumes with an elasticity of about 0.2. The impedance of travel time is high with an elasticity of −2.

Barrier effects are significant:

1. The effect of the Alps is noticeable; they reduce transport flows by 5 per cent.
2. The effect of barriers to eastern countries is also small; we measure an effect between −2 per cent and −5 per cent depending on the functional form of the equation. The reason is depressed income levels, which are more significant for the low level of spatial interaction than the Iron Curtain.
3. The Iron Curtain between east and west Germany only has a small impact of −2 per cent (if the factor 'east' of 2 per cent to 5 per cent is already accounted for). This result may be attributable to the special economic status of East Germany in the EC.

Table 9.1. Estimation of distribution functions for Germany, 1985

Model		1	2	3
Heterogeneity				
Population at origin	BEV85Q	0.949	0.859	0.950
		(19.85)	(22.17)	(19.87)
		LAM 1	LAM 1	LAM 1
Population at destination	BEV85Z	0.760	0.735	0.760
		(16.78)	(20.65)	(16.78)
		LAM 1	LAM 1	LAM 1
Area at origin and destination	FLAEQZ	−0.228	−0.266	−0.237
		(−6.08)	(−7.69)	(−6.09)
		LAM 1	LAM 1	LAM 1
Activity				
Income per employee at origin and destination	PKER85QZ	0.177	0.274	0.164
		(1.33)	(2.75)	(1.23)
		LAM 1	LAM 1	LAM 1
Metric				
Travel time	ZEITD	−2.375	−2.184	
		(−44.16)	(−55.52)	
		LAM 1	LAM 1	
Travel time in West Germany	ZEITDI			−2.383
				(−43.63)
				LAM 1
Travel time to foreign countries	ZEITDA			−2.330
				(−30.67)
				LAM 1
Correction				
Dummy for Alps	ALPZ	−0.054	−0.052	−0.065
		(−4.73)	(−7.57)	(−3.61)
Dummy for eastern countries	OSTZ	−0.048	−0.020	−0.052
		(−5.15)	(−2.85)	(−4.96)
Dummy for GDR	DDRZ	−0.019	−0.017	−0.032
		(−2.08)	(−2.52)	(−1.79)
Dummy for EC	EGZ	−0.107	−0.117	−0.150
		(−4.61)	(−8.17)	(−2.68)
Dummy for sea	MEERZ	0.043	0.037	0.042
		(5.25)	(7.84)	(5.13)
Regression constant	Constant	4.665	7.499	4.759
		(3.98)	(8.81)	(4.04)

Table 9.1. Continued

Model		1	2	3
Box–Cox transformations: uncond: [T-STATISTICS = 0]/[T-STATISTICS = 1]				
Lambda(Y)—Group 1	LAM 1	0.000 FIXED	0.077 [14.17] [−170.20]	0.000 FIXED
Lambda(X)—Group 1	LAM 1	0.000 FIXED	0.077 [14.17] [−170.20]	0.000 FIXED
Log-likelihood		−21365.7	−21269.5	−21365.3
Pseudo-R2: − (L)		0.999	0.999	0.999
− (L) Adjusted for D.F.		0.999	0.999	0.999
Average probability (Y = limit observ.)		0.000	0.000	0.000
Sample: − Number of observations		2000	2000	2000
− First observation		1	1	1
− Last observation		2000	2000	2000
Number of estimated parameters:				
− Fixed part:				
− Betas		11	11	12
− Box–Cox		0	1	0

4. Business interaction between Germany and its EC neighbours is severely hampered by barriers; we find a reduction between 10 per cent and 15 per cent. This, however, may also be a consequence of mode choice as, in this category of distances, the plane may be preferred to the car.
5. We had expected a reduction of flows through the sea barrier, which especially affects transport to Great Britain, Ireland and Scandinavia; in fact, we even find a small positive value of 4 per cent. It may be the effect of a systematic error in the estimation of sea trip times in the assignment model underlying the metric.

Note for reasons of interpretation that the GDR is also an Eastern nation and that some countries at the same time are EC members and are situated across the Alps.

If the impedance variable is split in the second model (column 3) we find no differences in structure and no improvement of the log-likelihood (column 3 against column 1); obviously, a structural break cannot be discovered by our approach.

Table 9.2. Estimation of distribution functions with spatial autocorrelation for Germany, 1985

Model		4	5
Heterogeneity			
Population at origin	BEV85Q	0.686	0.296
		(22.54)	(16.09)
		LAM 1	LAM 1
Population at destination	BEV85Z	0.468	0.206
		(19.54)	(11.58)
		LAM 1	LAM 1
Area at origin and destination	FLAEQZ	−0.123	−0.126
		(−5.46)	(6.61)
		LAM 1	LAM 1
Activity			
Income per employee at origin and destination	PKER85QZ	0.308	0.197
		(8.36)	(11.57)
		LAM 1	LAM 1
Metric			
Travel time	ZEITD	−2.265	−2.086
		(−61.29)	(−61.45)
		LAM 1	LAM 1
Regression constant	Constant	10.311	11.426
		(27.77)	(60.56)

Box–Cox transformations: uncond: [T-STATISTICS=0]/[T-STATISTICS=1]

Lambda(Y)—Group 1	LAM 1	0.084	0.083
		[14.99]	[14.95]
		[−163.41]	[−164.52]
Lambda(X)—Group 1	LAM 1	0.084	0.063
		[14.99]	[14.95]
		[−163.41]	[−164.52]

	4	5
Log-likelihood	−21319.7	−21424.9
Pseudo-R2: − (L)	0.999	0.998
− (L) Adjusted for D.F.	0.999	0.998
Average probability (Y=limit observ.)	0.000	0.000
Autocorrelation structure:		
− RHO	0.100	0.100
− A		0.990

9.4.2.4 Results from the Error Part

We thoroughly investigated the error structure of the model, especially the structure of spatial autocorrelation. Our hypothesis for the impact matrix is that residual errors are explained inside each spatial group, i.e. inside Germany and inside the foreign countries. Thus, the impact matrix always contains a '1' if a region of Germany is adjacent to another region in Germany or if a foreign region is adjacent to another foreign region, and a '0' otherwise.

Table 9.2 displays two models of spatial autocorrelation. In the first, an optimum with rho = 0.1 was found. All elasticities for population decline with respect to the initial model without dummy structure that is not shown here. We obtain a small improvement in the model's significance. We added a large distributed lag-factor of 0.99 (which was not estimated) in order to show the tendency of the impact model further reduces the elasticities. In doing so, we wanted to compute the effect of homogenization of the sample, as each observation obtains an infinite tail. We see that regional information still remains, although the model is not superior to the last one.

9.5 CONCLUSIONS

In our chapter, we have established new categories to define spatial interaction models, namely *metric* and *heterogeneity*. This allowed us to put forward a new model structure, which we used to estimate spatial interaction based on transportation flows for business trips in Europe, centred on Germany. We estimated 'barriers' through dummies, a split in the proxies for metric or the additive correction of the foreign part of our travel-time matrix, and autocorrelation (in place space, not yet in flow space). We conclude that our approach is successful in terms of the methodology used, but our data set has, especially in its foreign part, inconsistencies into which we will have to inquire further.

REFERENCES

Becker, G. (1982) *Ökonomische Erklärung menschlichen Verhaltens*, J.C.B. Mohr, Tübingen.

Blum, U. (1984) Regional production in border areas, in *Regional Research in an International Perspective*, Verlag V. Florentz, München, pp. 201–225.

Blum, U. (1987) *A Distributed Lag for Spatially Correlated Residuals*, Université de Montréal, CRT-Publication No. 512.

Blum, U. and L. Dudley (1991) A spatial model of the state, *Journal of Institutional and Theoretical Economics*, **147**(2), 312–336.

Blum, U. and M. Gaudry (1990) The impact of social security contributions on savings: an analysis of German households by category, *Jahrbuch für Sozialwissenschaft*, **41**(2), 217–242.

Blum, U. and F. Leibbrand (1995) *A Note on Spatial Interaction Models and the Underlying Theories*, Discussion Paper, forthcoming.

Blum, U., D. Bolduc and M. Gaudry (1990) *From Correlation to Distributed Contiguities: a Family of AR-C-D Autocorrelation Procedures*, Université de Montréal, CRT-Publication No. 734.

Blum, U., H. Gercek and J. Viegas (1992) High speed railway and the European peripheries: opportunities and challenges, *Transportation Research*, **26A**(2), 211–221.

Bröcker, J. (1984) How do international trade barriers affect international trade? in *Regional and Industrial Development Theories, Models and Empirical Evidence*, Elsevier, North-Holland, Amsterdam.

Cappellin, R. and P. W. Batey (1993) *Regional Networks, Border Regions, and European Integration* (European Research in Regional Science No. 3), Pion, London.

Cattan, N. and C. Grasland (1993) Migrations of population in Czechoslovakia; a comparison of political and spatial determinants of migration and measurement of barriers, *Trinity Papers in Geography*.

Christaller, W. (1933) *Die Zentralen Orte in Süddeutschland*, Wissenschaftliche Buchgesellschaft, Darmstadt.

ERTI (European Round Table of Industrialists) (1992) *Need for Renewing Transport Infrastructure in Europe*, Paris.

Gaudry, M. and U. Blum (1988) An example of autocorrelation among residuals in directly ordered data, *Economics Letters*, **26**, 335–340.

Gaudry, M., M. Dagenais, R. Lafferiere and T. Liem (1993) TRIO Model Types, Version 1, Université de Montréal, Centre de recherche sur les transports publication 904, Montréal.

Lösch, A. (1941) *Die räumliche Ordnung der Wirtschaft*, Gustav Fischer Verlag, Stuttgart 1962.

Ratti, R. (1993) How can existing border and barrier effects be overcome? A theoretical approach, in R. Cappellin and P. Batey (eds), *Regional Networks, Border Regions, and European Integration* (European Research in Regional Science No. 3), Pion, London, pp. 60–69.

Rietveld, P. (1993) Transport and communication barriers in Europe, in R. Cappellin and P. Batey (eds), *Regional Networks, Border Regions, and European Integration* (European Research in Regional Science No. 3), Pion, London, pp. 47–59.

Spiegel (1993) *Auto Verkehr und Umwelt*, Hamburg.

Thünen, J. H. von (1966) *Der isolierte Staat*, Fischer-Verlag, Stuttgart (first published 1826).

Part III

POLICY RESPONSES AND ISSUES

10 Private Sector Investment in Transport Infrastructure in Europe

DAVID BANISTER
University College London, UK

BJØRN ANDERSEN
Molde, Norway

SEAN BARRETT
University of Dublin, Ireland

10.1 INTRODUCTION

One of the basic political questions being addressed by decision-makers is the amount of infrastructure which should be provided—whether it should meet unconstrained demand or whether demand should be constrained— and how it should be financed. Most transport projects are funded by the public sector as they are large scale, involve a high risk and have long payback periods. It is only where the private sector has some degree of a monopoly position (e.g. in road bridges across fjords or estuaries) that real interest for some alternative forms of financing has been raised. Transport projects are notorious for their cost overruns, for technical deficiencies in construction and consequent high maintenance costs, and for optimism in their estimates of future demand—the Channel Tunnel project illustrates all of these problems as does the proposed link between Denmark and Sweden, and the Great Belt project between Zealand and Funen.

In the past, the public sector has been the main contributor, but with the increasing costs of new transport projects and with the desire of governments to reduce public deficits, new sources of capital are required. In 1985, only two of 22 OECD countries (Finland and Norway) had a budget surplus. While this increased to seven countries in 1990, the prediction for 1995 is that only Denmark, Norway, Australia and New Zealand will have a budget surplus. The preponderance of budget deficits among OECD countries was caused in part by 'considerable resources being channelled to enterprises in the public sector in the form of investment or to cover operating losses, or both. Thus the shedding of state-owned companies was seen

European Transport and Communications Networks: Policy Evolution and Change. Edited by David Banister, Roberta Capello and Peter Nijkamp. © 1995 John Wiley & Sons Ltd.

by many governments as a means of moving towards a more balanced budget' (Stevens 1992). In addition to reducing the budget difficulties facing the OECD countries as a whole since 1980, the other goals of privatization were to raise overall economic efficiency, enhanced competitiveness, curtail the state's involvement in the management and control of enterprises and to reduce the size of the public sector.

The most obvious means to raise finance would be to charge tolls on existing motorways. It has been estimated (Keating 1992) that tolls of 5 pence per mile (similar to current toll levels on French autoroutes) on UK motorways would raise about £2 billion per annum and its extension to other roads and to urban areas would raise £6 billion per annum. The key argument is whether the tolls are seen as a tax to increase exchequer revenues or a charge which can be reinvested in the transport system.

The level of private sector investment in transport in the past was high. Kelsey (1986) found the nominal value of transport companies quoted on the London Stock Exchange (LSE) in 1885 was £750 million (or about £250 billion at 1985 prices). However, in 1985 the value of transport companies quoted on the LSE was only £2.8 billion, one per cent of the 1885 figure! The loss of interest in the transport sector by private investors may be explained by changes in the financing of railway investment from international private sector capital flows to national public financing and the increase of road transport using public funds for infrastructure provision at the expense of rail transport. In the aviation sector in most countries, both airlines and airport infrastructure were until recently financed through the public sector, in contrast with the railways of the last century. The growth in road transport has meant that many firms in manufacturing operated their own transport fleets rather than buying in services from railway companies. The growth in own fleet operations was probably due to restrictive licensing in road transport. In addition to these issues is the question of property rights and ownership of the railways (Foster 1992), and more recently the ownership of the roads. Any uncertainty over the ownership of the (new and existing) infrastructure and for what period would increase the risk to the private sector.

The private sector has only really shown interest in terminals and interchanges as these are seen as being closer to the other sectors where private capital has been directed (e.g. the office and commercial sectors). Transport nodes allow development of associated buildings which can be sold or rented out, and it also allows internal space to be franchised. Airport terminals demonstrate this potential as some of the highest rents for floor space are charged. In the UK, the 15 largest airports will extend their retail floor space by 40 per cent to $56 \, m^2$ (600 000 ft^2 over the next seven years). However, the interest of the private sector in the links on the transport infrastructure has been far less.

This chapter presents some of the issues which European decision-makers must face concerning future investment in the links on the transport infrastructure. Some of the basic economic arguments are presented together with evidence from the recent involvement of the private sector in transport. The verdict is an open one as it seems that in competitive markets privately owned enterprises perform better than similar publicly owned enterprises, at least in economic terms, and there has been a substantial involvement in certain activities, but a reluctance in other activities. In the last part of the chapter, the possibilities of partnership between the public and private sectors are discussed, together with the means by which new funding instruments can be mobilized.

10.2 THE ECONOMIC CASE FOR PRIVATE SECTOR INVESTMENT

The historical evidence outlined above emphasizes the decline of private sector railways together with the legislative and administrative controls as the main reasons for the fall in market investment in transport. The infrastructure school of thought emphasizes certain features of the transport infrastructure which contrast with the directly productive activities financed through the market economy. Hirschman (1958) contrasts 'social overhead capital' and directly productive activities. He defines social overhead capital as comprising those basic services without which primary, secondary and tertiary productive services cannot function. In the widest sense it includes all public services from law and order through education and public health to transport, communications, power and water supply, and agricultural overhead capital (e.g. irrigation and drainage systems). The central part of the concept can be restricted to transport and power. Hirschman sees infrastructure projects as having the following characteristics:

1. they are an input to directly productive activities;
2. they are typically provided by public agencies or by private agencies under public control;
3. the products are supplied free or at regulated prices;
4. the products are not subject to competing imports;
5. production is characterized by 'lumpiness' (technical indivisibilities) and a high capital output ratio;
6. output may not be measurable.

In the Hirschman classification above, (3) would represent a major obstacle to market provision. The non-operation of the exclusion principle and the presence of free riders preclude market investment in a sector. The other characteristics such as capital lumpiness and non-measurable outputs would present less formidable obstacles to market provision. The Hirschman arguments would all argue for the continued involvement of the public sector in transport infrastructure provision.

Economic and technological changes since Hirschman defined infrastructure in the terms set out above have blurred the distinction between infrastructure and directly productive activity. Thus infrastructure can increasingly be provided in the market sector and a return to the high levels of private sector investment in transport, typical of the last century, can again occur. There are five changes which allow this to happen:

1. Pricing mechanisms are available for roads, seaports and airports and the exclusion principle may be applied. Smart card technology has reduced the transaction costs of road pricing.
2. Capital intensity and lumpiness are less of a barrier to private sector investment now than when Hirschman wrote. For example, the development of airports in the immediate post-war period coincided with the belief that only the State had the resources to undertake such large investments. This contrasts with the present situation when many governments experience severe constraints on public expenditures whereas private sector financial institutions have experienced a large increase in their supply of available funds for investment.
3. Administrative reforms have been instituted to establish agencies charged with devising pricing formulae for privatized utilities such as gas, water, electricity, telecommunications and airports.
4. Cross-border interconnectors for gas and electricity have made these products an internationally traded good. Reverse charges have brought competition in international telephone calls.
5. Output measurement techniques have been improved by research in areas such as programme budgeting, cost–benefit analysis, cost-effectiveness analysis, and research on factor productivity.

The ability to charge prices for transport infrastructure, the constraints imposed by the Public Sector Borrowing Requirement (PSBR) when private finance is more readily available, and the development of output and productivity measures have reduced the distinction between social overhead and directly productive investment.

Privatization has an important demonstration effect, and its perceived success is likely to lead to further actions. The arguments have included wider share ownership, income redistribution and reductions in trade union power. The empirical evidence is less convincing and privatization may not be the most suitable policy instrument for achieving each of these goals (Vickers and Yarrow 1988). However, the economic arguments may be more powerful. Stevens (1992) concludes that the property rights theory of the firm suggests that public enterprises should perform less efficiently and less profitably than private enterprises. In a private enterprise, both internal control—via the shareholders—and external control— through the discipline of the capital market—provide incentives to avoid inefficiencies. By contrast, public enterprises are not subject to the discipline of the capital markets, and internal monitoring is conducted by politicians who do not necessarily see their role as supervising the efficiency with which managers allocate resources. (p. 12)

Much of the evidence is inconclusive on whether the private sector is necessarily more efficient than the public sector. Many cost savings can and have been made in the public sector prior to privatization. This debate is still unresolved (Kay et al 1986; Millward 1986).

10.3   PROGRESS TOWARDS PRIVATE SECTOR INVOLVEMENT IN TRANSPORT IN THE 1980s AND 1990s

10.3.1   UNITED KINGDOM

In the 1980s several major transport companies were privatized (Table 10.1). The market capitalization of transport companies on the LSE (summer 1993) was £33.5 billion with 42 quoted companies. This is still far less than the figure quote in the introduction for 1885 (Kelsey 1986), but far higher than the 1985 figure. The market performance of these companies has been remarkable—the often quoted case of National Freight places its current market capitalization (1993) at £1351 million, some £1300 million higher than when it was sold in 1982. British Airways was capitalized in summer 1993 at £3.3 billion. In the previous decade it had been transformed from a poor performer in the high-cost European national airlines sector to a highly profitable airline with stakes in US, Australian, German and French carriers and significantly improved productivity.

Only two bus companies are quoted on the stock market (1993). National Express has a value of £81.6 million and Stagecoach has a value of £162 million. There have been important changes in the sector following privatization. Productivity, as measured by bus-kilometres per staff member, improved by 24 per cent and there have been substantial reductions in subsidies which are still some 22 per cent below their 1985/86 levels. These gains need to be balanced against the losses in patronage of some 30 per cent and the substantial real increases in fares, particularly in the period immediately following privatization (Banister 1992). Many National Bus Company subsidiaries were sold at attractive prices, particularly to management buy-outs. By undervaluing the assets of the companies, it was possible to find buyers and they were able to realize these assets and this released capital for further expansion (House of Commons Committee of Public Accounts 1991).

Table 10.1. Privatization in transport in the United Kingdom

National Freight Corporation—Road Haulage Operator	February 1982—Sold for £53 million to a consortium of managers, employers and company pensioners. The government paid back £47 million to the Company's pension fund to cover previous underfunding.
Associated British Ports—Ports and Property Development	February 1983—Part of equity sold, with £36 million raised, April 1984—Remainder sold by tender offer. £34 million raised.
British Rail	Some non-essential assets sold. British Rail Hotels sold in 1983 for £30 million. July 1984—Sealink Ferries sold to Sea Containers Ltd for £66 million. Sold to Stena Line in 1990 for over £100 million.
Jaguar—Luxury Car Manufacturer	Subsidiary of British Leyland. Sold in July 1984 for £294 million.
British Airways	Sold in January 1987 for £892 million.
Rolls Royce— Aeroengine Business	Bought by the Government in 1971. Sold in May 1987 for £1360 million.
British Airports Authority	Operates seven of the principal airports in Britain including Heathrow and Gatwick. Sold in July 1987 for £1280 million.
National Bus Company	Consisted of 72 subsidiary companies. Sale completed in December 1988 with gross proceeds of £323 million. Net surplus to Government after all debts paid of £89 million.

With respect to transport operating companies, the reduction in market investment in the sector up to the 1980s may be attributed to market prevention rather than market failure. Throughout Europe, road freight was controlled as a means of protecting the railways when the latter were in private and public ownership, and similar controls were placed on the bus industry. International aviation in Europe in the post-war era was based on colluding national airlines with price, output and market-sharing agreements rather than competing market-driven commercial airlines. National airlines achieved regulatory capture over governments and controlled the allocation of capacity ('slots') at major airports. Market entry was difficult except in the highly efficient European charter sector. Most of the traditional protections enjoyed by European national airlines remain in force and several new obstacles to contestability restrain market investment in the sector to levels below those which would be expected in a contestable market (Barrett 1993).

The success of privatized transport operating companies in areas such as aviation, road transport and shipping has been linked to factors such as the creation of a market in corporate control, the imposition of a bankruptcy constraint on producers and the links between contestability and privatization. The market in corporate control is created by the market in privatized transport companies. Investment analysts assess the performance of the companies as part of giving them a market rating. In state companies there are difficulties facing parliaments and public servants in ensuring corporate efficiency. The bankruptcy constraint controls the activities of firms in the market sector. By way of contrast the losses of state companies are borne as part of the PSBR.

The privatization of transport infrastructure differs from the privatization of the transport operating companies. Two major privatizations have taken place, Associated British Ports and the British Airports Authority (Tables 10.2 and 10.3). Ports are contestable markets in the contemporary economy because improved road and rail communications have made some ports substitutes for other ports. Ports have been operated by quasi-local authority bodies in order to prevent the abuse of monopolistic power by ports in an era when substitution of one port for another was largely impossible. Nationalization of ports also occurred as a by-product of the nationalization of railways. As private enterprises, ports have been readily sought after by investors and the performance of Associated British Ports has steadily improved since privatization. Some potential abuses of monopoly power may require regulatory intervention. For example, the European Union has sought greater access to slots at Holyhead for the British and Irish Line, a competitor of Stena Sealink which owns the port. Similarly, the Stena Line

Table 10.2. Associated British Ports

Associated British Ports comprises 22 ports including Cardiff, Fleetwood, Grimsby, Hull, Plymouth, Southampton and Swansea. It was formerly the British Transport Docks Board and was privatized in 1981 and 1984 for £70 million. The company made a loss in 1983/84.
The 1990 annual accounts show a turnover of £211.3 million comprising £176.4 million from port operations and £34.9 million from property transactions. Profits were £69.4 million of which £59.5 million came from port operations and £9.9 million from property.
Its market capitalization in summer 1993 was £766 million.

Table 10.3. British Airports Authority

British Airports Authority controls seven airports in Britain—Heathrow, Gatwick and Stansted in the London area, and Glasgow, Edinburgh, Aberdeen and Prestwick in Scotland.
Profits were £82 million in 1985 (prior to privatization in 1987), and £220 million in 1989. The sale price was £1280 million, and market capitalization was £1900 million in 1989 and £3700 million in 1993.

has taken the Danish State and the railways to the European Union for breaking competition rules by exerting monopoly control over several ferry links including Rødbyhavn–Puttgarten and Helsingør–Helsingborg.

British Airports Authority dominates the market in the south-east of England through its control of the major airports. Competition from Luton and London City Airport in the Docklands area is not a serious threat. The privatization of BAA took place as a single entity rather than as competing airports. The form of the privatization was favoured by the management of the nationalized company, and it was also seen as the means to maximize revenue from the sale. Although the pricing policy at the airports is limited by the RPI-X formula, profits have steadily increased since privatization in 1987 (Table 10.3). The economic explanation of these profit increases could be that under state monopolies economic rents are reduced by low productivity and weak management, whereas the more efficient market in corporate control after privatization leads to an increase in profits.

In addition to these major privatizations in transport, several transport infrastructure projects have been funded by the private sector. The main schemes include the Dartford Bridge on the M25 (Queen Elizabeth II Bridge costing £100 million), a new Severn Bridge (costing £300 million and due to open in 1996), the Channel Tunnel project (costing over £10 billion and officially opened in Spring 1994), the extension to the Docklands Light Railway and the proposed Channel Tunnel link. Other proposals have at present been delayed as joint funding between the private and public sectors has not been forthcoming—the Jubilee Line extension (£1.9 billion), the rail link between Paddington and Heathrow (£400 million), the Cross London Rail Link (£2.5 billion), the proposed second Forth bridge (£400 million), and several road proposals. The Jubilee Line extension has recently been approved (November 1993) with £98 million funding from the European Investment Bank and £300 million from the Canary Wharf Company spread over the next 29 years. The total private sector contribution amounts to about 10 per cent of the capital costs at 1993 prices.

Plans have now been published for Britain's first purpose-built toll road which will duplicate the heavily congested M6 motorway around Birmingham (Table 10.4). The M6 now carries up to 130 000 vehicles a day, and this level could rise to 200 000 vehicles a day in the next twenty years. This new toll route will duplicate the current M6, and the Government argues that this private facility will bring the benefits of private sector management, together with additional resources to the roads programme. However, it is unlikely that the private sector will be more generally interested in major investments in the infrastructure, provided that the vast majority of the motorway network remains untolled (Department of Transport 1993). To resolve this dilemma, the Government is now considering direct charging for the use of existing motorways (Department of Transport 1993).

Notwithstanding the above, Holland (1993) regards private sector invest-

Table 10.4. The Birmingham Northern Relief Road

Plans for the Birmingham Northern Relief Road (BNRR) were first announced in 1980. After extensive consultations, Draft Orders were published in 1987 for a 53 km motorway from the M54 at Featherstone to the M6 near Coleshill. A Public Inquiry was held in 1988.

In 1989 the Government announced a competition to design, build, finance and operate (DBFO) the BNRR as the first purpose-built inland tolled motorway in Britain.

In 1991 Midland Expressway Limited (MEL), a joint venture between Trafalgar House and Iritecna was announced as the winner and a Concession Agreement was signed in 1992. The revised preferred route was published in March 1992.

Draft compulsory purchase Orders authorizing the acquisition of land were published in 1993, and objections heard at Public Inquiry in 1994.

The total cost of the new road will be £270 million.

---

ment in roads in Britain as unsuccessful. 'One can say that the few schemes so far completed by the private sector and the further few in the pipeline, though individually large, are not of a scale which established a change in the programme size.' Holland states that

> the whole focus of the entrepreneurs has been to make money on the construction; and for this they have been willing to bear the obligation of operating the road until it has paid for itself and can be handed back to the Government. The country has acquired, and will acquire some infrastructure which might otherwise have been delayed by the Treasury, but the roads programme remains virtually controlled by an executive department of Government under Treasury discipline.

Holland proposes an alternative privatized system. What is required is 'the creation of a multiplicity of opportunities to invest in and manage road assets'. Two conditions are necessary for a privatized system. The period over which the road can be owned should be long and there should be a revenue base that is sufficient to attract entrepreneurs to the road sector.

There is a considerable debate in Norway and the UK on urban road pricing and toll roads—for a discussion of the situation in Norway, see Section 10.4.3. They have been debated in the Norwegian Parliament and are currently being debated in the UK through a consultation paper (Department of Transport 1993). Norway has experience of toll roads where environmental issues are perceived as important, and the revenues raised are used (at least in part) for further investment in roads and public transport. The UK Treasury has strongly resisted the demands for hypothecation in transport, but some transfer of funds seems to be crucial to the public acceptability of pricing— otherwise it will be seen as another transport tax. The Treasury may allow hypothecation, but reduce the transport budget by a similar amount, thus increasing Treasury revenue for expenditure in other areas. It is unlikely that pricing will substantially increase the transport budget, but it would allow some stability, which in turn may permit forward planning.

10.3.2   IRELAND

In Ireland, tolled routes are confined to two toll bridges on the east and west sides of Dublin. A bid of £100 million (123 MECU) was made by Manufacturers Hanover for a franchise to toll the M50 motorway on the west of Dublin. It was dropped because of opposition to tolls by local government representatives. The National Development Plan (1994–99) provides for the National Roads Authority to enter into joint venture agreements for toll-based private investment. The Plan anticipates £100 million in revenue from toll routes and this would be used to further accelerate road investment. The guidelines for toll-based private investment include: competition for each franchise; ownership of toll roads by the local authorities; reversion of franchises to the National Roads Authority; and a private sector investment of at least 20 per cent of the cost of the toll road. Since investment in national, regional and local roads will contribute to the generation of substantial economic activity (estimated at £1725 million), the projected toll contribution will be under 6 per cent. There appear to be doubts that the £100 million toll route revenue target will be attained. The total transport expenditure under the Plan is £2.6 billion, of which £1.1 billion will be financed by the Irish state and public bodies, £1.5 billion by the European Union, and only £14 million (0.5 per cent) from the private sector.

10.3.3   FRANCE

Construction of new motorways has been through letting concessions to semi-public bodies (Société d'Economie Mixte—SEM) which would then be contracted to build particular sections of road. Legislation has also been passed to allow private companies to bid for concessions to build certain roads. In all cases the State has implicitly supported the private sector by providing financial support through low or zero interest cash advances, guaranteed loans or the provision of related infrastructure. At present, France has seven autoroute SEMs, one private autoroute concessionaire (COFIROUTE), and two tunnel SEMs (Department of Transport 1993). The SEMs and COFIROUTE keep the revenues from the tolls which can then be used for road maintenance and the construction of new autoroutes.

In western Europe as a whole, there are some 40 000 km of motorway, of which 13 500 km are tolled. Over 90 per cent of the tolled motorways are in France, Italy and Spain. In 1991, the annual revenue from tolls varied from around £500 million in Spain, to over £1 billion in Italy and around £2 billion in France (Department of Transport 1993). Switzerland has imposed an annual charge of £13 per annum for all domestic and foreign vehicles using the national motorways. Each vehicle has to display a 'vignette', the price of which was fixed by legislation in 1984. In 1991 it raised revenue of about £90 million. The European Union has approved (June 1993) a maximum level of charges for freight vehicles and cars of 1250 Ecu per annum for the use of the European motorway system (Vreckem 1993).

10.3.4 NORWAY

The use of private financing of transport infrastructure was first permitted through the Road Act 1960, with a system of toll financing. These principles have been discussed in the Parliament in 1989, 1991 and 1993, and the following principles for using toll financing have now been established in Norway:

1. All investments must have local agreement—the municipalities and county councils in the areas concerned must agree to use of toll financing. In practice this means that most projects will originate locally, and that the local councils and local industry will be the driving forces for new projects.
2. The use of tolls can only be used for the defined road projects. But in 1991, Parliament agreed to a wider use, as there was a majority for changing the Road Act to use tolls for investments in railway projects if it could be documented that this was the best solution. This modification has not yet been carried through by further changes in the Road Act, and there is still considerable disagreement over this proposal.
3. From 1991, there is also an opportunity for the introduction of differentiated tolls by peak and off-peak in urban areas when the main motive is further investment and not 'traffic regulation'. To implement this proposal, there has to be local area agreement.
4. In the past a project must have at least 50 per cent toll road financing. This has been changed from 1991 so that tolls can also be used for lower levels of financing.
5. There is now the possibility for 'toll road companies' to use foreign loans to finance projects.
6. All projects must be carried through by 'toll road companies' owned by private investors, municipalities, county councils etc. These companies are responsible for financial planning and for charging proposals. When the project has been approved by Parliament, the role of the company will be to finance the project and to implement it.
7. The charges for toll roads have to be approved by the authorities. In Norway many projects are fjord-crossings relieving ferry connections and here the charges are normally set at the ferry fares plus a maximum of 20–40 per cent. The normal repayment period is 15 years, but in

exceptional circumstances 20 years. In addition if income does not fulfil expectations, a maximum 5 years extension can be granted.

8. There are normally no state guarantees for projects, but local authorities can give guarantees for their own local projects. If the maximum repayment periods are not enough, the local guarantees will be effective and the financial institutions will have to cover what is not covered by guarantees.

The recent policy review ended this spring (1993) with the Road and Road Transport Plan 1994–97. A more liberalized regime than before has been introduced. Some of the results of these changes result from the experiences of the late 1980s where many approved projects resulted in problems arising from:

1. increases in real interest rates;
2. reductions in levels of forecast traffic growth. This can partly be attributed to the optimistic forecasts from the local toll road companies;
3. increases in the levels of investment due to the underestimation of costs;
4. increases in maintenance costs, especially for tunnels. This has been an important factor in total cost overruns;
5. a high proportion of tolls being taken up by administration and the costs of toll collection.

It is important to note that a more liberal regime in the use of tolls has majority support in Parliament. This means that tolls can be used for railway investment in urban areas if this is the best overall solution economically, and tolls can be differentiated by peak and off-peak for investment purposes, but not for traffic regulation. Both these changes have not yet been implemented (December 1993).

10.3.4.1 The Proportion of Toll Financing of Total Investment in Roads in Norway in Recent Years

In Figure 10.1 the development in use of tolls for financing of roads is shown. The figure shows that in the period of 1985–86 about 25 per cent of the total investment in state roads were for toll facilities and the planned figure for the period 1990–93 was 36 per cent. In Table 10.5 the actual figures for 1990–93 are shown, with about 29 per cent of all investments in toll facilities.

The reason why the toll road investment is smaller than planned is due to the problems which many toll roads experienced in the beginning of the 1990s, and to more public money being available as a result of the anti-unemployment policy pursued by the Norwegian government.

In Figure 10.2 the actual amounts of toll-financed roads for each year in

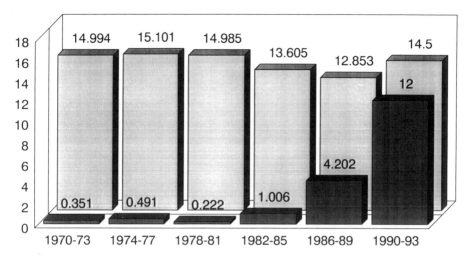

Figure 10.1. Investments in state roads in Norway 1970–1993: division between public investments and toll roads (1989 millions Nkr) (*Source*: St.meld.nr.46 (1990–91))

Table 10.5. Investment in state roads 1990–93 divided by public financed roads and toll roads (1993 prices)

	Planned investment 1990–93 in state roads (million NKr)	Actual investment state roads 1990–93 (million NKr)
Public investment	16 245	17 806
Toll roads	9 220	7 208
Total	25 465	25 014
Toll %	36 %	29 %

Source: Government data, St.Meld,nr.34 (1992–93)

the period 1984–1993 are shown and in Figure 10.3 the actual amounts of tolls paid each year are summarized. In 1993 it is expected that about NKr 1400 million will be paid in tolls.

10.3.4.2 Discussion of the Use of Road Pricing in Norway

During the spring of 1993 there was a thorough discussion of use of alternative financing for roads in Norway based on the Government's submission to the Norwegian Road and Road Traffic Plan 1994–97 (Government data,

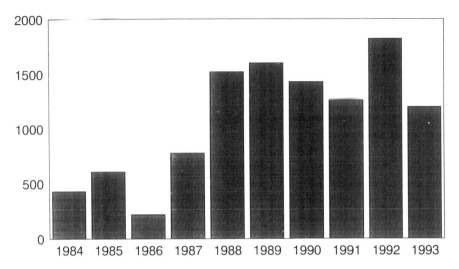

Figure 10.2. Toll roads built in Norway (million 1993 NKr) (*Source*: St.meld.nr.34 (1992–93))

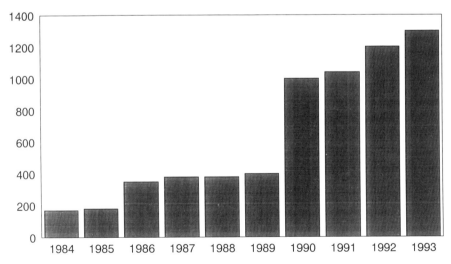

Figure 10.3. Toll revenue from road users in Norway, 1984–1993 (million 1993 NKr) (*Source*: St.meld.nr.34 (1992–93))

St.meld.nr.34, 1992–93). The Government submitted proposals and identified the following possibilities as *additional* forms of finance to assist public investment:

1. dedicated motor-taxes (including the annual tax for owning of cars and fuel taxes);
2. user charges through extra fuel taxes geographically located and toll charges;
3. road-pricing.

Dedicated motor taxes were rejected because it would create a system that was too unstable because of changing amounts of revenue which were dependent on the economic climate. This would be a very ineffective way of financing roads. The second alternative of geographically differentiated fuel taxes was also rejected because of the control problems between regions, even though it has been used in Tromsø. Toll charges have been the conventional means of attracting private finance or extra finance in Norway and this will be continued as described earlier.

As regards *road pricing*, the Government stated that in 1990 they started to study the future possibilities for use of road pricing in Norway including the potential need for such a system. The Government stated that there are at present several institutional problems that have to be solved before road pricing can be used. These include political processes, the use of revenues and the jurisdictional problems. The conclusion reached by the Ministry of Transport and Communications and the Government is that road pricing offers an interesting and useful means to regulate traffic, that the Government will continue to study the effects of such options, and that its use in Norway as a transport policy means that it will be judged on the basis of Norwegian and international studies.

The majority in Parliament shared the views of the Government, namely that road pricing provides an alternative in Norway, but that further studies were needed. It was also suggested that as a part of future studies, trials with road pricing could be used if there was local support. At present there is a 'wait and see' mood in Norway, and for the next four years at least, the system of toll roads will be used in preference to a wider system of road pricing.

10.3.4.3 The Results of the Large Toll Rings in Oslo, Bergen and Trondheim and Some General Comments on Toll Projects in Norway

In Norway, in addition to many toll projects in rural areas, particularly fjord crossings, three large urban toll rings have been established in the Oslo area, in Bergen and in Trondheim (the three largest urban areas in Norway). The results of these toll rings are worth commenting on in some detail.

The Oslo project was established on 1 February 1990 after considerable debate, especially locally in the Oslo area. The resistance to the project was strong, even though the local authorities had favoured the solution as being the only means to quickly resolve *some* of the road capacity problems in the area. The main object of the project was to raise money for road building, for investment in public transport (rail) and for better provision of pedestrian and bicycle facilities. The project follows the principles of the Road Act. One of the main objections to the project was the *high costs* of running of the system (revenue collection, etc.). The investment in the road and underground tunnel with roads 'Fjell-linjen' was NKr 250 million and the yearly income was estimated at NKr 600 million.

The system consisted of a charge for each car (NKr 10 per small car, now NKr 11) and season tickets with an unlimited number of journeys (per month, per ½ year and yearly). The electronic system was established in December 1990 to replace the manual system in operation between February and December. There was some concern over queues, and that was the main reason for the introduction of the season-ticket system. A new system of prepaid passes was introduced in 1991 and this has replaced the original season-ticket system. There are 19 stations located 3–8 km from the city centre. Approximately 210 000 cars pass every day (towards the city) which amounts to 40 per cent of all car trips in the Oslo region.

The results of the toll ring have been monitored and the main conclusions are the following:

1. The toll ring has resulted in about a 12 per cent reduction in car trips in the Oslo region from 1989 to 1990. The general decrease continued in 1991. However, this change can only be attributed in part to the toll ring as unemployment figures increased and as fuel prices were raised during the same period.
2. The effect of the toll ring itself on the total car traffic is larger than the expected reduction of between 5 and 10 per cent.
3. The toll ring has not contributed greatly to reducing rush-hour traffic, but there has been a reduction in leisure trips. There are clear changes in travel patterns as people tend to change their geographical travel pattern towards seeking destinations on 'their own side' of the toll ring.
4. The public transport has not generally benefitted from more traffic, but the toll ring has contributed towards helping public transport maintain its market share and mitigate the recent downward trend in public transport demand. However, the public transport share of trips crossing the toll ring showed an increase of about 10 per cent.
5. It is difficult to test a general hypothesis that people using season tickets travel more than they did before the introduction of the toll ring. People using season tickets seem to travel less than before by reducing their general travel, but they still travel more than people without season tickets.

6. It was found that the percentage of people strongly opposed to the toll ring has reduced (1989 to 1990). Studies in 1991 show the same tendency, so the public acceptability of the toll ring is increasing.
7. In the autumn of 1990, 34 per cent of all households in the region owning a car had bought a season ticket, the majority for one year. Almost 50 per cent of ticket holders had received financial support from their employer.

The Bergen project was the first urban project to be established. Its intention was also to raise capital to improve the main roads in the area (from 2 January 1987). The toll road has six stations. The toll-ring project has been a success and the public seems to have accepted the project and the revenue from the project has also been higher than expected. The principles used are based on the Road Act and the later changes made by Parliament.

Some of the early results of the Bergen toll ring can be summarized:

1. On average toll-road users having a pass use the ring 1.7 times per day.
2. The share of pass users is approximately 55 per cent, but 70 per cent of users in the morning peak use a pass.
3. The toll ring reduced the traffic by 6–7 per cent the first year compared with what might be expected from the free use of road.
4. In Bergen, there has not been any increase in public transport and some bus companies also lost traffic in 1986 compared with 1985. There has been no positive effect on public transport resulting from the establishment of the toll ring.

The Trondheim project is the newest of the large urban projects, established through recent decisions in Parliament (1990). The plans consist of road building, public transport, and safety and environmental projects costing NKr 2.5 billion. The project has faced severe problems due to the late start in charging tolls, increases in the investments needed for road building, and this in turn has led to higher financing requirements. The traffic forecasts of the projected toll ring also seem to be overestimated, especially the expected use of new toll road to the east of Trondheim (a local road exists for diverted traffic).

In general, several of the new toll projects in Norway have had problems with creating enough revenue to pay interests and loans. This can be attributed to several factors, in particular to:

1. Overestimation of traffic demand. The official forecasts for traffic from 1989 to 2000 have been reduced from an increase of 40 per cent to an increase of 25 per cent. The economic climate and increased unemployment has also added to the lower levels of traffic demand increases;
2. The real rate of interest up to 1993 has been much higher than estimated;

3. People have been creative in making use of seasonal tickets and thus making the average fare lower than expected.

A good example of the problems experienced is the so-called Ålesund tunnels, which consist of three submarine tunnels connecting the town of Ålesund with surrounding islands including the airport. These tunnels are privately financed and the toll company is bankrupt and has been taken over by the banks with debts of NKr 900 million. The main reasons for this situation are the overestimation of traffic by 20–30 per cent as the toll charges were set too high, and because the charges were expected to be set at 1.4 times the previous ferry fares, but the actual levels are lower due to extensive use of discounts. This is a problem for many schemes, as the elasticities of demand are high. The market is not large enough to sustain the fare levels to secure the necessary revenue to pay for costs of administration and collection and interest and loans. As *time savings* account for 50–80 per cent of total benefits in the cost–benefit analysis, when comparing tunnels with ferries, drivers do not necessarily recognize these benefits as the time savings are often internalized by car users in their travel decisions. Consequently, they continue to use cheaper, slower ferries.

10.3.4.4 Privatization in Scandinavia

The process of privatization found in Great Britain is very different to that found in the Scandinavian countries. Up to the change in Government in Sweden (1991), the Labour Government led a policy of non-privatization, but corporatization. This changed in 1991 when the Conservative Government embarked upon a vast programme of privatization, including selling the shares in the SAS parent company, plans for privatization of the Swedish telecommunications industry, and the privatization of many other companies.

In Denmark the situation was different, and the Conservative Government (up to 1993) privatized 45 per cent of bus operations in Copenhagen through tendering which explicitly excludes public participation, prepared the privatization of Copenhagen Airport and the telecommunications industry, and sold off several banks and insurance companies. The Labour Government (1993) has held a more pragmatic view, but has signalled a continuation of the programme for financial reasons.

In Norway no real privatization has taken place outside some industrial firms, mostly 'lame ducks'. This has been a clear policy of different Labour Governments preferring corporatization instead of privatization. In the 1980s, the Conservative Governments proposed privatization of the state railway travel agencies and bus operations, but this did not get support in Parliament. The proposal for selling 49 per cent of the telecommunications terminal equipment subsidiary of Norwegian telecommunications was withdrawn when there was a change of government in 1990.

Real privatization in Scandinavia has been a mixed bag, and outside Denmark there has not been any real progress in this area. However, there has been major progress made in introducing toll roads in urban areas to raise revenue for further investment in roads and public transport. Most of these schemes have been organized by the public sector, but there is no reason why the private sector could not become more fully involved. However, where private sector involvement has taken place (e.g. the Ålesund tunnels), the returns have been mixed as the high prices necessary to recoup the capital costs have been resisted with car drivers using the cheaper, slower ferries or not travelling at all. This is a problem in thin transport markets where existing levels of demand are low and the risks to the private sector are high.

10.4 PARTNERSHIP POSSIBILITIES

Several different strategies seem to have been developed to reduce levels of public expenditure in transport. Privatization of transport companies is one means to reduce pressures on public sector budgets and reduce the public sector borrowing requirement (Section 10.3). However, when the transport infrastructure itself is considered, the options available are less clear. In the past, transport networks have been seen as a public good and a long list of problems have been presented as to why it would be difficult to use private-sector funding, except in certain situations or where very clear guarantees were given.

The main concern to the private sector is the risk of investing large amounts of capital in projects where there are:

1. long periods between the start of the investment and the financial returns to investors;
2. irreversibility of investment or where there are substantial sunk costs. It is costly to start a project and even more costly to withdraw;
3. financial returns do not flow until the whole infrastructure is completed;
4. political influences on the production of goods and services;
5. long amortization periods when loan repayment periods are often over a much shorter term.

Quite naturally, the conservatism of developers would mean that investment would be concentrated on buildings and transport terminals/interchanges rather than in the infrastructure itself. There are lower risk projects available with shorter payback periods and greater certainty (until recently) over the rates of return.

In addition to the general questions of risk, transport projects have in the past proved difficult to assess. They often have:

1. substantial cost overruns;
2. levels of demand have been difficult to predict accurately;
3. been considered as a free good at the point of delivery;
4. aroused considerable public opposition to their construction and this has often delayed their implementation, sometimes after substantial investment has been made in scoping studies (e.g. environmental impact assessments).

High levels of risk and uncertainty mean that the private sector has been reluctant to get involved, despite having the resources available. Investment funds currently have large sums of capital available, and it would seem natural to direct some of this cash into transport infrastructure.

The key requirement facing governments in many European countries is to establish the means to bridge this funding gap. Both the public sector and the private sector have important roles to play in the construction, renewal and maintenance of the road, rail and air infrastructure. Over the next decade it is likely that new means of financing will be established, yet the public sector cannot withdraw completely. Much of the funding will still remain the responsibility of the public sector, but the role for transport planning must be to facilitate private sector and joint ventures through advice, predictions, land assembly and accelerated public inquiry procedures. The public sector still has a key role to play:

1. where the market fails and intervention takes place for accessibility, distributional and equity reasons, and where transaction costs are high;
2. where there are significant externalities involving the use of non-renewable resources, land acquisition, safety and environmental concerns;
3. where transport interacts with other sectors, such as the generation effects of new developments and priorities given for regional or local development objectives;
4. where transport has national and international implications, such as promoting a capital city (e.g. London or Paris) as a world city or maintaining high-quality international air and rail links.

However, all these roles are essentially passive and the more important position for the public sector must be to promote a partnership between the public and private sectors.

10.4.1 PUBLIC SECTOR ACTIONS

There is a series of positive actions which can be taken by the public sector to facilitate interest from the private sector. As most transport infrastructure investment is long term, there must be a stable policy framework and a set of planning procedures within which the private sector can operate. In addition, the public sector could raise capital through special funds which are seen as ancillary budgets remaining under Parliamentary control. Only the balance would be shown in the general budget. Access to these funds would take the form of a competition between the various government departments or there could be a competition within departments for the available budget.

Alternatively, earmarked charges could be raised from toll roads (as in Norway) to finance further investment. Some economists (e.g. Buchanan 1963) have favoured earmarking as a means to compartmentalize fiscal decisions and to allow individuals to participate in public expenditure decisions. A new impetus has been given to the hypothecation arguments with the realization that greater transparency and public acceptability of increased taxation is essential if substantial sums of public expenditure are to be allocated to transport infrastructure. The acceptability of environmentally efficient taxes on motorists will depend upon part of the revenues raised being used to fund less environmentally damaging modes of transport. If long-run changes in demand are also desired then the decisions concerning hypothecation must also be transparent, otherwise most motorists will continue to use the car and pay a higher price.

A second major contribution that can be made by the public sector is to facilitate the complex processes of raising the substantial sums of capital required for transport infrastructure projects. These include raising capital:

1. Through loans from the European Investment Bank and the European Union's European Coal and Steel Community, the Regional Development Funds and the new European Investment Fund.
2. Through transport bonds and other long-term investments (such as pension funds) which are extensively used in Japan.
3. Through tax incentives to the private sector by making their capital contributions tax deductible. This might allow pension funds and financial institutions to become involved in road construction as an investment opportunity (suggested for road building in Norway with funds set aside in development areas. This proposal was not accepted by the Government).
4. Through employment taxes (as in Paris and other French cities) or a tax on petrol (as in Germany and the USA).
5. Through user charges from tolls and road pricing.

More controversially, the public sector could guarantee loans to the public or the private sectors, thus accepting a substantial part of the risk. Considerable debate is currently taking place on this issue as the balance of the risk is still with the public sector and not the operator of the system, for example where the government (public sector) underwrites the loans to the railways (public sector). Much of the French TGV routes have been funded in this way. Under French transport policy, the French Railways (SNCF) are responsible for developing the railways and they have had a high credit rating as the government guarantees the loans. This means that interest rates are lower than commercial rates (typically by 1 per cent), and high rates of return are not essential (as they would be on equity risk capital). This allows SNCF to run at a loss and pay no corporation tax. A private sector package would have taken much longer to set up (Gerardin 1990). In the longer term, the TGV system will make a profit and may repay a substantial part of the amortized debt, but in the short term SNCF has substantial debts. A similar company in the private sector with similar levels of debts (FF 100 billion, 1987) would have been placed in liquidation.

However, these semi-public undertakings in both the road and rail sectors in France have allowed revenue funding to switch between projects. The cash flows generated by the infrastructure for which the loans have been completely repaid have been used to finance further extensions to the network—what Gerardin (1990) calls the overspill principle, similar to that used in the last century to finance the last great expansion of the rail network.

10.4.2   PRIVATE SECTOR PROJECTS

As outlined in Section 10.3, there are particular types of infrastructure projects which the private sector are prepared to finance and operate with minimum levels of guarantees (e.g. the Queen Elizabeth II bridge on the M25 around London). These low-risk projects are comparatively small in scale and place the private sector operator in a monopoly position. Interest is particularly high where there are congested conditions on the existing infrastructure and where anticipated growth in demand is high. If a reasonably lengthy ownership period can be negotiated, then the expected payback is substantial. In this case the private sector plans, designs, builds, operates, owns and finances the project (Table 10.6). The role of the public sector is secondary, and limited to the promotion of supporting actions (e.g. enabling

Table 10.6. New partnership possibilities between the private and public sectors for road investment

Function	Traditional model	Partnership possibilities			Private sector model
		1	2	3	
Planning	Public	Public	Public	Public	Private
Design	Public	Public	Private	Private	Private
Construction	Private	Private	Private	Private	Private
Operation	Public	Private	Private	Private	Private
Ownership	Public	Public	Public	Private	Private
Finance	Public	Joint	Joint	Private	Private

legislation) and negotiations on the length of the ownership period. This needs to be balanced against the expected growth in traffic and the levels of charges to be paid by the users. Successful negotiations on time, traffic growth and user charges means that the project will go ahead.

In most cases, one or more of these conditions are not met and as a consequence the private sector has been reluctant about making any firm commitment. A further complication is the ownership of the infrastructure after the period guaranteed to the private sector. Reversion to the public sector may mean that the route no longer has a toll on it, but maintenance costs and reconstruction costs are likely to be substantial (particularly on tunnel infrastructure). This means that the public sector may incur substantial costs at some time in the future.

The most notable exception to all the above is the Channel Tunnel project which has been funded by the private sector. This is Europe's largest transport construction project, costing some £10 billion, and the length of the ownership period by the private sector is 55 years. It has been completely designed, constructed, owned, operated and financed by the private sector. The very substantial risks are with the private sector over a very long period of time, yet the rewards may also be substantial if predicted levels of demand are met.

10.4.3 JOINT PROJECTS

It is in partnership between the private and public sectors where most potential lies. The public sector can anticipate the growth in demand which results from the growth in the economy, rising income levels, and from new developments. It can also assist in the land assembly process and in the public inquiry so that time from project inception to completion is minimized. In certain situations, it can also help with the costs of construction of the actual infrastructure, but it is the private sector which will manage and run the facility, including setting the levels of tolls or fares.

There are several different approaches which can be used. The land could remain in public ownership with a contract between the public and private sectors. This concession would be granted through a tender or franchise, and the funding would be the responsibility of the private sector with some public contribution in the form of loans and loan guarantee (Table 10.6, partnership possibility 1). The private sector operator would set the charges, but constraints on the quality of service would be set by the public authority granting the concession (e.g. minimum levels of service and safety standards). Maintenance of the infrastructure would also remain in the public sector, and public money would only be used to ensure that the project actually takes place. This means that the risk to the private sector has to be equivalent to other investment opportunities. One possibility here is to restrict the public sector contribution to those factors which reduce the negative externalities (i.e. the public sector would pay for environmental improvements), but even this is controversial if the polluter pays principle is used. It is often cheaper to guarantee loans (Section 10.4.2) than to contribute directly. Rather than have a toll for the use of the facility, the public sector could pay a fee or shadow toll to the private sector for every vehicle using the road (Button and Rietveld 1993). Such an arrangement would allow a scheme proposed by the public sector to be built by the private sector and operated as a free (at the point of use) facility. The risk to the contractor is in the demand forecasts as they would only be paid a fixed rate for the actual numbers of vehicles using the facility. At the end of the contract period, the public sector might have the option to buy the road.

Development gains can also provide an important incentive to the private sector. In the past, planning permission has often been granted subject to certain conditions being met. These conditions have involved road construction, particularly in locations where development pressures are substantial. A different option would be to give the developers the rights to develop land around the road or rail infrastructure which that developer had financed. Land values at new accessible locations, principally at road interchanges or rail terminals/stations, rise substantially (Stopher 1993), and the potential for development is considerable.

In this case the public authority would acquire more land than is actually required for the construction of the infrastructure. Either this land could be sold to the developer with the profits being used to finance the construction of the infrastructure by the public sector, or an agreement could be reached with the private sector that it takes the responsibility for construction and acquires the land. In this second alternative, the private sector has the associated land development rights which are similar to the air rights being granted in urban areas over new station developments. The private sector could either carry out the full development or build the infrastructure and sell on the associated development rights.

10.4.4   IMPLICATIONS OF PRIVATELY FINANCED ROADS FOR THE FINANCIAL AND THE CONSTRUCTION SECTORS

The traditional division of road transport investment and construction between the public and private sectors is shown in Table 10.6. With mixed or partnership systems, there is a range of new opportunities with the private sector taking on many more roles than the traditional one of construction for a single client, the State. In Table 10.6, various degrees of private sector involvement have been presented (Section 10.4.3) for road investment. In a totally privatized system and in partnership possibility 3, the private sector would serve many clients, the user. Investors would be exposed to risks such as overestimation of demand, design standards for a higher level of service than customer willingness to pay, cost overruns, and changes in interest rates. Privatized road investors would presumably resemble the developers that now exist in the property sectors. Private sector investors have responded to the investment opportunities of privatized infrastructure such as ports and airports.

Holland (1993) states that conditions for attracting major private sector equity to road finance are a long, even indefinite period, over which the road is owned by the investor, and a revenue base that is sufficient to attract entrepreneurs to the sector. The revenue stream from road investment would be based on the direct assessment of the facility by users. Currently, user benefits from roads are assessed by models such as COBA which impute values for benefits such as work and leisure time savings, accident cost reductions and fuel cost savings, net of taxation. The COBA model values leisure time savings at a quarter of work time savings and bases the latter on labour costs. Small time savings are treated in the same way and are valued at the same level per unit of time as large time savings. The users of tolled routes might not value time savings as specified in COBA and investors in these roads would have to develop estimates of willingness to pay for time savings. The actual time savings to the individual on any particular trip will not be the same on each occasion, but it is likely that the toll will be the same. So the user of the toll road is paying a fixed price for a variable time saving.

The accident cost savings from higher levels of road investment are based on the virtual elimination of head-on accidents by dual carriageways, and junction accidents by grade separation. These benefits are currently imputed for cost–benefit analysis rather than sold to the user as they would be under a toll road system. The safety benefits may not be widely understood at present, but could be an inducement to users when entrepreneurs market toll roads. Fuel savings are a relatively small part of the user benefits of road investment in the COBA model. The financial assessment of road investment would overstate the value of fuel savings because a substantial part of the savings is a reduction in sales tax which is a transfer payment rather than a resource cost.

It will be necessary to persuade investors and fund managers of the earnings potential of road investment. The gains from privatization of road investment, based on privatizations in other sectors, might be expected to be a more efficient use of labour and capital, and the choice of a level of service which is market-led. At the macroeconomic level the gains would be a reduction in the public sector deficit, the tax burden and the size of the public sector.

Less ambitious partnership possibilities involve joint funding between the private and public sectors. In this situation the public sector would maintain ownership of the infrastructure, but would lease it to the private sector over a lengthy period of time and share revenues, or pay shadow tolls for the number of vehicles using the road (Section 10.4.3). In most situations the planning of the road would remain in the public sector as delays resulting from the public inquiry, the environmental impact assessment, forecasting, consultation and land acquisition procedures would all mean an unacceptable delay to the private sector. The private sector can be fully involved in the design of the road, and in the construction and operation of the infrastructure as a toll facility.

The three partnership possibilities (Table 10.6) are consistent with the clearing-before-awarding approach proposed by Fielding and Klein (1993). Auctioning a prepared project reduces the uncertainty on time and costs as construction can begin immediately. The private sector has to make a bid to complete the final design and to finance, construct and operate the project. Fielding and Klein also argue that competition in the bidding would be enhanced and the post-contractual administration costs would be reduced. The clearing-before-awarding approach should ensure that:

1. The risk to the private sector is reduced and thus bids would be expected to have a lower rate of return.
2. The costs to the public sector for inquiries, environmental impact assessment and land acquisition may be lower than they would be to the private sector. This would again reduce risks and sink costs.
3. The whole process involves both the public and private sectors in appropriate roles. Public sector involvement in the approving and awarding phases means that the proposals may be less vulnerable to political tampering and all the necessary procedures are followed in full.

10.5   CONCLUSIONS

There do seem to be substantial opportunities for the private sector to become more involved in the process of planning, design, building, financing and operating the links on the transport network as well as investing in and operating the nodes on the network. It is likely that progress must be made in partnership between the public and the private sectors. The combination of both sectors (Table 10.6) substantially reduces the front end risk and the likelihood of final cost overrun. It clarifies the difficulties of estimating the payback to the private sector, and the necessary period over which it would accrue. The financial rates of return could be substantial and revenue flows in the short term are possible. The public sector also needs to take a much more active role in promoting the project and steering it through the planning and design stages.

The main difficulties presented in the first two parts of the chapter have been overcome, but other factors must also be considered.

1. Risk sharing between the public and private sectors is necessary.
2. The free rider problem has to be resolved.
3. New transport infrastructure needs to be seen as part of the development process, not separate from it.
4. Consistency and stability in financing needs to be maintained as political and commercial horizons are often short term, whilst major infrastructure decisions are long term.

With joint projects, the private sector would recover their costs through user charges, but a balance needs to be sought between the private and public sectors in terms of their risks and rewards.

Competition between the private sector would be ensured through competitive bidding for franchises, and the public sector would specify its own contribution including the possibility of taking a non-controlling equity stake in the project. Some of the options here have been presented in the section on funding sources, and Fielding and Klein (1993) have developed some of the means by which roads can be franchised. A similar scheme has been adopted in Norway for the new Oslo area airport and for the new high-speed rail link between Oslo and the airport. Both will be built by joint stock companies owned by the state, but the opportunity will remain for future private participation. The projects will be funded by private banks through loans that will be repaid through charging for use of the infrastructure.

A second, less integrated approach to partnership would be through the public sector leasing road space, rail space, or rolling stock from the private sector. The effect of this strategy would be to reduce the capital outlay by the public sector, and allow financial allocations to go further as only the current leasing payments would be set against allocations. This practice is already in operation in the private sector where many airline companies have reduced their exposure in the market by leasing aircraft. In each case the principal risk remains with the owner of the facility and the operator only takes the risk for the leasing period.

In conjunction with each of the above possibilities, the government could set up new private companies or quasi-private companies (e.g. the SEMs in France) to design, build, finance and operate new transport infrastructure. These companies would have the right to raise capital on the national and international markets, they could reinvest the revenues obtained from the charges for using the infrastructure facility, and they would also have responsibilities for maintaining the infrastructure. There would be some debate over whether these companies were completely independent of government, or whether government would underwrite some of the risk, thereby obtaining lower rates of interest on the capital markets (as in France).

As industry has become increasingly global in scale with large multinational corporations, the involvement of the private sector becomes important as very few public enterprises operate beyond national boundaries. Privatization has provided one mechanism for transport enterprises to diversify and to become powerful operators in world markets (e.g. British Airways). Conversely, many public enterprises are protected by national governments from international competition. Capital for investment has to be raised through national governments or through international loans (e.g. from the European Investment Bank). This approach has been very successfully used by the French state-owned companies which have borrowed on both the national and international markets at low interest rates as the risk has been underwritten by the national government.

An alternative would be to privatize transport enterprises and allow them free access to the national and international financial markets. One of the main limiting factors, namely the linking of an individual company's borrowing ability to the macro-economic government policies determined by the PSBR, would be broken. Conversely, private enterprises could take minority stakes in public enterprises. This financial input from both national and international companies permits a wider financial base and allows public-sector priorities to be promoted in conjunction with private-sector objectives.

This chapter has attempted to open up the debate. It has not come up with a solution to the problem of financing transport infrastructure, but it has raised some of the principal issues which must be resolved. The private sector has a strong tradition of commercial, office and residential development, and it has successfully moved into the transport sector to take over the development of terminals and interchanges. The challenge set here is to determine whether private-sector portfolios can be further extended to include transport links. If the continued growth in travel demand is accepted, the European transport infrastructure needs substantial investment in new and upgraded roads and railways. Public budgets are limited and there must be a commercial opportunity for the private sector to enter the market.

REFERENCES

Banister, D. (1992) The British experience of bus deregulation in urban transport: Lessons for Europe, Paper presented at the *Spanish Regional Science Association's Seminar on Urban Transport Problems*, Madrid, June, and Working Paper 5 in the series produced by the Planning and Development Research Centre, University College, London, p. 29.

Banister, D. (1993) Charging systems for the use of urban infrastructure: Possibilities and realities. Paper presented at the *ECMT 97th Round Table on Charging Systems for the Use of the Urban Infrastructure*, Paris, 4–5 November, p. 34.

Barrett, S. (1993) Air transport markets, in D. Banister and J. Berechman (eds) *Transport in a Unified Europe: Policies and Challenges*, North-Holland, Amsterdam, pp. 91–124.

Buchanan, J. M. (1963) The economies of earmarked taxes, *Journal of Political Economy*, **71**, 457–469.

Button, K. and P. Rietveld (1993) Financing urban transport projects in Europe, *Transportation*, **20**(3), 251–265.

Department of Transport (1993) *Paying for Better Motorways: Issues for Discussion*, London: HMSO, May.

Fielding, G. J. and D. B. Klein (1993) How to franchise highways? *Journal of Transport Economics and Policy*, **27**(2), 113–130.

Foster, C. (1992) *Privatisation, Public Ownership and Regulation of Natural Monopoly*, Blackwell, Oxford.

Gerardin, B. (1990) Private and public investment in transport: Possibilities and costs, *Proceedings of the ECMT*, Round Table 81, Paris, pp. 5–32.

Hirschman, A. (1958) *The Strategy of Economic Development*, Yale University Press, New York.

Holland, C. (1993) Paying for roads, *Transport*, May–June.

House of Commons Committee of Public Accounts (1991) *Sale of the National Bus Company*, 9th Report of the House of Commons Committee of Public Accounts, Session 1990–91, No. 119, HMSO, London.

Kay, J. A., C. Mayer and D. Thompson (1986) *Privatisation and Regulation: The UK Experience*, Clarendon Press, Oxford.

Keating, G. (1992) Toll tales on the highway to prosperity, *Independent*, Wednesday, 16 December, p. 21.

Kelsey, A. (1986) How the stock market sees transport, *Journal of the Chartered Institute of Transport*, February.

Millward, R. (1986) The comparative performance of public and private enterprises, in J. A. Kay, C. Mayer and D. Thompson (eds) *Privatisation and Regulation: The UK Experience*, Clarendon Press, Oxford.

Stevens, B. (1992) Prospects for privatization in OECD countries, *National Westminster Bank Quarterly Review*, August, pp. 2–22.

Stopher, P. (1993) Financing urban rail projects: The case of Los Angeles, *Transportation*, **20**(3), 229–250.

Vickers, J. and G. Yarrow (1988) *Privatisation: An Economic Analysis*, MIT Press, Cambridge, MA.

Vreckem, D. van (1993) European Comunity Policy on Taxes and Charges in the Road Transport Sector, Paper presented at the *Joint OECD/ECMT Seminar on Internalising the Social Costs of Transport*, Working Document 4, Paris, November.

11 Franchising Alternatives for European Transport[1]

BJØRN ANDERSEN
Molde, Norway

11.1 INTRODUCTION

Franchising as a future regulatory regime in public transport has been proposed as an intermediate solution taking into account both the market-oriented approach and an approach based on social-effectiveness goals. One claim made is that franchising can eliminate the weaknesses of full deregulation (i.e. competition in the market) as has been experienced in Britain outside London, and it can also eliminate the weaknesses of regulatory regimes based on a system of monopoly by opening up the market for competition. In Table 11.1 the system of franchising is defined and compared to other regulatory regimes.

However, the notion of franchising is not clear, and it must be defined more precisely. In this chapter the term franchising will mean a regulatory regime where an organizing authority or other public body can be classified according to Table 11.1.

With franchising, an organizing authority or other public body enters into *contract* on operations or operators/builders of public transport either by arbitration or competitive tendering; and where the operators/builders can be either privately owned, publicly owned or mixed private/public ownerships or partnerships.

When competitive tendering is used, there is *competition for the market*. Competition for the market can be seen as an alternative to *competition in the market*, i.e. full deregulation as a regulatory system.

In Section 11.2 we will try to give an overview of the extent to which the various franchising systems are used in Europe and the extent to which competitive solutions or arbitration are used.

[1]The research on which this paper is based is financed by the Nordic Council of Ministers/ Nordic Transport Research Committee project 'Organizational Change in Scandinavian Transport' and the Royal Norwegian Research Council/NORAS project 'Changing Working Conditions in European Public Transport'.

European Transport and Communications Networks: Policy Evolution and Change. Edited by David Banister, Roberta Capello and Peter Nijkamp. © 1995 John Wiley & Sons Ltd.

Table 11.1. Regulatory regimes in relation to market-decided costs and service-levels, and administratively decided costs and service-levels (routes)

Market-decided service levels

	Competition in the market and supplementary tendering. 'Threatened competition'	Private monopoly (protected)	
Market-decided costs			Administratively-decided costs
	Competition for the market (Tendering/ franchising)	Public monopoly (protected)	

Administratively decided service-levels

Source: Cox and Love (1991).

11.2   AN OVERVIEW OF THE FRANCHISING SITUATION IN EUROPE TO-DAY

Franchising systems can be found in many European countries, but there are three regions where there has been substantial use of franchising systems:

1. the Scandinavian countries
2. France and
3. Great Britain (London).

In addition, there are franchising systems in Switzerland (Zürich) and Spain (Madrid), but in this paper the focus is on Scandinavia and France. London is well documented elsewhere (Banister *et al.*, 1992a; and Andersen 1993b).

The use of franchising in France is not new, but important developments have taken place since 1982. The enacting of new legislation for public transport in France in 1982 (LoTi) laid the foundations for the recent important developments concerning infrastructure building and the creation of public/private partnerships, especially for new rail-based transport systems.

In Scandinavia the development started with new public transport policies in Sweden (enacted in 1985 and brought into force in 1989), Denmark (Copenhagen 1990, outside Copenhagen 1994), Norway (1991) and Finland (1991). The latter two countries have not yet started using competitive tendering. It will be in use from 1994/95.

The systems in Scandinavia and France are different in organization. Scandinavia has chosen to use total tenders or franchises, implying that the operator also owns the rolling stock and facilities for maintenance. In France the most common type of franchising arrangement is an operating franchising where a public transport organization owns the rolling stock and other infrastructure, and leases this out to the operating companies of which the dominant ones are private. Recently, we have also seen the creation of private/public partnerships especially as regards designing, building, operation and financing of railway/tramways for urban use. It is therefore interesting to compare the two types of system and especially to study the disadvantages and advantages of franchising systems in light of the two different types of system.

This chapter will also discuss the importance of European Community rules regarding 'purchasing of services', i.e. the so-called directives on public procurement in the excluded sectors, of which transport is one. This is important because European Community policy in this area will influence the future public transport regulatory regimes in Europe. This is so because there will be compulsory rules for the public procurement of services which came into force in July 1994 where the purchase of public transport is based on an agreement and not on legal licences.

11.3   THE FRANCHISING APPROACHES: AN OUTLINE OF DIFFERENT SYSTEMS, ADVANTAGES AND DISADVANTAGES

The franchising systems discussed in Section 11.1 can be of several different types. Normally, they are divided into systems based on arbitration or systems based on competition. The latter are termed competitive tendering for the market.

Irrespective of whether we use arbitration or competition there are, however, different systems which can be classified as follows:

1. *Total franchises* include both the operation and provision of necessary infrastructure/rolling stock. This means that the franchisee is also responsible for capital costs.
2. *Operations franchises* exist where the franchisee operates the system, but with rolling stock and infrastructure provided by the franchisor, normally a public authority responsible for the planning and financing of public transport.
3. *Management franchises* occur where a public body is responsible financially for both operation and for rolling stock/infrastructure, but where an outside franchisee provides the necessary management competence for the operation of the system.

4. *Planning franchises* operate where there is public operation and infrastructure and rolling stock are owned by the same authority, but where the planning of the system is done by a franchise.

Of the four different systems the two first are the most common. In bus operation in Scandinavia total franchises are found, whilst in France the operations franchise is the most common system. Each type of franchising system has its advantages and disadvantages, and it is appropriate to study some of these in general before we look into the different systems in more detail.

Competitive franchises (competition for the market) have two different purposes or aims namely, the control aim and the efficiency aim. The *control aim* is the system which places franchising in *the administrative planning camp* of regulatory regimes. It is a *planned system* where different types of transport authorities define route structures, fares and the overall structure of the public transport system. The control role can therefore be said to reflect the 'social-effectiveness' view of public transport as opposed to the market view leading to competition in the market.

The *efficiency role* on the other hand will be used through competition for the market to create a more efficient public transport system, both as regards internal efficiency (x-efficiency) and allocative efficiency.

These two roles can be conflicting. The public transport system which results from a control point of view or from a social effectiveness point of view, may not be the same as the one that would give the most efficient public transport system. This is because the operator's freedom to adjust the supply of public transport will be affected by what the public authority decides that the supply shall be through planning.

Berechman (1993) calls this trade-off between control and efficiency a form of institutional failure built into the franchising model.

We can summarize the main problem areas of franchising systems as:

1. control over the winner of the franchise;
2. organization of an effective franchising process;
3. internal efficiency and allocative efficiency problems with franchising.

Williamson (1986), Demsetz (1968), Posner (1975) and Hensher (1988b) have all discussed these problems in relation to the criteria for allocation of franchises, the problems during the franchising period and the absence of competition under renewal of franchises (incumbent advantages).

Many of the problems encompassed by franchising can create barriers to competition which may eliminate the advantages of competition. The literature has been especially occupied with the question of fixed facilities/rolling stock/infrastructure and the influence on competition, i.e. fixed facilities and barriers to competition. This is dealt with in Banister et al. (1992b).

Another important factor which must be taken into account when considering franchising are the transaction costs. These are the costs of carrying through the process of franchising. One of the arguments against franchising is that there are heavy costs associated with franchising especially concerned with defining tendering criteria and the tendering process, and the evaluation of tenders. These transaction costs can in certain circumstances outweigh the efficiency gains.

On the question of barriers to competition, the most important factor has been the question of capital equipment and associated capital costs. It has been strongly argued that total tenders favour the incumbents. This is particularly true at the time of renewal of franchises. If heavy capital costs are involved, this favours the use of operation franchises like those in France. The disadvantages of franchising are different to those concerned with competition in the market, but on the other hand there are advantages through franchising as some of the disadvantages, that competition in the market has had, can be avoided (see Britain outside London), Banister et al., 1992a. Thus we have to accept that franchising is a planned system, but created with the aim of taking account of some of the social-effectiveness goals associated with a good public transport system.

11.4 THE EUROPEAN COMMUNITY PUBLIC PROCUREMENT RULES AND THE CONSEQUENCES FOR USE OF TENDERING

The European Economic Space Agreement has been signed between the EC and the European Free Trade Association (EFTA). This agreement came into force in 1994 and creates a 'common market' in industrial products, including transport, i.e. the EFTA countries are a part of the Single Market created by the EC which came into force from 1 January 1994. The EFTA countries have in that respect accepted EC rules concerning transport and public procurement in force for the Single Market of the EC.

The European Economic Space Agreement will mean that there cannot be any discrimination on the basis of nationality concerning public procurement, and rules have already been passed for buying of goods and for works contracts in the transport sector (European Community 1990). There has also been submitted a proposal for services (European Communities 1989), and this was accepted by the Council of Ministers of the EC in June 1993 (European Communities 1993). It has significance also for public transport in Europe. The new rules came into force from 1 July 1994. These new rules of public procurement mean that a process of *transparency* in public procurement of services, including land transport, has been created; there are clear rules as to how this procurement process can be undertaken and also as to what criteria to use to choose seller (price or economically most favourable contract); and the three different types of buying process-tendering—restricted tendering or buying by arbitration—can be used.

The rules do not prohibit public production of public transport. The rules will not apply for a transport authority providing all its public transport internally (own resources), but they will have to be applied in a situation where there is a transport authority buying services from *outside operators* or where the authority decides to use tendering when the process between seller and buyer is based on an agreement. The new rules will most likely result in a movement towards the use of tendering in many countries not using it today, and it will also speed up the process in countries where tendering is partially used today. This is because tendering will be the only system of securing non-discrimination according to the article in the Treaty of Rome (articles on competition and on transport), and, as noted above, it is compulsory when the process between seller and buyer is based on an agreement.

In the event of cases where the operators hold a licence or are given operating rights on the basis of administrative rules, the new service Directive does not apply. However, the EC Commission is working also with these forms of service-contracts, the so-called 'licence contracts'.

11.5 THE FRENCH FRANCHISING APPROACH

11.5.1 A GENERAL INTRODUCTION AND OUTLINE OF THE FRENCH SYSTEM

French public transport both in urban areas and in rural areas is organized through a model where an organizing authority (l'autorité organisatrice, OA) is central. This model was created through the Act on Internal Transport (loi d'orientation des transports interieurs – LoTi) of 30 December 1982. The OA has the following responsibilities in urban areas: the creation of the urban transport perimeter, i.e. the area for which the OA shall function; the organization of public transport (routes to be implemented, choice of operators and technical operating methods, the fares, contracts with companies, setting and financing of subsidies for capital construction and operation); the creation and management of infrastructure and equipment assigned to transport, the regulation of transport activity and monitoring of its application, and the development of information of the transport system. In short, this means that the OA is responsible for the planning of the public transport system and for overall route structure and fares policy.

There are *five different* forms of possible *legal structures* for OAs: a commune (municipality); intercommunal syndicates; districts; urban communities; and mixed syndicates.

Their different status is based on certain fundamental attributes including the extent to which the body has been set up voluntarily, the degree of rigidity of its structure, the attribution or not of mandatory competence, and the extent of its fiscal powers.

The most important syndicates are those with a single or multiple objective, i.e. whether they have responsibility for public transport or another single activity, or whether they have more than one duty, e.g. public transport and other service functions. In 1988 there were 12 500 syndicates with a single objective (SIVU) in France. There were 2000 syndicates with multiple objectives (SIVOM), where public transport could be one of several services undertaken. There were also 750 mixed syndicates, where there is an association between departments, urban communities, districts, communal syndicates, Conbus and commerce and industry, agricultural or trades associations, etc. The OAs for public transport in urban areas are shown in Table 11.2 (CETUR 1990).

The relationship between the OAs and the operating of public transport in France is organized through a system of franchising where companies undertake the operation of public transport, where the OA is the *owner* of infrastructure, equipment and rolling stock *in most instances* (see below concerning private/public partnerships). This means a system of operating franchises or a system of operating through public administration (régie) by the OA. This is used where a municipality is an OA (e.g. in Marseilles). The 1982 Act (LoTi) sets a minimum framework within which urban public transport can be operated for both 'régie' and 'agreement', and for the agreement type of operation, the LoTi defines the following minimum requirements. The agreement must be of fixed duration and it is recommended that it includes measures to be taken at expiry date concerning rolling stock and installations or staff. The recommended period of agreement is 5–10 years. The agreement must specify the general content of the services provided (routes to be served; route standard, comfort, etc.) the conditions applicable to operation of the services (timetable, periodicity, etc.); the financing conditions of the service; especially

Table 11.2. The distribution of organizing authorities in France, 1988

Communes	42
Communal syndicates	56
Districts	19
Urban communities	6
Mixed syndicates	10
Paris	1
Total	134

the fares, the renumeration of the operators, financing of capital investment in rolling stock and installations; the obligations of both parties to users, especially concerning information; and the methods to be used for monitoring the use of funds committed or guaranteed by the organizing authority.

In addition, it is recommended that the condition should provide for situations where the interruption of the services takes place before the expiry of the agreement; the reasons for the renegotiation of the general terms before the due date; the responsibility of each party in setting of fares. This is especially important when the operator takes a high commercial risk and where he is given a right to set fares or part of the fares policy (e.g. discount policy).

Recently, there has been a clear tendency to increase the commercial risks for operators by using competitive tendering instead of arbitration for securing new agreements or by changing from public administration (régie) to agreement.

As regards operators, the private sector plays an important part in urban public transport provision, but in a different form to many other countries in Europe, i.e. through the system of operations franchising where they rarely own transport equipment. In France there are three nationwide groups or holding companies of which two are private and one is semi-public. The three are

1. VIA Général de Transport et d'Industrie, VIA GTI owning VIA TRANSEXEL (urban transport).
   VIA TRANS-CAR (inter-urban transport) and VIA-TRANSETUDE providing research and consultancy.
   VIA GTI is owned by Compagnie de Navigation Mixte;
2. TRANSCET S.A., a subsidiary of Société Européenne de Transport Public, TRANSDEV, operating also inter-urban public transport and also owning the subsidiary, PROGECAR. This operator is semi-public, owned by the group Caisse des Dépôts et Consignations (C3D) which is state controlled;
3. CGEA-CGFTE groups controlled by Compagnie Génerale des Eaux, a multinational service company.

In addition to these three large groups, there are several other smaller groups. These three groups operate through different forms of ownership such as through their own subsidiaries, or through mixed economy companies, i.e. owned jointly by OAs and operators or through public administrations (régies) where the operators have management contracts. CETUR (1990) gives the following information: in 1988 TRANSEXEL operated through 35 subsidiaries, 8 mixed-economy operators and 1 régie. TRANSCET had the following distribution: 12 subsidiaries, 13 mixed-economy companies and 2

Table 11.3. The organization of urban public transport in France

Organizing authority (OA)		
Owner of	*Commune*	*Aim*
– infrastructure	– intercommunal syndicate	
– rolling stock	– district	PLANNING
– installations, etc.	– urban community	AND
in most instances	– mixed-syndicate	FINANCING
	decides whether to operate through	
	FRANCHISING AGREEMENT (operating agreement)	RÉGIE
	Private operators	*Public operators*
examples:	– private operator or subsidiaries	– i.e. *own* operation through a public
TRANS-	– mixed-economy	administration	OPERATION
EXEL	companies		AND
TRANSCET			MANAGEMENT
CGEA-	– régie through	
CGFTE	management contract	

régies. CGEA-CGFTE had 16 subsidiaries, 2 mixed-economy companies and no régies. The system in Paris is different and is based on the operation of two state-owned companies, RATP (bus and underground) and SNCF (railways). This is not dealt with in this chapter.

In Table 11.3 the structure of the French system is described in more detail. The following conclusions can be drawn: a clear separation has been made between the planning and financing of public transport and the operation of public transport. Orientation through franchising contracts has been based on the organizing authority providing infrastructure and rolling stock. Strong private participation in operation has taken place through three strong nationwide public transport groups, but often in joint ownership with OAs through mixed-economy companies or purely management contracts.

11.5.2 NEW FORMS OF PRIVATE//PUBLIC PARTNERSHIPS IN FINANCING/OPERATING OF NEW PUBLIC TRANSPORT PROJECTS

In Section 11.5.1 the general system of public transport planning and operation in France was outlined. In this section we study in more detail the development concerning building, operating and financing of new public transport systems in France, where interesting forms of private/public partnerships have developed. The private franchising of building, operating and financing of public transport systems began to appear in the mid-1980s (Amsler 1993). The reason for this new development is summarized by Amsler (1993) as follows: a new economic environment and a general trend towards privatization; the political will in France to develop new public transport systems; a shortage of local resources especially as regards more rapid transit systems; for medium-sized cities (200 000–600 000 inhabitants); and an interest from the financial sector and new ideas for services especially from banks. In Table 11.3 an overview of projects is given as categorized by franchising characteristics.

The two first private franchising projects were signed in 1988 and comprised designing, building, operating and financing of so-called VAL-Lines, one in Paris (ORLYVAL at Orly airport) and one in Toulouse. The Paris system was abandoned because of low levels of traffic and transferred to the OA in Paris-STP which entitled RATP to maintain and operate the line (Table 11.3). In Toulouse the system is operated through a mixed syndicate and the franchising agreement awarded to a private consortium headed by C3D for 30 years, the bank owning TRANSCET (Section 11.5.1). There is also public money invested in this project, while the Paris project was purely privately financed.

After these two initial projects, three more projects have been signed up— Rouen, Strasbourg and Paris Charles de Gaulle airport. However, seven more projects have been studied for development through franchising schemes and are in different stages of progress (Table 11.4). The interesting part of the franchise system in France is *how* the partnership between private and public has been organized, and especially the involvement of the local public authorities in the finance and operation of the projects. This has obviously influenced the success of the projects and dictated how the *internal partnership structure* of the franchise company can and has been structured (Table 11.4).

The partners in France are normally five, namely the banks, operators, suppliers of rolling stock and equipment, civil contractors and other engineering contracts.

The French view is that there must be found a *balanced share of responsibilities* and *risks* between the different parties, in particular between the public franchisor and the private promoter. Amsler (1993) summarizes the conditions which have to be met by the franchise company: the financial resources provided by the supplier, the civil engineering contractor and the bank; an interest in the long-term success of the project, which concerns the bank and operator (and to a degree, the supplier too); technical knowledge on the part of engineers, the supplier and the civil engineering contractor. It is important to note that each consortium lasts for 25–30 years, the life of the franchise.

The role of the operation authority (OA) (Amsler 1993) is in the overall

Table 11.4. New private–public relationships in public transport in France

Area franchise characteristics	Partners in the the franchise company	Leading partner	Missing partner	Status of project
Brest	Bank Operator Engineer	No	Suppliers Civil contractor	Cancelled
Bordeaux	Bank[1]	Bank	Civil contractor	Under negotiation
Paris–Orly	Bank Operator Supplier Others[2]	Manufacturer	Civil contractor	Franchising interrupted
Reims	Bank Supplier Civil contractor Engineer Operator	No	No	Cancelled
Paris–CDG	Bank Operator Supplier	No	No[3]	Under construction
Rouen	Bank Operator Supplier Civil contractor	No	No	Under construction
Strasbourg	Franchised through the operator	Franchisor	All others	Under construction
Toulouse	Bank Operator Supplier	Bank	Civil contractor	Under construction
Grenoble	Bank Manufacturer Civil contractor	Manufacturer Contractor	Operator Engineer	Under construction
Caen	Bank Manufacturer Civil contractor	Manufacturer Contractor	Operator Engineer	In negotiation

Source: Amsler (1993)
[1]Other partners are designated by the public authorities (supplier/operator).
[2]Others: Air Inter/RATP.
[3]Civil contractor in charge of the franchisor.

design of the project; the definition of the level of service to be offered during the lifetime of the franchise; the agreement on the share of risks/responsibilities between partners and on the legal structure of the franchise company; the possible partial involvement in the project through financing or participation in the operating company; the sharing of the revenue risk related to integration of the system into a global network (both traffic and fares), the ability to control the proper development of the contract during the construction and operation phases, and the ability to revise, at fixed intervals, the application condition in view of political and economic development.

This division of power between the OA and the franchise company can be compared with the system advocated by Fielding and Klein (1993) for franchising of highways in USA, where they advocate a system of division of responsibilities encompassed by the 'clearing before awarding' concept. This process makes it easier for the franchise company to evaluate the risks of the project. The new developments in France are interesting and they should form an important part of the future public transport system in France, but this is dependent on the success of the projects now under construction or in negotiation (Table 11.4).

11.6   THE SCANDINAVIAN EXPERIENCE

11.6.1   INTRODUCTION

In Scandinavia the late 1980s and the beginning of the 1990s have resulted in regulatory reforms for regional and local public transport. In the four Scandinavian countries (Denmark, Finland, Norway and Sweden), competition for the market has been introduced or will be introduced. The process of change in Scandinavia has been going on for many years and has been a gradual development. Denmark, Norway and Sweden in the late 1970s and the beginning of the 1980s transferred responsibility for regional and local public transport to regional level either directly to county councils (Norway) or to transport organizations owned and financed by regional and local councils (Denmark and Sweden). These regional organizations still have responsibility for public transport and have powers to create an integrated public transport system for their areas with uniform fares systems and fare levels, and with a politically defined quality of service for public transport.

During the 1980s the Scandinavian countries again moved towards more competitive regimes, but unlike Britain they opted not for competition in the market (full deregulation), but competition for the market through tendering. (Andersen 1992; Andersen 1993a,b). The aims of the change of regulatory regimes in Scandinavia are the same for all countries, namely to establish a more effective public transport system through opening up the market for competition, but at the same time securing the benefit of an integrated public transport system. The solution used is tendering. The reason for this is that all Scandinavian countries have a clearly stated transport policy which aims at providing public transport as an essential social service. However, there has been a necessity to reduce public expenditure in all countries and this has led to a solution of competition for the market rather than in the market. In Denmark there has also been a clear ideological reason for opening up the market for tendering in Copenhagen, as this has resulted in a demand for the privatization of 45 per cent of bus provision in the metropolitan area through tendering.

In the other three countries the need for reducing public expenditure and creating more efficient public transport systems has been the main motive behind the reforms outlined. In Sweden, Denmark and Norway the process of change has been based on the same fundamental principles of creating what can be called *buyer and seller roles* in public transport, with the opportunity for using competition as a means of obtaining the necessary provision of public transport.

The basic characteristics of the systems are described in Table 11.5 with public authorities being responsible for defining what and how much public transport to buy and the fare levels to charge, and public and private operators for the task of operating the system. The advantages and disadvantages of this system were much discussed when Britain changed its regulatory regime and will not be discussed in detail here, see Banister (1985), Beesley

Table 11.5. The buyer–seller relationship in Scandinavian public transport. Responsibilities and aims

Responsibility:		*Aims:*
Planning of public transport, fare system and level, marketing terminals	BUYER Transport authorities or county councils directly	*Allocative efficiency* i.e. securing the right level and distribution of public transport
		Contracts, based on tendering or arbitration
Responsibility: Operation of the public transport system	SELLERS Publicly and privately owned transport operators	*Aims:* *X-efficiency* i.e. public transport provision atminimum cost per unit

and Glaister (1985), Gwilliam et al. (1985), Hensher (1988b) and Section 11.3 of this chapter.

The organization of the new system in practice, where there are publicly owned operators or public operation that has been a part of a previous monopoly regime, has been a vital one in Scandinavia. Principally, this problem has been dealt with sensitively in order to create transparency and an atmosphere of mutual trust. This suggests that the buyer role must be strictly separated from the operator role if the previously publicly owned operation shall continue in one way or another. In Table 11.6 the main characteristics of all the Scandinavian tendering systems are described. The table shows us that there are significant differences in the systems, but still some main features that are similar in all countries. The following conclusions can be drawn:

1. With the exception of Copenhagen, the decision to use tendering rests with the local authorities, i.e. it is not compulsory by law.
2. In Denmark and Sweden where tendering has been used, there is a system of minimum cost contracts where transport authorities are responsible for the revenue side, i.e. they have the risk for income.
3. Tendering periods are normally 3–4 years except in Norway where the minimum period is 5 years.
4. Sweden and Norway have opened up services outside the bus sector for tendering to cover rail (Sweden) and shipping services (Norway) and ferries (Norway), on an experimental basis.
5. Certain types of conditions concerning buy-back of buses, transitional periods of tendering introduction or the use of accepted employment contracts exist in the countries, so that a fully competitive market is not created.

The introduction of tendering in the Scandinavian countries can be said to have developed at different stages, from agreement on the rules of working (Norway, Finland and Denmark outside Copenhagen) to extensive use of tenders in Sweden and Copenhagen. The results of the tendering system so far are dealt with below. The general conclusion from Scandinavia points towards more use of tendering. A similar conclusion can be made with respect to the development of tendering in the European Community as a whole. This must be seen as a clear alternative to the British full deregulation.

11.6.2   SWEDISH RESULTS TO DATE

The use of tendering started in Sweden in 1990. The first acts had been passed in 1985 and the 'Transport Policy Decisions' of 1988 confirmed the process towards increased competition in public transport. The situation in

Table 11.6. Summary of tendering in Scandinavia

	Sweden	Denmark Copenhagen	Denmark outside Copenhagen	Norway	Finland
Legal basis	Yes, through acts of 1985 and 1988 transferring licences to transport authorities	Yes, through act of 1989 demanding privatization of 45% of bus provision in Copenhagen through tendering	No, through an agreement between organization of transport authorities and operators' organization	Yes, through act of 1991	Yes, through act of 1991
Tendering compulsory	No	Yes, for 45% of service provision	No	No	No
Max./min. tendering period	Not defined, normally 3–5 years	Max. 8 years, normally 4 years	4 years for those opting for tendering through agreement	Min. 5 years	Not decided
Tendering system	Min. cost contracts	Min. cost contracts	Min. cost contracts	Not decided, net subsidy contracts based on arbitration	Not decided, both min. cost and net subsidy contracts today
Percentage of operation tendered	*Bus:* About 50% of buses in scheduled transport spring 1993. *Rail:* Three lines in Stockholm. Use of tenders for local railways. Tenders for government subsidized rail traffic (100%)	30% per April 1992. 15% new + 20% old tenders, i.e. 35% from April 1994, 45% in total from same date (discussions on 100% tendering through opening up for authority's own services to compete)	Nothing except local bus in one area. Five transport authorities decided to use tendering from 1994, 4 not yet decided and 2 continuing with arbitration	Not yet in operation. One area in Oslo in 1991 (3 lines). Can be used both for bus and shipping services in regional and local transport And on experimental basis for ferries	Not yet in operation. Helsinki from 1994 starting with regional routes
Conditions attached to tendering	No specific conditions	Compulsory use of wage agreement in force	Compulsory buy-back of buses when changing to tendering, first time not later. Wage agreements	Compulsory use of wage agreements in force. Five years transitional period if tendering is used to avoid buy-back clauses	

Sweden in October 1992 indicated that only two transport authorities had not tendered any of their traffic by autumn 1992. However, there are great variations in the percentage tendered and if we take this into account, about 50 per cent of the total number of buses in scheduled transport had been tendered. This percentage will gradually change as some of the larger cities in Sweden (Stockholm, Gothenburg and Malmø) will start in 1994, having allowed its municipally owned operator a long period of adjustment.

The results of tendering in Sweden are in many respects interesting. Andersen (1992, 1993a) deals with the savings in more detail, but the following conclusions can be drawn (all savings refer to costs in minimum-cost contracts):

1. The savings in costs have been between 5 and 15 per cent in the first rounds up to 1991.
2. The new rounds conducted in 1992/93 have given much larger savings, especially in medium-sized towns and where municipally owned operators have been forced to compete through tendering. In areas where there has been re-tenders, the savings from the first rounds have been sustained and increased. In one area, Kronoberg, the saving in the first round was 10–15 per cent, but has been increased to 20 per cent through re-tendering in 1992 for operations starting in 1993. The increased saving can be attributed to more experience with tendering both on the buyer and the seller side.
3. The savings in Stockholm for 20 per cent of bus traffic and three suburban rail-lines are 4 per cent. But one must here take into account the fact that large savings were taken out in the authority-operated buses before tendering started, in order to be able to compete (15 per cent cost reduction before tendering in own account operation). The saving in the Gothenburg area for 30 per cent of city traffic put out to tender in 1992 for operation from 1993, was a reduction of 45 per cent in costs from 1989 to 1992.

During the autumn of 1992 and spring of 1993 many new tenders have been carried through with interesting results:

1. Several municipal companies lost out completely and had to close down their activities, e.g. Jønkøping and Halmstad. These losses are due to high cost levels.
2. In *three towns* in Northern Sweden (Västernorrland) three municipal operators lost to Swebus (a nationwide state-owned operator), but the authorities changed the rules during the process and made the winner negotiate with the municipal companies about taking over employees, buses and plants. If no agreement could be reached, the municipal operation should continue, but with the winner's price. The conclusion of the

process has been that Swebus has taken over one town (Härnøsand) and two continue (Sundsvall and Ørnskøldsvik), but at the winner's price. Whether they will survive depends on their ability to cut costs (about 20 per cent). This experience shows that there is much to be said about competition rules and the performing of the process, but this is under change in Sweden now, where a new Competition Act will be introduced this year (1994).

3. Both in northern Sweden and western Sweden several new *bus pools* have been created, for both municipal operators and private operators. Their main task has been to compete with the large operators, Swebus and Linjebuss. The municipal pool won nothing, but some of the private pools won tenders. These pools consist of smaller operators not being able to compete for larger 'packages' in their own right. One could discuss whether such bus pools can be said to be cartels restricting competition, but in Sweden they have been accepted as creating rather more competition.

The situation in Sweden today shows that municipal operators face a serious competition problem unless they can do something about their costs. An example reveals this. In one town, one of the private operators calculated that it needed 19 employees in maintenance and administration. The incumbent municipal operator used 54. However, the competition rules between private and public operators are not equal, and if the public operators succeed in the future, they have to be allowed to compete in other areas than their own. At present this is not possible even though some have competed both through bus pools and in Stockholm. To allow this, the Government will most likely demand an *arms-length relationship* through the creation of limited companies. But private operators fear that municipal operators will be at an advantage through their owners' demanding lower rates on return than private operators have to earn on their investment.

11.6.3 DENMARK – COPENHAGEN

Through the passing of an act of reorganization of public transport in the Copenhagen area in 1989, *three important* issues were taken up. *First* the operation was reorganized through the creation of two different departments of the Transport Authority (HT), one responsible for planning and the buying of the necessary public transport provision in the Greater Copenhagen area, and the other department operating the authority-owned buses in the area. The *second* proposal changed the governing system of the transport authority in the area, slimming down the board and changing local responsibilities, and the *third* was a proposal for privatizing through tender 45 per cent of the bus operations in Greater Copenhagen (15 per cent from 1 April 1991, 30 per cent from 1 April 1992 and 45 per cent from 1 April 1994). In addition there have been tenders for several other service lines in the area, in addition to the legally binding tendering.

Up to the present three tenders have been carried through: the first tender around August 1989, 5 per cent, the second around October 1989, 15 per cent, and the third around March 1991, 10 per cent.

The fourth round was published in March 1993 and includes the last 15 per cent of the original tenders plus 20 per cent of old tenders (rounds 1 and 2), in total 35 per cent of the bus traffic in Copenhagen has been put into operation from 1 April 1994.

The results of the tendering process in the Copenhagen area are as follows.

1. Through tendering, the price per vehicle hour has been reduced by about 9 per cent from 1990 to 1992 through the change from own operation to use of private tenderers. There has been a decrease in HT's own operation cost of about 5 per cent per vehicle hour, which includes the different structure of lines, rush hours, etc.
2. The experience shows that a clear strategy for tenders was necessary and also clear rules for choosing of tenderers. If price alone was to be used as a criteria, the Swedish operator Linjebuss, a nationwide operator in Sweden, would have won all tenders in the third round.
3. The monopoly tendencies are obvious and over time there will be increased concentration in Danish bus operation.
4. The experience in Copenhagen shows that new operators came into the market, both scheduled operators, non-scheduled operators, goods transport-operators and one foreign operator (Linjebuss), though there have been several other foreign bidders interested.
5. HT has found that the optimal tender period is 4 years and that price differences are very small indeed by increasing the tendering period from 4 to 5 years (0–2 per cent). This is an interesting result as one of the arguments of the operators was that the prices would decrease if the tender period increased. The results of increasing the period from 4 to 5 years show that this is generally not true. But it may be that much longer periods of 8–10 years are required before any significant change takes place.

The experience in Copenhagen has been interesting in many respects and included one important *controversy*, namely the privatization process combined with tendering. This means that HT's own operation has had to be decreased as the given result was to be 45 per cent of bus traffic privatization through tendering. The fourth tender in Copenhagen was published on 15 March 1993, consisting of 15 per cent new tenders (the last part of the original 45 per cent given by the act of 1989) and 20 per cent re-tenders expiring.

The tenders started on 1 April 1994, and the date for submission of tenders was 18 May 1993. The board of HT has taken its decision on allocation of tenders after consultation with the Minister of Transport as regards HT's own participation in the fourth round. HT was denied participation due to the fact that the act of 1989 did not allow it. The results of the fifth round is that cost savings are sustained. Linjebuss has increased its share to about 25 per cent. Of the 11 original private operators only one is left.

11.6.4   DENMARK OUTSIDE COPENHAGEN, NORWAY AND FINLAND

In Denmark outside Copenhagen, changes in the agreements between operators and public transport authorities took place in 1994, and is now an opportunity for tendering. Of the 11 authorities outside Copenhagen 6 have opted for tendering, and 5 have chosen renewal arbitrated contracts based on standard-costing, though 3 of them will tender some of their routes. Through acts in 1991, both Finland and Norway can now make use of tendering.

In Finland the process has not fully started yet, but it has been decided that there will be tendering in the Helsinki area from 1995. In Helsinki the regional routes in the suburban areas will be the first services to be tendered.

In Norway new regulations in addition to the act have had to be produced. These regulations came into force on 1 January 1994.

11.6.5   SOME AREAS OF CONCERN IN THE SCANDINAVIAN USE OF TENDERING

During the process of tendering in Scandinavia, some areas of concern have arisen that have to be tackled in the future if competition for the market is to be an alternative to other solutions like full deregulation. The areas of concern include the following.

1. The problem of concentration in Scandinavian public transport: is it a threat to sustainable competition?
2. The problem of using minimum-cost contracts with a transport authority responsible for the revenue side of public transport. This may lead to a lack of incentives for operators as regards passengers and income.
3. The problems of capital cost and the use of tendering, especially in rail tendering and ferry and shipping tendering: can operating tenders be used to avoid barriers to competition?
4. The problems of creating x-efficiency in an environment where there is no competition in the factor markets for wages, where there are different

buy-back clauses for buses or operating tenders, and where vehicles and plants are owned by the tendering authority.

As regards concentration, the question arises as to whether the market is contestable, in the sense that if large operators take out a monopolistic rent, new operators will be able to enter the market. This will be dependent on economies of scale in bus operations and barriers to entry.

The question of economies of scale in bus operation has been dealt with by many researchers including Lee and Steedman (1970), Koshal (1970), Viton (1981) and Windle (1988). Most of these studies conclude that there are constant returns to scale between unit costs and bus miles in urban bus transport. This coincides with Norwegian results from the 1960–70 period (Andersen 1968). In the discussion on economies of scale there are other important aspects of economies referring to economies of densities and economies of networks, but they are not covered here. These are dealt with in Windle (1988). The important questions on productivity in privately owned and publicly owned operators are covered in Hensher (1988a).

The empirical research done in the USA, Australia and UK suggests that there are economies of scale, but the experience found in Scandinavia, and especially in Sweden during the change over to the use of tendering, suggest that there may be economies of scale under certain conditions. The conclusions are similar in Scandinavia as in other countries for some of the input factors, but the large national operators claim important economies of scale in the purchasing of buses, the purchasing of fuel, the purchasing of insurance, finance and administration.

Both large Swedish national operators claim economies in these respects to be substantial, and this was one of the reasons why they succeeded in tendering. They based their strategy on local operation (without certain economies of scale) and centralized purchasing (with economies of scale). The important factor binding these two together was their *leadership philosophy*—the administration of the operators combining small-scale operation with large-scale purchasing.

One of the most significant features of the systems in Scandinavia where tendering has been introduced is in the use of minimum cost contracts, leaving the responsibility for the revenue side to the transport authority. This question has been dealt by Andersen (1992), Tough (1992) and Hensher (1988a). The problem here is the lack of an incentive for the revenue side and the means to increase the number of passengers. In Scandinavia this is seen as necessary if cost savings are to be sustained. Cox and Love (1991) have discussed the alternative of 'threatened competition' used in Australia. In Scandinavia there is now work going on in trying to establish some kind of *minimum cost contracts with incentives*. There are several problems here:

1. How do you create a system of incentives where accountability is taken care of, both regarding an increase in passengers and the quality of the public transport system, and at the same time giving the operators an increased possibility of acting on their own?
2. Can you preserve a system which depends on the authority with uniform responsibility for fares and transport planning, and which has opportunity for more operator initiative?
3. How do you monitor a system with more flexibility? Is a system of quality monitoring necessary for introduction of incentives and if so, how do you create it?

The really big problem must be to find a solution to how you separate gains created by the operators in their own right, gains created by the system as a whole and external factors. As shown in Sweden, tendering has also been used for rail traffic (e.g. Stockholm and transport authority railways) and for government subsidized rail traffic. This has been accomplished through operating tenders where equipment has been leased from the transport authority.

The problems with barriers to entry are apparent here and could give the incumbents obvious advantages over new entrants. This has also been one of the main concerns that has resulted in the use of operating tenders and in giving the new entrants an opportunity to lease vehicles. Unlike the situation in Britain where competition in the market (White 1990) has also created competition in factor markets, the situation in Scandinavia is different, both in Denmark and Norway, where there are legal requirements to maintain the prevailing wage contracts within the public transport sector. There have been provisions for transferring of employees to new operators (Sweden and Denmark) and there has been much discussion on whether or not there should be buy-back clauses for vehicles.

All these rules inhibit competition in the factor markets and the tenderers thus have to compete on the utilization of personnel, on procurement, on administration and maintenance costs, etc. The important question becomes what savings can be achieved and whether these savings can be sustained. Experience from the last tenders in Sweden suggest that savings can be sustained, but definite conclusions cannot yet be reached as only a few tenders have come up for renewal. One indication of sustainable savings is however the extensive changes in operators in the re-tendering process. (Even though the two large national operators have increased their market shares, they have also had to make extensive changes to tenders and regions.)

Even though there is no real direct competition in factor markets, indirectly it does exist, as there is a process towards equalization of tariffs for the different types of employees (municipal, state and private). This indicates that municipal employees are not able to compete unless they compete on equal wage levels and structure including social costs. (At present the difference between private and municipal operators is about 10 per cent in Sweden.) There is a clear tendency towards more equal tariffs and this process has been speeded up by the many losses of tenders suffered by municipal operators in 1992/93.

11.6.6 SOME CONCLUSIONS IN SCANDINAVIA

The following preliminary conclusions can be drawn from the Scandinavian experience:

1. The cost saving by using tendering has been considerable both in Denmark and Sweden, between 10 per cent in rural parts of Sweden and up to 20–30 per cent in urban parts operated by municipal operators. In Copenhagen the savings have been around 10 per cent.
2. There are signs that cost savings can be sustained when services are re-tendered, but there are two main obstacles, namely the use of minimum-cost contracts and the lack of competition in factor markets.
3. There is also a fear that increased concentration in the market may also reduce savings in the longer run due to re-monopolization and the creation of new barriers to entry. The two nationwide operators in Sweden with economies of scale in purchasing and an efficient system for yardstick competition between operating units may be seen as particular threats. The real danger may, however, be overstated, as the contestability of the market is obviously there.
4. The future for municipal operators who are able to cut costs will be bleak: they may either lose out in competition or be sold to private operators for financial reasons. In any event it is necessary to come up with clear rules for competition between publicly and privately owned operators. This regards both rate of return and pricing policy in publicly owned operators with operation both in competitive markets and sheltered markets. The problem of cross-subsidization between the tendered regional/local market in Sweden and the unsubsidized rather monopolitic interregional market is worrying.
5. In all Scandinavian countries there is a need for strengthening of competition policy and competition authorities. This must be done whether or not the Scandinavian countries become members of the EC. Today the powers of the competition authorities are too weak.
6. New methods of combining operator incentives with authority control over important parts of fares and routes have to be found. Some kinds of incentives combined with strict quality control may be the answer. Here much work remains to be done. The threatened competition model may be an alternative.

In any case the change in regulatory policies in Scandinavia have shown that it is possible to combine a system of social effectiveness goals for public transport with both allocative and x-efficiency in operation, namely through a system of competition for the market—tendering. In years to come when Denmark, Finland and Norway all actively start using tendering combined with more use in Sweden, the answers to many of the problems will be found. Tendering seems to be a better alternative than full deregulation and competition in the market, because the social-effectiveness goals which are obviously still important in public transport can be maintained.

11.7   IS FRANCHISING THE SOLUTION FOR EUROPE? SOME CONCLUSIONS

This chapter has discussed whether franchising is the best solution for European public transport as regards the regulatory regime. Franchising has been investigated in several contexts, principally through the use of franchising for the operation of public transport systems (both total franchises including operations and infrastructure investment or pure operations franchises) and through the use of franchising as an alternative for investment in public/private partnerships, especially for rail investments.

The use of various forms of franchising will be an effective means of solving the need for further efficiency measures in public transport operations without choosing the most radical alternative—full deregulation. This allows the politicians to take 'social effectiveness' into account, and at the same time reduce the pressure on public budgets. Franchising also permits the future demands from the European Community regarding public procurement procedures to be followed, and thus ensures non-discrimination between nations and individual firms in the area of public transport.

Of the above-mentioned reasons for using franchising, the most important one is undoubtedly the future securing of 'social effectiveness' aims concerning equity and the environment through the political control of future route systems and fare levels and structures. It is obvious from the situation, outlined here, that in most European countries the British solution of full deregulation is not a real alternative. If one wants an alternative to public monopoly or private monopoly, then there are few alternatives to franchising.

However, one must be aware of the fact that franchising as a regulatory regime also has its disadvantages. These include conflicting aims between control and efficiency, an implication of the fact that franchising is an administratively planned system; barriers to competition created by the system of franchising, especially through the fact that the incumbent in a franchise will have some advantages on renewal of the franchise in market knowledge, and in owning or leasing rolling stock and often infrastructure; and the increased concentration in public transport, again creating barriers to competition. This particularly relates to the tendency of nationwide operators found both in Sweden and France towards cooperation instead of competition between these nationwide groups.

New forms of public/private partnerships have developed concerning investment in public transport. This will realize projects that would not have been possible to finance with public budgets, and even if competition will be reduced, there can be no doubt that these solutions are favourable in the long run, especially through bringing private capital into the sector.

The experience of public transport franchising found in Scandinavia, France and Great Britain (London) is favourable. Efficiency gains per produced unit have been between 5 and 20 per cent, in some cases even more. In rural areas the gains have been between 5 and 15 per cent and in urban areas in Scandinavia between 10 and 20 per cent. More important, the gains seem to be sustained when the franchises are renewed, and even increased in some cases, especially where municipal operators are involved. The choice between total and operations franchising arises where operations franchising is a good solution to the problem of investment in rolling stock and other infrastructure. In this respect the publicly owned transport authorities act as an owner of rolling stock and infrastructure and lease it out to operators. Opportunities for the solution of investment in new public transport investment exist, especially in rail through public/private partnerships franchising, responsible for design, planning, building, operating and financing the project. As explained above, many countries seek new solutions to budgetary problems, but are not willing to choose radical solutions, implying that the social effectiveness perpective on public transport is lost. In this situation there are few options other than franchising.

The countries in southern parts of Europe and Germany will gain from opening up markets for franchising solutions. In these countries public monopolies still dominate public transport and if account is taken of the experience from other countries, the opportunities for efficiency gains are obviously present without giving up public control of the public transport system, which seems to be important in many continental countries. Choosing a combination of public control and competition for the market allows efficiency gains to be realized without having to accept some of the negative experiences of full deregulation found in Great Britain outside London.

REFERENCES

Amsler, Y. (1993) New public–private relationship, in *Urban Mass Transport in France, Public Transport International*, **3**, 42–46.

Andersen, B. (1968) *Prognose for driftskostnader og kapitalkostnader for kollektive naertrafikkselskap i årene 1968–80 (Prognoses of operating and capital costs in urban bus operation)*, TØI, Oslo.

Andersen, B. (1992) *Organisatoriske endringer i samferdselssektoren i Skandinavia. Privatisering, anbud og omlegging av jernbanepolitikken. (Organizational changes in the transport sector in Scandinavia. Privatization, tendering and change of railway policy)*, Nordic Council of Ministers, in Norwegian.

Andersen, B. (1993a) Reform in local public transport. Some evidence from Scandinavia, in D. Banister and J. Berechman (eds) *Transportation in a Unified Europe: Policies and Challenges*, Elsevier, Amsterdam, pp. 249–290.

Andersen, B. (1993b) Endrede rammebetingelser i europeisk kollektiv-trafikk (Changed operating conditions in European Public Transport) MRDH, Oslo, in Norwegian.

Banister, D. (1985) Deregulating the bus industry, the proposals. *Transport Reviews*, **5**, 99–103.

Banister, D., J. Berechman and G. De Rus (1992a) Competitive regimes within the European bus industry. Theory and practice. *Transportation Research*, **26A**, 167–178.

Banister, D., J. Berechman, B. Andersen and S. Barrett (1992b) Access to facilities in a competitive market: The North European experience. Paper presented at the *ESRC Seminar on Entry Barriers in Transport*, University of Liverpool. Also in *Transportation Planning and Technology* **17**(3), 341–348.

Beesley, M. and S. Glaister (1985) Deregulating the bus industry in Britain. A response, *Transport Reviews*, **5**, 133–142.

Berechman, J. (1993) *Public Transit Economics and Deregulation Policy*, SRSUE, North-Holland, Amsterdam.

CETUR (1990) *Urban Public Transport in France. The Institutional Approach*, Direction des Transports Terrestres (DTT), Paris.

Cox, W. and J. Love (1991) International experience in competitive tendering, paper for *Int. Conf. in Privatization and Deregulation in Passenger Transport*, Tampere.

Demsetz, H. (1968) Why regulate utilities? *Journal of Law and Economics*, **11**, 55–65.

EC (1989) Directive on provision of services in energy, transport and telecommunications, Brussels, 90/531/EEC.

EC (1990) Directive on public procurement in energy, transport and telecommunications, Brussels, September.

EC (1993) Report of the Council of Ministers, Brussels, October.

Europeiske Faellesskaber (1989) Forslag til endring af direktiv 90/531/EØF om fremgangsmåderne ved tilbudsgivning indenfor vandog energiforsyning samt transport og telekommunikation.

Europeiske Faellesskaber (1990) Rådets direktiv af 17. sept. 1990 om fremgangsmåderne ved tilbudsgivning indenfor vandog energiforsyning samt transport og telekommunikation (90/531/EØF).

Fielding, C. J. and D. B. Klein (1993) How to franchise highways, *Journal of Transport Economics and Policy*, **XXVII** (2), 113–130.

Gwilliam, K. M., C. A. Nash and P. J. Mackie (1985) Deregulating the bus industry. The case and against, *Transport Reviews*, **5**, 133–142.

Hensher, D. A. (1988a) Productivity in privately owned and operated bus firms in Australia, in J. S. Dodgson and N. Topham (eds), *Bus Deregulation and Privatization*, Avebury, Aldershot.

Hensher, D. A. (1988b) Some thoughts on competitive tendering in local bus operations, *Transport Reviews*, **8**, 363–372.

Koshal, R. K. (1970) Economies of scale in bus transport II, some Indian experience, *Journal of Transport Economics and Policy*, **IV**(1), 29–36.

Lee, N. and I. Steedman (1970) Economies of scale in bus transport I. Some British municipal results, *Journal of Transport Economics and Policy*, **IV**(1), 15–27.

Posner, R. A. (1975) Theories of economic regulation, *The Bell Journal of Economics and Management Science*, **5**(2), 335–359.

Svensk Lokaltrafik (1992) Mer att vinna än någonsin i årets upphandlingar, *Svensk Lokaltrafik*, June, 8–15 in Swedish. (More to win than ever by tendering).

Svensk Lokaltrafik (1993) GL først ut med incitamentsavtal. Svensk Lokaltrafik nr.2.93. 10-11 (GL first by incentive agreements) in Swedish.

Sveriges offentliga Utredningar (1993) Økad konkurrens på järnvägen. SOU 1993:13 (Increased railway competition), in Swedish.

Tough, S. (1992) *A Comparison of Minimum-Cost and Minimum Subsidy Public Transport Tendering Methods*, T3G, University of Westminster, mimeo.

Viton, P. (1981) A translog cost function for urban bus transport, *Journal of Industrial Economics* **29**, 287–304.

White, P. (1990) Bus deregulation: a welfare balance-sheet, *Journal of Transport Economics and Policy*, **xxiv**, 311–322.

Williamson, O. E. (1986) *Economic Organization. Firms, Markets and Policy Control*, Harvester Wheatsheaf, New York, pp. 258–297.

Windle, R. J. (1988) Transit policy and the cost structure of urban bus operations, in J. S. Dodgson and N. Topham (eds) *Bus Deregulation and Privatization*, Avebury, Aldershot.

12 Future Positions and Strategies of Dutch Shippers and Carriers

STEF WEIJERS

Ministry of Transport, Rotterdam, The Netherlands

12.1 INTRODUCTION

Are the steps which Dutch transport companies are considering for their future activities in line with the new shifts in the strategies of big shippers, regarding their in- and outbound transport flows within and through the Netherlands? Is there any parallelism, or are strategies of shippers and carriers going in different directions?

These questions are central to the research programme at the policy research unit which is a part of the Dutch Ministry of Transportation. Research concentrates on the question of whether things are going right or wrong, with regard to the Dutch transport industry. Before advising policy departments, it is necessary to know what is happening, and why things are evolving the way they do.

Compared to the situation a few years ago, there are now signs that there is substantial movement in the positions of carriers, as well as in positions of shippers.

12.2 ACTUAL SHIFTS IN STRATEGIES AND POSITIONS OF CARRIERS AND SHIPPERS

In this chapter attention is given to developments in regard to carriers. In 1993, three issues were central in the transport industry. The first one was European integration, the second logistics, and the third information technology. All three are still important, but at least in 1993 they shared one common feature: their perspective for a brand new future.

European integration and the disappearance of Europe's internal frontiers has raised the amount of transport flows substantially. It raised the already high rate of international activities of Dutch transport, particularly as Dutch hauliers feared competition of hauliers from low-wage countries. Predictions at the time differed from one another not in growth versus decline, but in the expected degree of growth.

European Transport and Communications Networks: Policy Evolution and Change. Edited by David Banister, Roberta Capello and Peter Nijkamp. © 1995 John Wiley & Sons Ltd.

Secondly, logistics was said to open a new perspective for carriers. Repacking, labelling, invoicing, tracking and tracing and even assembling and finishing of goods were expected to establish a steady position very quickly in the activities of carriers. Shippers were expected to out-source transport activities and logistic activities. So carriers were expected to take over management of transport more and more.

Thirdly, new carriers were supposed not to restrict their activities to the simple transport of goods from one point in the country to another, but complex European transport networks would distinguish these new carriers from their predecessors. More particularly, this kind of networking asked for a thorough adaptation of information technology (IT). Electronic data interchange (EDI) between carriers and shippers has the potential to conquer the world more quickly than one normally would expect.

Another, the transport industry, and with it much research, was quite optimistic about the European market for transport services, about opportunities for logistic services, and about transport networks, including IT networks.

From 1990, the spread of information technology (IT) in the transport industry was examined on several occasions. The first enquiries showed a very low degree of adaptation to IT by the transport industry. Maybe it was too early. As stated earlier, many people were expecting that the degree of use would rise very quickly. So the enquiries were repeated in 1991 and in 1992. The last investigation showed that 1 per cent of all Dutch road hauliers are applying electronic data interchange (EDI), or at least claimed to apply EDI. But 44 per cent of all Dutch road hauliers had never heard of EDI.

Moreover this investigation showed that 1.8 per cent of all Dutch road hauliers claim to apply electronic position-fixing systems; 2 per cent pride themselves on black boxes; 2 per cent have route-planning systems, and 6 per cent have on-line planning systems. As can be seen from Figure 12.1, two simple forms of IT are rather common. These concern the administration of salaries (for 62 per cent of all companies) and accountancy support systems. Hardly any difference appears to exist between small firms and the average Dutch haulier—the small ones only apply IT a little less. More complex forms of IT, like fleet management systems and EDI, feature at the lower levels of use.

These findings were a bit disappointing, but this chapter does not go into the reasons for this slowness of IT growth. The problem is said to be lack of money, but there is another more structural reason: any serious form of information technology contradicts the nature of current Dutch road haulage (Weijers, 1992).

These investigations into the spread of IT were not supported by common sense. But that was not all. Europe 1992 promised a massive growth in the transport of goods. Despite this prediction, profits in the transport industry

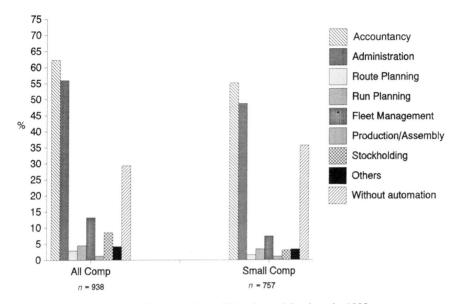

Figure 12.1. The degree of automation of Dutch road haulage in 1992

became more under pressure than expected. In general, profits of carriers are under pressure of shippers, in particular with respect to transport rates. Behind this phenomenon, shippers are themselves under pressure from other shippers. But ultimately in this relationship between shippers and carriers, carriers are basically the balancing item, the closing entry, especially in periods of recession. The situation was further complicated as many new carriers are entering the market simultaneously, especially in road haulage. More often, truck drivers who had been fired start their own company, on low rates with a cheap second-hand truck.

Moreover, the number of bankruptcies and redundancies increased substantially, especially in 1992. For example, unemployment rates in road haulage have risen by 43 per cent (1992–93). These tendencies are quite contradictory to the general picture a few years ago. The developments in the Dutch transport industry after 1992 appeared to be much more contradictory than predicted. Not only did IT appear to be adopted more slowly than expected, but also the growth of the European transport market appeared to be more critical than the common analysis had suggested.

However, there was still the new logistics, of which our colleagues in the policy departments and the carriers themselves were very optimistic. As an illustration, three years ago I spoke with the president of a road haulage company in The Netherlands. He owned 200 trucks. In discussions on new demands of carriers, he said that he was completely convinced that logistics were his future. He said 'I will start again. It is better to sell my trucks, and fire my truck drivers. I will buy a new warehouse, I will contract new people and procure new equipment.' 'If you can help me with a suggestion, you are welcome,' he added. This may be an extreme example of the attitude towards logistics, but it illustrates the common feeling at the time.

But at our research department, the same stories were heard about the same companies. Sometimes, the feeling was that these new logistics were wishful thinking, rather than reality. For example, NedLloyd recently opened a new dedicated warehouse for IBM in Amsterdam. It was promoted as a new step in logistics. But now, it appears to be a normal warehouse, certainly not a warehouse with value added logistics (VAL). The haulier mentioned above, who wanted to sell his trucks and fire his truck drivers, is still a road haulier.

So on the side of the carriers, we notice contradictory movements which are taking place both with respect to market growth and pressure on profits: contradictory regarding a quick technical development of IT and a slow growth of its adaptation, and contradictory regarding new logistic options that create a picture which deviates from day-to-day practice.

At the same time, there seemed to be some movement on the side of the shippers. There was some evidence that shippers are shifting their strategies in a different direction. For example, there are simultaneous processes of globalization and localization; there is a concentration of distribution activities resulting in a growing number of European distribution centres and a deconcentration of distribution activities, resulting in an enforcement of regional distribution centres. These tendencies are referred to again later.

In summary, what has been learned about the potential for new developments like global sourcing or outsourcing certainly did not hold in all cases. Day-to-day practice seems to be much more differentiated than suggested by the optimistic view which was generated only a few years ago.

12.3   RESEARCH QUESTIONS

The evidence about the changes in strategies and positions of shippers and carriers is contradictory, and it is unclear what structures could be identified. To help in this search, a series of complementary research projects was set up.

Firstly some research projects were established on shifts in the strategies of shippers. A study of logistics structures of big shippers, who are important for Dutch carriers, was started. Its central question was determining whether different forms of differentiation were developing on inbound and outbound logistics structures; what were its driving forces, what were its barriers, and what its enablers? This study was conducted by NEA-zuid and the Cranfield Centre for Logistics and Transportation, for the Policy Research Unit AVV. Results were published in January 1994.

This work can be compared with a parallel study by Arn Raaymakers and René Buck. This study was conducted by Buck Consultants International, for the Dutch Projectburo Integrale Verkeers en Vervoersstudies. This research unit has resulted from a cooperation between the Dutch Railway Company, the National Physical Planning Agency, and the Policy Research Unit AVV. The study investigated the Dutch and Belgian automotive, copiers, and final chemicals/polymers sectors, and is focused on the identification of spatial effects of shifts in production, distribution and procurement structures. Suppliers, shippers and carriers have all been interviewed. During the last five years, several studies have been conducted on new strategies of carriers. Most studies had one or more common characteristics, but it was difficult to compare their results.

A theoretically based comparison of these studies was made and this resulted in an interesting method to establish the current positions of transport companies, and to identify their strategic options for the future. The possibilities for strategic development of transport companies differ largely between transport markets. The draft of this study makes it possible to differentiate between different transport markets. To find evidence for their ideas a survey was carried out which enabled a comparison to be made between the 1990 opinions and expectations of Dutch road hauliers in 1993. This study was conducted by INRO-TNO and the Dutch PTT Post for the Policy Research Unit AVV. It was published in December 1993.

With each of these research projects, it was expected that an answer would be obtained to the question of whether the steps which Dutch transport companies are considering now for their future activities follow the strategies of big shippers regarding inbound and outbound transport flows in and through The Netherlands. The preliminary results are summarized in the following section.

12.4 RESEARCH RESULTS

12.4.1 FUTURE LOGISTICS STRUCTURES OF SHIPPERS

In three branches of the freight industry, Buck Consultants International investigated new strategies in logistical concepts and their spatial implications. This included an investigation of positions of their suppliers and logistic services companies. The branches covered in the research involved passenger cars, medium- and high-volume copiers, and polymers and final chemicals. The following companies in The Netherlands and Belgium have been interviewed, together with some of their suppliers and logistics service companies: Nedcar, Ford-Genk, Honda Europe, Nissan Europe, Rank Xerox, Océ, NRG, Canon Europe, Kodak, Dow Chemicals, AKZP Fibers, DSM Polymers, BASF, Synbra and the Bredase Polystyreen Maatschappij.

This research shows that there is no uniform change to be expected in the patterns of current logistics of leading companies in these branches. However, in some cases the relations between producers, suppliers and logistics service companies will change substantially. This change relates to both inbound and outbound flows of goods.

On the inbound side of production facilities, strong hierarchical relations are developing between suppliers: main suppliers are located in part near production sites, co-suppliers are located at a certain distance, and jobbers are located anywhere in the world, mainly in regions with cheap labour conditions. For example, the electronic cables for Nedcar used to be made by a single supplier in Western Brabant in The Netherlands. In order to retain the status of supplier and to capture the status of co-supplier, this Dutch company now buys its wires at the production department of a big prison north of Prague in Czechia. So these wires are not made by blue-collar workers, but by men in striped prison clothes.

Although at a general level future developments may not affect inbound transport flows severely, this growing hierarchy between suppliers might affect transport movements substantially at the local level. Shorter transport movements will take place between main suppliers and producers; longer movements between jobber and co-supplier; and more transport movements because of the increase of just-in-time delivery procedures.

At the other end of the production process, concentration of production activities in Euro-hub-sites is at stake, such as launching satellites for European Distribution Centres, and the direct delivery of goods to customers and dealers. Although not in the car sector, nor in copiers or in the final chemicals sector, leading companies expect the global position of Europe to change. Inside Europe transport flows may shift substantially.

At this moment, many large copying machines are delivered out of production plants, all along different national distribution centres. Direct (physical) delivery, to customers at a relatively short distance from the production plant, is expected to replace this system, without any interference from the national distribution centres. The daily or weekly numbers of medium-sized and big copying machines produced are large enough to justify a European distribution centre.

In the car industry, outbound flows are voluminous enough to create European distribution centres, or at least centres on an international level. Both features are identified—centralization on a European level, and centralization or decentralization on a level between the European and the national level.

So, trends are going in different directions: more centralization versus less centralization. Differences are occurring not only between branches of industry, but also between companies within the same branches.

All leading companies examined are involved in a process of reducing their numbers of logistics service companies. In fact, many shippers claim to prefer contracting just one logistics service company for the whole of western Europe. The problem is that they cannot find companies offering such a service. So, most shippers are striving to contract just one company for each country or each region. For every other region, another company is contracted.

For example, the Dutch carrier De Rooy is carrying cars for Nedcar to Spain. For Nedcar this is quite profitable, because this carrier is receiving orders from another car shipper to transport cars from Spain in the direction of The Netherlands. And, one step further, it really would be interesting for Nedcar if its carrier were also transporting cars from Italy to the Benelux countries, or from Switzerland, etc. Such a carrier could reach a substantially higher capacity for its transport lines, and this offers an opportunity to proportionately reduce haulage rates. This research indicates that carriers fail in their network capacities.

Besides network facilities, carriers can offer another kind of service to shippers. These are activities called value added logistics (VAL). Recently VAL has become a topic for research involving the leading companies in The Netherlands. If someone would like to trace VAL, these companies should be a proper subject of research. Buck Consultants International asked in all three sectors (cars, copiers and chemicals) to what extent VAL-activities have already been implemented. Despite all optimistic debate on this subject, only in a few cases had VAL actually been put into operation.

Some VAL-activities were noticed in the automotive industry, but only in inbound transport flows. In some cases storage and administration of components was accomplished by carriers. But activities like waxing the car are still performed by companies themselves, or by dealers, not by logistics companies. Companies feared the extra cost. Moreover, the companies feared this would bring one more link in the logistical chain. Among copiers, outbound activities were sometimes positively performed by service companies. For example, copying machines are made customer-ready by logistic services companies, with such activities as the inclusion of manuals. Among copiers, hardly any activities concerning the inbound flow of goods are performed by carriers. In this sense copiers reflect an opposite trend to the automotive sector.

This is the overall picture. But there appears to be quite a lot of variation. Some companies do not outsource VAL activities at all, while others are experimenting with VAL both on the inbound and outbound sides. Sometimes carriers take over VAL-activities from shippers. Such a step appears to make high demands upon carriers. It presupposes that the carrier has an overview of the possible differentiated desires of customers. And more than once this has appeared to be too much. This was the case of a Japanese shipper of copiers. He outsourced all Dutch national distribution activities to a carrier a few years ago. Currently the shipper is managing Dutch distribution itself again, only outsourcing its activities.

In a certain sense, shippers do want to outsource VAL activities. But they do not seem entirely convinced of the success of this operation. At least two barriers drive them in the other direction. The first barrier is in the limited possibilities of carriers, or at least in the idea that carriers dispose of only restricted abilities (knowledge of specific products). The second barrier is that shippers fear that outsourcing will be at the cost of the transparency of their cost structures and their customer service. A third barrier could be the fear of disturbing labour relations. Outsourcing can imply a severe reduction of wages, due to differences between different collective agreements. So outsourcing could result in an industrial dispute.

The limited character of current VAL activities is not only due to carriers, but also to shippers themselves. Some shippers have had some bad experiences. But the overall picture is that shippers are trying to find out how far they can go in outsourcing VAL. This sometimes includes going too far.

12.4.2   THE RESEARCH ON FUTURE LOGISTIC STRUCTURES

Research on changes to the structure of logistics systems is being undertaken at NEA and Cranfield (UK), and there are four basic issues for investigation. The first issue is the growing competition in the 1980s. It is traced back in particular to the decline in the power of brands as a differentiating factor between products, which has taken place because, contrary to the 1970s, customers now can buy high-quality products which come down from a range of manufacturers, not only from single-brand manufacturers. This implies that service to customers has increasingly become vital in competition. And this in turn implies that logistics become more central in the activities of a manufacturing company. So hauliers are anxious to ensure that logistics are organized as effectively as possible.

Their second point is the integration and rationalization in European logistics. From the 1970s channel integration developed, combining transport, warehousing and inventory management. Driven by the growing competition in the late 1980s, a few companies began to integrate activities across natoinal borders, searching for better economies of scale. For example, Bosch–Siemens is now rationalizing its Nordic warehouses. Bosch–Siemens used to organize its warehouse operations on a national basis: one warehouse in each of its three Scandinavian countries. Now it organizes its warehouse operations out of one Nordic location to all three countries.

A third point is the current deregulation in road freight transport. Deregulation has taken place on a European level, and in most countries on a national level as well.

Their last point is information technology (IT), which could allow a far greater transparency of the logistical chain. Currently, many problems occur in the systems themselves and in their management. But once these problems can be solved, logistic systems can develop rapidly.

So starting from these trends, developments of logistic systems are analysed. Attention has been paid to its driving forces, enablers and barriers. Twenty-four leading companies in six branches of industry have been interviewed. This research shows several shifts in logistic structures of the selected companies. Here, the identified branch patterns will be presented and these include:

1. changes on the supply side—inbound logistics structures;
2. changes in the level of production—manufacturing logistics structures; and
3. changes in distribution structures—outbound structures.

Figure 12.2 shows a differentiated overview of branch options for logistic systems. The figure shows how products flow in three separate logistics systems. First in manufacturing, where products may be transported trans-nationally, out of one country towards other countries. The current, traditional way is to organize product logistics multinationally, each product being produced in its own country. Finally production and product logistics can be organized on a local basis—some speciality chemicals are organized this way. Similarities and differences appear to exist between different branches. Six industry sectors are analysed for the way they organize logistics in production:

1. automotive
2. business equipment
3. channel management retailers
4. fast-moving consumer goods
5. food and agriculture products
6. speciality chemicals.

On the level of manufacturing, many similarities appear to exist between branches. The picture is clear: transnational organization of production is the current option, for some companies already for a long period—illustrated by an 'O'. For others, it is seen as a new option—marked by a '↑'. The multinational option is declining, marked by a '↓'. The local option did not come across—its mark is a '–'. Differences between branches occur in many different ways on the inbound side, and especially on the outbound side in the lower part of the figure (Figure 12.2).

Starting from their sourcing unit, goods may flow directly to the production site. This is the one extreme. The other implies a hierarchical way of collecting on a European and a regional level, before goods are finally delivered to the production site. In between these extremes, the collection of goods may occur on a European, national or a regional level. This last formula—regional collection—did not come across. Direct sourcing appeared to be a current option in all channel management and retail

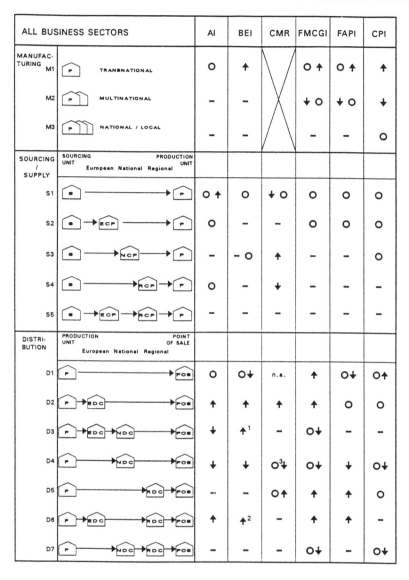

ALL BUSINESS SECTORS		AI	BEI	CMR	FMCGI	FAPI	CPI
MANUFAC-TURING M1	TRANSNATIONAL	O	↑	✕	O ↑	O ↑	↑
M2	MULTINATIONAL	–	–	✕	↓ O	↓ O	↓
M3	NATIONAL / LOCAL	–	–	✕	–	–	O
SOURCING / SUPPLY	SOURCING UNIT — PRODUCTION UNIT European National Regional						
S1	■ ——→ ⌂	O ↑	O	↓ O	O	O	O
S2	■ → ECP ——→ ⌂	O	–	–	O	O	O
S3	■ ——→ NCP ——→ ⌂	–	– O	↑	–	–	O
S4	■ ——→ RCP → ⌂	O	–	↓	–	–	–
S5	■ → ECP → RCP → ⌂	–	–	–	–	–	–
DISTRI-BUTION	PRODUCTION UNIT — POINT OF SALE European National Regional						
D1	⌂ ——→ POS	O	O↓	n.a.	↑	O↓	O↑
D2	⌂ → EDC ——→ POS	↑	↑	↑	↑	O	O
D3	⌂ → EDC → NDC → POS	↓	↑[1]	–	O↓	–	–
D4	⌂ ——→ NDC → POS	↓	↓	O[3]↓	O↓	↓	O↓
D5	⌂ ——→ RDC → POS	–	–	O↑	↑	↑	O
D6	⌂ → EDC ——→ RDC → POS	↑	↑[2]	–	↑	↑	–
D7	⌂ ——→ NDC → RDC → POS	–	–	–	O↓	–	O↓

Source: Interviews with 22 large producing and retailing companies in west Europe

Symbols:
↑ increasing
o no change
Manufacturing locational structures
Transnational = Specialized, complementary, multi-sited
Multinational = Multi-domestic, not complementary
National/local = Domestic, single or multi-sited
Notes:
1. increasing use of wholesale retailers, using their own NDC
2. RDCs run by retailers
3. multipurpose stockholding NDCs

n.a. not applicable
↓ decreasing
– not identified as relevant during study
Sectors
AI = Automotive Industry
BEI = Business Equipment
CMR = Channel Management Retail
FMCGI = Fast Moving Consumer Goods
FAPI = Food and Agricultural Products
CPI = Chemical Products

national collection tends to replace regional collection of goods. Collection may imply trans-shipment or warehousing.

The picture really is divergent on the distribution side. Options may vary between direct delivery and linked delivery on a European, national and regional level, or combinations. This picture is not clear from an overall point of view. But on a branch level, all studied companies show quite similar options and a shift in a similar direction. Let us take the branch which shows the most unclear pattern: fast-moving consumer goods. All different options were identified in the examined cases. But the branch picture points to an increase of the combined EDC/NDC options, at the cost of the other two NDC options (NDC and NDC/RDC). Another example is speciality chemicals, where direct appears as an option. Both EDCs and NDCs are options, but not in combination. The other branch pictures show similar parallels between options of companies. These results clearly point to the importance of a branch approach.

The general picture is far more complex on the distribution side than on the supply side. Combinations of collection activities have not been identified. Supplying goods seems to have been rationalized much more than distribution activities. This research shows that carriers on the supply side are not the same as those in distribution. For a carrier who has been contracted at the supply side, offering a freight loop service appears to have been quite decisive for selection. This is not the case in distribution activities. In this realm, either network services are asked for, or warehousing specialists are contracted, or regional specialists are used. So this general picture also points to different demands on carriers, depending on the question as to whether they operate at the supply side of the shipper, or the distribution side. This brings us to the positions and strategies of carriers.

12.4.3   REASSESSMENT BY CARRIERS

In 1990, 60 managers of road haulage companies were interviewed by Vermunt (INRO-TNO). This enquiry was restricted to road hauliers who employed between 50 and 500 people, and who were mainly involved in general cargo activities. This category of cargo can be defined as requiring no special treatment for reasons of weight, volume, appearance or conditions. Most Dutch road hauliers deal in general with cargo. They were asked to express their views on the strategic position of their company at the moment, and on the future development of their company. In 1993, three years later, the same managers were asked to answer similar questions. Half of the group responded. The 61 enquiries of 1990 appeared to be a repre-

Figure 12.2. An overview of options for logistics systems

sentative sample of all medium-sized hauliers in general cargo. Moreover, the 28 enquiries of 1993 appeared to be representative of the enquiry of 1990 on all three strategic directions: 54 per cent of the 1990 sample expected to internationalize, against 57 per cent of the 1993 sample; 31 per cent of the 1990 sample expected to extend their services, against 32 per cent of the 1993 sample; 22 per cent of the 1990 sample expected to develop more network activities, against 25 per cent of the 1993 sample. This made it possible to compare their positions in 1990 to those in 1993: did their predictions actually come out or not? The repeated enquiry established to what degree predicted strategic options had been modified.

Managers were asked to answer questions firstly on the general scope of their companies (number of employees, turnover, fleet). A second set of questions was about the geographical scope of the haulage company. Thirdly, questions were asked about true transport activities, including their internal preparatory activities and proceeding to the actual rendering of additional services. A fourth set of questions was about additional final services to customers performed by the road haulage companies.

In 1990, the sample companies each employed on average 116 people. At the time, they expected to grow on average to 146 employees in 1995. In fact, they employed on average 119 people in 1993. That is not a spectacular shift. It only seems to show that the hauliers had been too optimistic in 1990. In 1990, only one haulier expected that his number of employees would diminish. But in 1993 nearly 50 per cent of all hauliers appeared to have cut their staff, most of them by 10 per cent or more. In proportion, a few companies had seen a substantial growth in their number of employees. Nevertheless nearly all hauliers expect a substantial increase in the number of employees in the years to come. Although less pronounced, figures about turnovers and numbers of cars are also in the same direction (Figure 12.3).

This example of the size of the staff shows two things. First, hauliers were too optimistic in 1990. Second, behind this general stable development of employment—from 116 to 119 employees—a substantial shift in positions took place on the level of individual companies. There was a dramatic decrease in the case of many companies versus a substantial increase for others. The same pattern can be recognized for changes in turnover and fleet size.

In 1990, Dutch hauliers in general cargo generated 50 per cent of their turnovers inside Dutch borders, and nearly 50 per cent on the European market. Only a few per cent were generated by global activities. These figures have remained stable. In 1993, 50 per cent of their turnover was made from activities inside the country, and 50 per cent outside. Global activities had shrunk further, from 4 to 2 per cent. This is not a spectacular result, but let us look at it in more detail.

In 1990, nearly 50 per cent of the hauliers expected that their national activities would rise; the same expectations were expressed with regard to

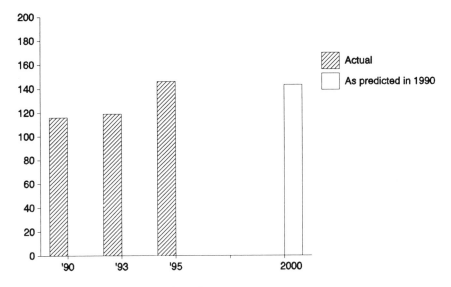

Figure 12.3. Average number of employees

European activities. Only a few companies expected a decrease, mainly of national activities. In fact, more than one-third had a greater share of national activities in 1993; the same counts for European activities as measured in proportion to their turnovers. But at the same time, just as many companies had contracted their national activities; the same appeared to be the case for European activities. In half of all cases, shifts appeared to be more than 10 per cent in terms of turnover. Again we had to conclude that hauliers had to adjust their 1990 expectations, and that quite a lot of hauliers noticed a quite substantial shift, this time not only in size, but also in geographical scope (Figures 12.4 and 12.5).

Companies were compared according to their form of organization, both in 1990 and 1993. In 1990, half of all hauliers in general cargo organized their transport services without any form of trans-shipment. This implied transport from just one point in the country to another, or to a third point or more, without the organization of any trans-shipment. One third of the companies had one trans-shipment site at their disposal. One fifth commanded several sites (Figure 12.6).

Again these figures have been remarkably stable. In general, no shifts seem to have taken place between 1990 and 1993. However, many individual companies experienced large changes in their networks. Nearly 30 per cent of the companies appeared to have developed a network on a higher level. And nearly 15 per cent had withdrawn to a lower level. So nearly half of all companies changed the character of their transport organization.

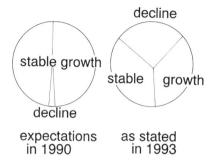

Figure 12.4. Changes in geographical scope

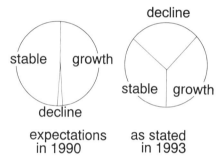

Figure 12.5. Changes in European scope

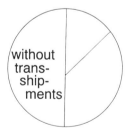

Figure 12.6. Kinds of transport organizations

A distinction has been made between the management of activities in transport and logistics on the one hand, and performance of these activities on the other (Figure 12.7). In 1990, the Dutch hauliers interviewed received 90 per cent of their turnover from performance activities in transport or storage. Management was performed minimally. And again hardly any

Figure 12.7. Kind of delivered services (shares measured as per cent of turnover)

general change in 1993, but in more than 50 per cent of all cases a substantial change had occurred. Three quarters of the hauliers obtained their turnovers only from transport, both in 1990 and 1993. But nearly 50 per cent of them decreased their proportion of executive transport, and about 40 per cent raised this share, at the cost of storage, trans-shipment or management.

Hence, the overall picture is that no big changes had taken place at first sight, and that there were vigorous adjustments compared to 1990. But that is not all. There are also some signs of a general shift in strategies.

Together with these questions on their actual kind of services, hauliers were asked about which kind of activities will focus their future strategies. In 1990 more than 50 per cent of them wanted to develop more fulfilling activities, as many hauliers wanted to perform more services than just transport for shippers: 40 per cent wanted to take over management activities from shippers, and 20 per cent wanted more international activities. In 1993, only one-third of them wanted to develop more fulfilling activities, but as many as two-thirds of them were aiming for a bigger range of services. A similar number now want to alleviate management activities of shippers. These more extreme figures point to a more pronounced strategy. This does not necessarily imply a better deliberated view, but it shows that clearer choices are being made.

A similar phenomenon can be identified in investment. The average number of investment items has decreased from 4.1 items in 1990 to 3.4 in 1993. This suggests that investments tend to be more concentrated. Although the vehicle fleet remains the biggest investment item, more hauliers invest in EDI, quality programmes and management of warehouses. Fewer hauliers invest in buildings and in management of transport, but those that do invest have nearly doubled their investment share.

12.5   CONCLUSIONS

Three research projects have been discussed in this chapter, two on shippers, and one on carriers. It is now appropriate to conclude this survey with some general and comparative remarks.

1.   It is important for carriers to develop in line with shippers. Moreover, liberalization of cabotage and deregulation on a European level and a domestic level is making the transport market more open. So in order to survive competition, hauliers have to be more service oriented towards shippers than they have been in the past.

2.   The Cranfield/NEA-research indicated a market improvement of service performance from shippers towards customers, because of new relationships in competition. So one could conclude that the more a manufacturer is oriented to customers, the more logistics will be seen as crucial. This could imply that the more logistics are seen as crucial, the less they will be outsourced to carriers, at least the management part of it. If not, carriers will have to meet very high standards of managing logistics for shippers. Regarding operational activities by carriers, new requirements develop. Either it becomes important that a carrier operates on a transnational level, or he has to expand his capacity, or he has to perform better-fitting services. Combinations of options are possible, but complex and expensive.

3.   At first sight the results of the inquiry on hauliers seemed not to be surprising. Apart from individual cases only small overall changes could be identified. But going into more detail resulted in interesting findings. Recent changes have been impressive for a very large number of individual companies. In 1990, expectations were high about the possibilities for (a) internationalization, (b) logistic services, and (c) network development. Contrary to this general positive outlook, many hauliers experienced contrary movements between 1990 and 1993. Hauliers had to adjust their expectations quite strongly, sometimes by expanding, but often in a more restrictive sense. Several hauliers had to lower their ambitions regarding internationalization, the extension of their supply of services, and their range of network activities. A 'positive' development in one direction was often combined with a 'negative' one in another direction. A 'negative' adjustment does not necessarily imply that it is a bad solution, and the converse can be said for 'positive' adjustments.

A pattern could not be identified in the changes experienced by the companies, but there were some interesting signs. The researchers found out that:

(a) Dutch hauliers, who operated in 1990 mainly on a national basis, seemed to dispose of a higher level network in 1993.

(b) Hauliers who did not dispose of trans-shipment facilities in 1990 tended to raise the level of their network; they also tended to extend their supply of services, but they tended to shrink their geographical range.

(c) Finally, the more strongly hauliers were focused on international activities in 1990, the more extremely they tended to develop in all three directions—that is to say, in both the positive and the negative sense.

Again, the number of the surveyed companies is too small to draw certain conclusions, but trends seem to go in the following directions:

(a) The more international, the greater the risk.
(b) The less trans-shipment, the more possibilities for extension of services and the less possibilities of going international.

4. Many of the predictions of hauliers did not work out. This poses the question of whether circumstances have changed, or the level of charges were underestimated. The answer is both. An economic recession set in, and there are some indications that carriers did not adequately consider the implications of new developments in Europe and logistics with respect to European integration: a strong pressure on profits could have been expected to result from growing competition. Knowledge of earlier forms of deregulation in the United States and in Australia has helped here, so, in part, the pressure on profits could have been expected. The effects of the overall economic recession have been more difficult to predict with respect to networking—scale appears to be a big problem. Developing a more complex network requires an enormous organization, and an enormous amount of investment. Current Dutch carriers are not big enough to develop in this way.

With respect to logistics, it is difficult to assess whether the current economic recession was favourable to the amount of logistical orders for carriers. The recession will have put higher pressure on rates of logistical services. But in the research projects there is some evidence that carriers experienced some difficulties in managing logistical services. For example, in the copiers branch, recycling is an important phenomenon. Lack of knowledge of different product types is a severe handicap for a carrier in this logistical area. The growing competition was predicted in part, but the required scale for networking, and especially the required skills for logistics and IT seem to have been underestimated.

5. A general conclusion is that shippers are undergoing a fundamental process of reassessment. Many shippers are reconsidering far-reaching forms of outsourcing. The VAL example in the copiers branch and the IBM/ NedLloyd example are illustrative. Loss of transparency of the rendered services and of cost structure seems to be a barrier for shippers, as is fear of disturbance in labour relations. In the end, there seems to be a general reconsideration among both carriers and shippers. Both parties have withdrawn from their primary goals.

6. Finally, our general question was whether the actual strategic steps of Dutch transport companies are in line with the transport effects of the strategies of big shippers. Taking all three research projects together, it can be seen that carriers appeared to develop in all three ways which were asked for by shippers:

(a) geographically
(b) regarding expansion of network activities
(c) regarding expansion of logistic services.

This does not mean that the answer to our central question is yes. Maybe a carrier is doing too much in one field of activity and too little in another. Carriers are not only advancing in one or more of the three directions, but they are also retreating from the same three directions. The answer as to whether advancement or withdrawal is the proper strategy can be found in their respective transport markets.

The NEA/Cranfield research shows clearly that some branches ask carriers to be a serious sparring partner in IT or EDI. Other branches do not. For example manufacturers in chemicals used to contract medium-sized road hauliers because of their traffic dominance. But these same companies lack the skills and financial resources to implement sophisticated EDI systems. Up till now shippers have financed the total individual EDI projects with carriers. But they are no longer prepared to do so.

The Buck Consultants International research showed that many shippers would like to contract a—not yet existing—Pan-European logistic service company. The NEA/Cranfield research showed that such a carrier would be too expensive and too complex. Moreover, for the shipper this contracting out could imply loss of control.

The existence of more than one Pan-European logistics service company would be quite profitable for shippers. Because, competition between them would put transport rates under pressure. But, given the importance of logistical services, one could seriously doubt whether shippers would contract such a Pan-European service company. In fact, shippers ask for carriers who are specialists in transport in one country, or who are specialists in warehousing, or in trans-shipments. The INRO research showed that these kinds of carriers in general cargo are available in The Netherlands. Logistics, especially their management, seems to be too crucial.

REFERENCE

Weijers, S. (1992) *Dutch Road Haulage, Questions on its Organisation and Information Technology*, Utrecht University, Utrecht.

13 Decision-making for Trans-European links—The Danish Case

KAI LEMBERG
Copenhagen, Denmark

13.1 INTRODUCTION

The approach in this chapter is sociological and politological. It discusses the decision-making process concerning fixed transport links between Sealand and Funen (Great Belt), between Sealand and Scania in Sweden (Øresund, the Sound), and between Lolland and Fehmarn in Germany (Fehmarn Belt) as parts of a superior European road and rail network.

My major objective is to examine the way in which investment decisions for these 'missing links' are made concerning choice of mode of transport, type of construction and time of implementation, especially order of succession, and to demonstrate which economic and political actors have dominated the decision-making process.

The structure of the chapter is as follows. The problem of transport barriers in Denmark and their possible elimination by substituting ferry lines by bridges or tunnels is discussed in Section 13.2. In subsequent sections important actors in this field are described: Round Table of European Industrialists (Section 13.3), Scandinavian Link (Section 13.4) and the EU Commission (Section 13.5). In Section 13.6 a politological theory of attitudes and powers in the decision-making process is presented. Section 13.7 contains conclusions on the nature of the decision-making process for these links.

13.2 THE PROBLEM: TRANSPORTATION BARRIERS AND THEIR ELIMINATION

Denmark is a country of islands plus one peninsula: Jutland. From time immemorial we have been accustomed to a situation in which all transportation by land was interrupted by sounds and belts. To overcome this a number of ferry services connect the Danish provinces as well as Denmark with the neighbouring countries.

During this century these ferry services have been adapted to convey

European Transport and Communications Networks: Policy Evolution and Change. Edited by David Banister, Roberta Capello and Peter Nijkamp. © 1995 John Wiley & Sons Ltd.

motor vehicles and (for some main lines) trains, and have been modernized, so that loading and emptying of ferries run quite smoothly; there is—except for a few holidays (around Easter, starts of school holidays, industrial holidays and Christmas)—no long queuing with excessive waiting time for drivers who have reserved a place on-board; and the crossing time has been reduced, but remains, of course, longer than an alternative driving time or riding time by train passing over a bridge or through a tunnel.

Many drivers consider the interruption of a long car or truck journey a relaxing breathing space and use it for a meal or a coffee break, which for journeys beyond a few hours' time would take place anyhow, at some other location. But for business undertakings any prolongation of a trunk or goods train journey, including queuing and waiting, is a loss of time and money, unless it fits in with an obligatory time of rest for the truck driver.

This problem of transportation barriers has a special importance for the Scandinavian countries, situated in a peripheral geographical position in Europe. The actual prolongation of travel time, the higher costs of travelling by ferry than by road or trail, and the psychological barrier because transport users consider ferry services a complication—all means an intensification of the peripheral problem. So, for several important straits and fjords, the Danish ferry lines have been replaced by bridges or tunnels. But the three most important connections have remained ferry lines. The construction of fixed links for these three connections—the Great Belt, the Sound and the Fehmarn Belt—are considered part of the creation of a superior, fast European road and rail network.

In traditional transportation policy it is presumed that:

1. The realization of these (and other) large transportation infrastructure investments and their mutual time priorities must depend on whether they constitute economical use of resources or whether (conversely) their societal profitability is too low to justify their realization.
2. The construction of large transport infrastructure like these connections will create considerable economic growth.
3. These evaluations can be made by the use of objective, quantitative calculations according to cost/benefit methods and similar economic methods, including also environmental and other non-economic issues, converted into economic terms.
4. National superior links must be given priority over international links, independently of their relative profitabilities.
5. The decisions about these investments are made by a majority of politicians in the Danish Parliament (Folketing) in cooperation with the Swedish and German Parliaments (i.e. according to the rules of representative democracy).

By examining these hypotheses against actual incidents concerning the three superior European links, this study might contribute to an understanding of the national and international economic—political processes behind the decisions on the largest transport infrastructure investments ever made or planned in Denmark.

For decades discussions have been going on concerning the construction of bridges and/or tunnels for road and/or rail transport instead of ferry services at the Great Belt and the Sound. Advantages and disadvantages in the prolonged discussions on establishing fixed links instead of ferries for transportation between east and west Denmark (Great Belt) and between the Scandinavian peninsula, Denmark and the European continent (the Sound, Fehmarn Belt) have followed lines of thought as indicated below, according to a huge number of official reports, political debates, newspaper articles, etc.:

Adherents of fixed connections for road and rail instead of ferry routes stress the importance of establishing motorway links independently of railway connections to secure a free choice for the transport user. They are primarily representatives of trades and industries oriented towards import, export or long-distance home market sales, plus car drivers, road hauliers, oil companies and motor vehicle trades and industries. Ideologically, most are either conservative or liberalists with a strong belief in the advantages of economic growth and a market economy with free competition—or they are social democrats with strong labour union interests in the motor industry or connected trades.

The crucial argument for combined fixed road and rail links is the alleged strong impact on economic growth through increased international competitiveness, increased optimism in trades and industries and for the Sound Bridge an expected synergy effect by joining Greater Copenhagen with the highly urbanized South Scania (including Malmoe, Lund, Helsingborg, etc.) into a Danish–Swedish metropolitan agglomeration (Ministry of the Environment 1993). The fixed combined connections mean time savings for goods transport, car driving and train passengers, and subsequently lower costs and increased competitive power. This is of particular importance because of the general tendency towards internationalization of trades and industries, liberalization of world trade and especially the Single Market within the European Union.

Secondary arguments are to obtain independence of weather and semi-monopolistic state railways and ferries. The motor lobby stresses the importance of free choice of mode of transport, optimal flexibility and the pleasure of individual car driving as a contrast to train journeys or the use of train wagons carrying cars. Trade union people and (other) social democrats in particular emphasize the large employment created during construction and the expected employment effects of increased economic growth.

Generally, the combined bridge devotees claim that the bridge will show a high transport economic profitability and imply no state expense, as full user payment is taken for granted. A railway tunnel (also with wagons for motor vehicles), they claim, will prove unprofitable and will therefore never be realized.

Finally, maritime environmental damage can be completely avoided by the technical design of the bridge, and air pollution and other environmental damage and nuisance will be minimal and outweighed by a cessation of ferry pollution.

Opponents of combined road and rail bridges are not correspondingly easy to define. They include non-drivers, people travelling less (except by charter flights), ecologists and environmentalists, left-wing and most centre party voters, town planners, etc. Most of them originally preferred continued ferry services, but now most prefer railway tunnels. For the Sound link some have advocated a tunnel between Elsinore and Helsingborg instead of between Copenhagen and Malmoe.

Their main arguments are ecological, environmental and economic. A railway tunnel will cost about 60 per cent of a combined link and will be more economical, energy-saving and environmentally neutral or even positive, because it will encourage public transport and discourage car and heavy truck driving. Alternatively, modernized ferry services can be low-polluting, energy-saving and fast, especially direct superferries between Sweden and Germany.

The heavy extra capital investments in combined links constitute a waste of resources that could be better used for alternative purposes: environmental improvements, public transport, energy savings, etc., and the larger capital investments will cause a heavy strain on the weak Danish capital market and necessitate large foreign loans.

The opponents reject the alleged long-term stimulating economic effects on Danish trades and industries. This theory has no empirical support from existing large bridges, and the same stimulus in Denmark should also occur to Swedish and German trades and industries—but not everybody can gain from the marginal direct transport cost savings. The fixed links in the superior European transport networks might just as well make Hamburg the future commercial capital of North Europe. So, the long-term employment effect is dubious, and the large construction employment will be counter-balanced by loss of ferry employment. In the social economic cost/benefit calculations, the economic value of the very large accumulated time savings for huge numbers of individual car travellers for leisure, holidays and personal visits is highly questionable. The whole economic growth argument is a postulate and is, furthermore, in deep conflict with the superior global goal of sustainable development and a large-scale reduction of energy consumption in the industrialized world.

Opponents of the combined links lay stress on the negative environmental effects of the motorway bridge, which will stimulate lorry and car traffic, especially local traffic between Greater Copenhagen and Malmoe–Lund, thereby causing increased noise and air pollution by emission of CO, NO_x, SO_2, soot, heavy metals and CO_2. Catalysts in cars will reduce these emissions except CO_2 from each car, but more cars will run, and more complete combustion means increased CO_2 emissions, which is the main factor behind the global greenhouse effect. In addition, the motorway bridges will reduce traffic safety, cause vibrations and cut through urban quarters and landscapes. Conversely, a railway tunnel (even with wagons for motor vehicles) will improve environmental conditions on land, because it will give preference to public transport and subdue motor vehicle traffic.

Another important type of environmental damage from a bridge, as opposed to a bored tunnel, is its diminution of the water-flow, influencing the salinity and the water renewal in the Baltic Sea and causing disturbances in the bottom of the local waters. Both types of effects, even marginal, can disturb the maritime ecological balance and thereby the fisheries. These problems have played an increasing role in the debate for and against a Sound Bridge. In May 1994, the Swedish special Water Court has commissioned new model testings and calculations before their final evaluation of the level of such damages.

Finally, fixed links (especially at the Great Belt) will close down several existing ferry routes and concentrate long-distance traffic at a single link, which will aggravate the peripheral situation of provinces like North Jutland and South Norway. The competitiveness of maritime transport to and from the ports of Copenhagen, Malmoe, Gothenburg and Oslo may also decline. A bridge with concentrated international traffic is vulnerable to terrorist and warfare attacks and to maritime collision accidents.

13.3 ROUND TABLE OF EUROPEAN INDUSTRIALISTS

The second oil crisis in 1979 and the increased technological and economic competition from the United States and Japan caused shock waves through big business in Europe and a state of pessimism. The division of the European market into a large number of national markets, each with its own regulations and protective measures against foreign competition, and the slow process of economic integration among the 12 countries of the European Economic Communities (now the European Union) was detrimental to the large multinational concerns in Europe—within and outside the EC—and to expanding dynamic industries.

So, at the initiative of the (then) Swedish managing director of Volvo, Pehr Gyllenhammar, a small number of leading European industrialists started to act as a think-tank. Among them were top leaders from the car concerns Daimler, Benz, Fiat and Volvo, the tyre factory Pirelli, and the electronics concerns Philips, Asea Brown Bovery and Siemens. Gradually others were invited to join. In 1983 they established the organization The

Round Table of European Industrialists (originally 17 top leaders, but by 1990, 43—all men of course) had become members. They represented the largest European industrial concerns, including also Royal Dutch Shell, Petrofina, Pilkington, Nestlé, Hoffmann la Roche, Bosch, Thyssen, etc. Only invited firms can join the party. Two Danish members have been invited: Poul Svanholm (Carlsberg–Tuborg breweries) and the shipowner, overseas trader and offshore oil producer Mærsk McKinney–Møller. Other Scandinavian members are (in addition to Gyllenhammar) Curt Nicolin (Asea ABB), Torvild Aakvaag (Norsk Hydro) and Kari Kairamo (the Finnish Nokia Oy).

The first report, published 1984 by European Round Table, called *Missing Links* (Round Table 1984), indicated the high priority the group paid to European transport infrastructure, predominantly motorways, and concentrated upon north–south links between Scandinavia and the European continent (the Sound and Fehmarn Belt), Scotland–England (via the 'Chunnel')–France, transit lines through the Federal Republic of Germany, using transalpine tunnel links (Figure 13.1, left). This map included fixed links crossing the Sound and Fehmarn Belt, but no Great Belt Link. Another map illustrated in more detail 'a high quality road and rail connection' Oslo–Copenhagen–Hamburg, including bridges or tunnels across the Sound, Fehmarn Belt and Great Belt (Figure 13.1, right).

The European Round Table analysed the problems of (western) Europe: the split into many nations, the too slow process of implementing high technology in European industries, the multitude of national regulations, the lack of capital, the slow process of international goods transportation and several missing links of a superior, coherent European rail network for high-speed trains and motorways for fast car and truck traffic.

The crucial weak point in solving these problems was, according to the European Round Table, the decentralized decision processes, state interventions in the plans and decisions of private business, lack of cooperation between the big industrial concerns, and the slow and bureaucratic functioning of political democracy.

To overcome these difficulties and bring about new dynamic economic development, the Round Table members started an intense cooperation with the EC Commission, especially through former commissioner Etienne Davignon and the then president Jacques Delors. The Philips executive director Wisse Dekker prepared a sketch entitled Europa 1990 for the organization of what became known as The Single Market of the EC (now the EU) by 1993—an operational vision of a European Communities superstate without internal borders, duties, controls, import restrictions, etc., and with a breaking down of all barriers inconvenient to industrial development and expansion. This principle was presented as 'free movements of goods, people, capital and ideas'.

The technological solution of the European problem would be, then, a

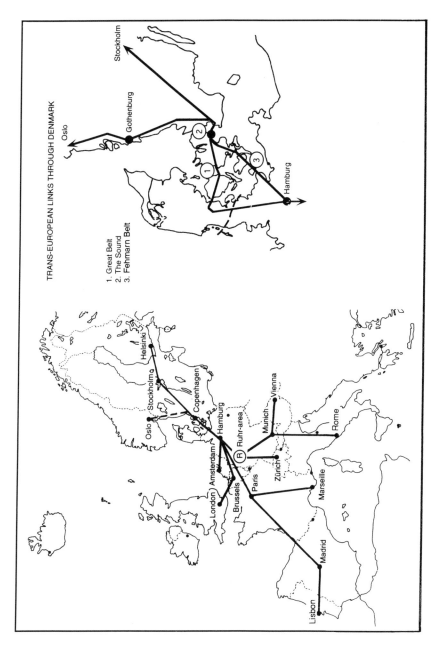

Figure 13.1. Trans-European links through Denmark

fast development of high technology in production, steering of production and distribution by computers, large investments in fast road and rail transport, and a new 'logistic revolution', i.e. steering of the flows of goods, information and values between suppliers, producers and consumers to reduce the amount of capital that must be bound in stock. To influence the decision-making as strongly as possible, the European Round Table worked in close collaboration with the EC Commission and with transnational associations of trades and industries, employers' associations, some trade union top leaders and politicians from relevant parties: conservative, liberals and social democrats.

In a publication 1988, entitled *Needs for Renewing Transport Infrastructure in Europe* (Round Table 1988), the European Round Table warned against

> the threatening breakdown of the physical movement of people and goods(!)

and stressed

> the strong impact of transport services on economic efficiency, the growing inadequacy of Europe's transport infrastructure

And the need for new and radically improved infrastructures for transport and telecommunications. Trades and industries and the private transport sector are up against an increasing political

> power of those who are not directly or indirectly concerned to delay or even veto projects. ... It seems to be important to prevent a similar development from occurring at the European level.

The real problem, claims the Round Table, is not lack of financial resources, but the political process of decision-making, planning and coordination. Against this the European industrialists claim the necessity to develop an overall concept of a European transport infrastructure, harmonizing transport policies between EC member countries and revitalizing the role of the European Community.

In Europe (as opposed to the United States) transport infrastructure is primarily a government responsibility, based upon goals of equity and non-discrimination, but with resulting

> lack of balance between (transport) supply and demand ... the absence of an efficient pricing and financing mechanism, neglect of economic efficiency; lack of transparency in planning and decision-making and a vulnerability to strong vested interests in public administration and private sectors.

The worst bottleneck is national sovereignty; further inefficiencies in the decision-making process are due to:

> the involvement of environmental groups, local residents and others affected by infrastructure projects. ... Vested interest groups tend to oppose competition and a systems approach to infrastructure planning—being rather effective in forging public opinion in their favour.

Table 13.1. Obstacles in infrastructure decision-making (Round Table)

	F	D	I	S	DK	N	SF
Lack of finance	M	W	M	W	M	M	M
Bureaucracy	W	M	S	M	M	M	M
Environmental problems	W	M	M	M	M	M	M
Lack of space	W	M	M	W	M	W	W
Political problems (greens)	W	S	M	S	S	M	W
Political problems (party politics)	M	M	M	S	M	M	W

F = France; D = Germany; I = Italy; S = Sweden; DK = Denmark; N = Norway; SF = Finland.
S = Strong   M = Medium   W = Weak.

The publication indicates the important obstacles in infrastructure, decision-making and the (bleak) state of decision-making procedures in a table, shown as Table 13.1.

A number of case studies are described, including the Eurotunnel (the 'Chunnel'), European High-Speed Train Network—and the Scandinavian Link, which 'could serve as a model for an organization at European level' for transport infrastructure planning. The Round Table notes (with silent regret) that no single owner of the proposed transport infrastructure is foreseen (the three large bridge/tunnel projects, etc.), but is content that the Scandinavian Link Consortium has the role of a unitarian designer. Among the basic requirements for successful projects, the European Round Table stresses

> involving industrialists and bankers in the early stages of planning and decision-making and establishing economic conditions for a profitable operation of the project.

Environmental interests must not be ignored or underestimated.

> However, they should put limits and constraints on needed infrastructure projects, not block them.

13.4   THE LOBBY ORGANIZATION SCANDINAVIAN LINK

In February 1984, the five Nordic prime ministers recognized a privately established working group of top leaders in industries and banking—including directors of Swedish Volvo and Asea, Finnish Nokia, Norwegian Hydro and Danish Carlsberg Breweries plus one trade union boss: Georg Poulsen from Danish Metal Workers—to present experiences and ideas for economic cooperation between the Nordic countries. Volvo's top director Pehr Gyllenhammar was appointed chairman. A group report in 1985 (Arbetsgruppen 1985) recommended the construction of a system of motorways and fast railway lines linking Norway, Sweden and Denmark with the European continent, including 'fixed connections over Great Belt, Øresund and Fehmarn Belt as fast as possible' (and a motor road to a new established Volvo factory near Gothenburg) plus effective ferry connections between Finland and Sweden. The group declared that Nordic manufacturing industries were ready to build these lines. A subsequent attempt by the group to be affiliated to the Nordic Council of members from the five Parliaments failed, and they decided to continue work as a private consortium 'Scandinavian Link' with close ties to the European Round Table. The consortium was organized in four national sections, the shareholders of which were 55 of the largest industrial, trading, transport and financial companies in Denmark, Finland, Sweden and Norway, with director Curt Nicolin (Asea) as the chairman.

After a first period of silent contacts with politicians, Scandinavian Link addressed the general public at large by a number of reports, partly elaborated by officials 'borrowed' from the Ministry of Transport (Scandinavian Link 1987), analysing and recommending a complete transport system for a high-speed railway and motorway network to be finished by the year 2000, including the three large bridges.

To the leaders of big business (but hardly to most politicians), it was evident that the Sound and Fehmarn Belt links were the important links in a European perspective as the international transit traffic through Denmark. The Great Belt link was of only secondary interest, as a supplementary link with a 150 km detour between Copenhagen and Hamburg compared with the Bee Line over the Fehmarn Belt. But psychologically the lack of a Great Belt bridge would in Denmark act as a strong barrier against political acceptance of the two other links. Scandinavian Link consequently used their money and energy primarily in promoting the Great Belt Bridge, which was in itself less interesting.

Conversely, it also was evident that the Fehmarn Belt connection was the least politically popular link in Denmark, considered to be primarily a transit link (and an advantage to German, Swedish and Copenhagen businesses). Consequently, Scandinavian Link has kept a low profile on this part of the total transit infrastructure programme and they have not even included it in their cost/benefit calculations.

Behind the practical operational goals of the Scandinavian Link of fast, secure and reliable transports to be delivered just-in-time are some more fundamental, societal goals. Generally speaking, the goals of Scandinavian Link are economic and quantitative, concerned with economic growth, rationalization and efficiency, large-scale transport, and belief in the market mechanism with a minimum of public intervention and restrictions. In addition, the transport user should have a free choice between road and rail transport (and pay accordingly), and the Nordic countries should be integrated into the EC.

These attitudes are marked by the interests of the participating trades and industries in an improved competitive position in the European market through a better transportation system. Ecological and environmental interests are recognized, but as secondary aspects with some public relations value.

Having succeeded in establishing a large political majority and support from trades and industries, employers and trade unions for the Great Belt Link and the Sound Link, Scandinavian Link dissolved its secretariat in Copenhagen. The job was done; politicians had taken over, following the Round Table/Scandinavian Link concept.

13.5 THE COMMON CONTEXT OF EUROPEAN INDUSTRIALISTS AND SCANDINAVIAN LINK: THE SINGLE MARKET OF THE EUROPEAN COMMUNITIES

The 'historic decision' of the EC Commission in 1985 to establish a single interior market comprising the 12 member countries was the starting signal to extensive studies under the leadership of the Italian economist Paolo Cecchini on the realization of the Single Market from 1992 and its economic consequences. The resulting report in 1988 (CEC 1988) in 16 large volumes investigated the economy of an EC market of that type, with free interchange of goods, services, people, capital and establishment of business firms, and with complete elimination of the frontier lines between the member states.

A prerequisite for a development along these lines is that the business leaders of western Europe must integrate with the political leaders, and that national governments must refrain from limiting trade with goods not observing essential claims from national states. National prescriptions and standards, 'especially on the protection of health, safety and the environment' are explicitly mentioned as obstacles to the realization of the Single Market, because they distort competition and cause enormous costs to trades and industries, predominantly to the car-producing industry. Other obstacles are differences in taxation (VAT, taxes on cars, cigarettes, wines and alcohol, etc.), the suspicion from national tax agencies towards multinational companies, discriminating impediments against imports, and social security elements in labour market legislation.

The point for goods transportation in EC member countries is to substitute national products by imported EC products (through increased specialization) and to substitute third countries' products with EC products. Both of these endeavours will tend to increase European goods transportation by road and rail (and possibly to reduce overseas maritime transport).

The general philosophy of the Cecchini Report is based upon a firm belief in the market mechanism, in large-scale management and specialization; in a consensus of interests between all classes in society; in the necessity to involve business leaders in political decision-making; in continued mastery of free competition (despite semi-monopolistic markets) without national state intervention; and in harmonization of conditions for production, trade, labour and capital markets, etc., within the EC.

These goals have been further stressed by the Commission (Group Transport 2000 Plus 1990) and in the Maastricht Treaty (European Communities, 1992) and in the Treaty widened to taxes and finance, currency and banking, social policy, education, foreign policy and defence, etc., by the plans for converting the EC into an economic, monetary, political and military Union. The transport infrastructure plans for high-speed trains, motorways, bridges and tunnels crossing the Sound and the Fehmarn Belt are high-priority plans for establishing superior trans-European road and rail networks of the type emphasized in the Maastricht Treaty.

13.6   A POLITOLOGICAL THEORY OF ATTITUDES AND POWERS IN THE DECISION-MAKING PROCESS

Comparing the more general goals underlying the practical, operational goals (on infrastructure investments, removal of frontier barriers, integration of business top leaders in political decision-making, etc.) of the EC plans for the Single Market, the Round Table of European Industrialists and Scandinavian Link make it obvious that they have a lot in common. There are nuances in the emphasis of some issues; but they are all strongly economic growth oriented, and they follow traditional lines of conservative–liberal politics with some concessions, especially for the Scandinavian Link, to environmental and safety aspects, and to regional considerations.

The primary goals described in the preceding sections for the three actors mentioned at the European scene are of an economic nature and can generally be measured and compared in money terms, using evaluation techniques of the cost/benefit type, such as calculations of discounted net capital value, internal real rate of interest, and break-even traffic levels. The calculations include some non-economic or social aspects (advantages or disadvantages) in comprehensive evaluations, trying in different ways to transfer these into economic terms. It seems that these methods have not proved convincing, as the values of a number of qualities and nuisances are largely a matter of political judgment. The cost/benefit methodology has been stretched far beyond what is comprehensible and scientifically acceptable.

Concepts like income redistribution, equity, employment and especially ecology and environmental quality are theoretically, as well as practically, impossible to translate into money values in a way acceptable to everybody. Briefly expressed, the conflict between efficiency goals and equity goals, and between accessibility and environment can only be solved politically.

Politically, the opposite of *right-wing (conservative) politics* is different degrees of *socialist and social reformatory left-wing politics* with *liberal politics* in between, but closer to right wing. As far as transportation is concerned, right-wing policies put emphasis on the transport user's free choice between modes of transport; economic efficiency, including time savings; private car driving and private road hauliers, reduced taxes on cars and petrol, etc. Left-wing policies usually restrict private car driving and parking and give preference to rail transport compared to road transport, put a heavy weight on equity and equality considerations, accept public (semi)monopolies in transportation, maintain or increase taxes on cars and petrol, etc.

Scientific methods of transport economics, energy theory, ecology and environmentalism can secure a certain consistency of goals and elucidate consequences of projects and policies; but they can never identify objectively true or false goals.

Right- and left-wing politicians may, however, well agree on putting the main emphasis on economic growth, efficiency, centralized decisions and large production units—but probably with differing views on the distribution of wealth, the level of employment and the size of the public sector.

However, during the last decades a new dimension of conflicting ideological attitudes have become topical: the conflict between economic growth attitudes with emphasis on efficiency, centralization, material wealth, hard technologies, expert steering and representative democracy decisions on the one hand, and 'green attitudes' with emphasis on ecology, environment, decentralization, non-materialistic qualities of life, soft technologies, public participation and self-management on the other hand. This two-dimensional quality of political/ideological attitudes is illustrated in Figure 13.2.

Economic growth attitudes may be right wing (conservatives, christian democrats, etc.), centre (liberals, most democrats, green parties) or left wing (socialists, communists, etc.), and correspondingly, a 'green attitude' may be combined with a conservative, a liberal, or a socialist/social reform view. A liberal attitude is in some, but not all, countries combined with some ecological and environmental concerns; and green politicians often oppose socialism as well as (hard-boiled) capitalism, placing themselves in between, near the *Y*-axis in the diagram.

The EU, the European Round Table and the Scandinavian Link consortium are in the 'north-east quadrant' of the diagram: economic-growth oriented and right wing.

In collective decision-making on issues of a political nature a hierarchy of powers exists between the different actors at the economic and political scene of society—depending, of course, on the type of society, the cultural, political and religious traditions, social classes, etc. For most advanced industrialized countries, including those of west Europe, the following power layers can be identified, as illustrated in Figure 13.3. The diagram starts with the strongest power layers and continues downward to the weakest or powerless.

278

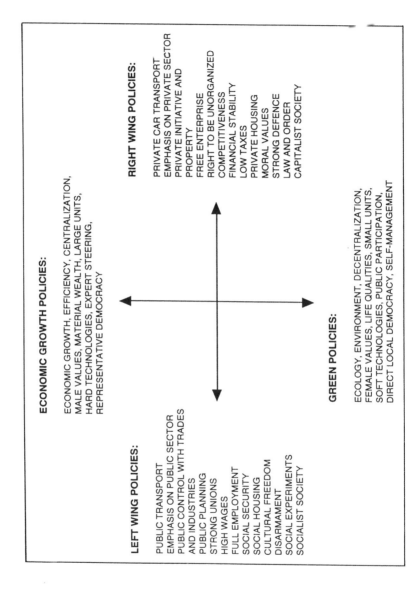

Figure 13.2. The two-dimensional politological diagram

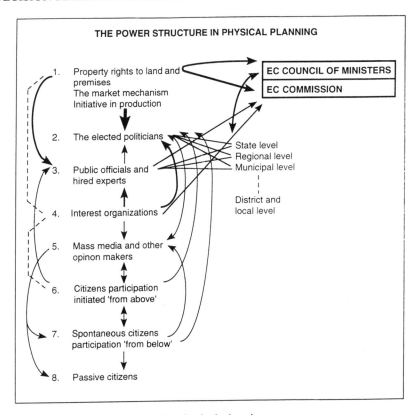

Figure 13.3. The power structure in physical planning

These layers of power groups are, however, in several ways combined, interrelated, and influencing each other. Fat arrows indicate typical strong influences, for instance from property owners and business leaders (layer 1) and interest organizations (layer 4) on elected politicians (layer 2) and public officials (layer 3). Other influences are weaker; some may go both ways in the decision-making process as between elected politicians and mass media (layer 5).

Politicians and public officials operate at several levels, from federal (in some countries) or state level to regional and local (municipal) level

For EU member countries the European Union acts at supranational level, over and above the (nationally) elected politicians. In the diagram EU is (for simplicity) represented by the Council of Ministers and the Commission, being the primary decision-makers, even if other EU institutions are relevant in certain matters.

European Round Table and Scandinavian Link can be described as belonging to power layer 1. They are primarily—in cooperation with the EU

Commission—concerned with influencing politicians (power layer 2), partly through public officials (3), and have close contacts with organizations (4). Only secondarily, when projects become political issues and, therefore, publicly known, does the Scandinavian Link become interested in mass media (5) and through them (and directly) relevant groups of the population (6, 7 and 8).

A consequence of this politological approach with the power structure and the diagram of ideologies is that the very idea of looking for 'the right evaluation' or just 'the best evaluation' of the impacts of a transport project must be rejected. Instead, we must accept that evaluations are not only complicated and burdened by uncertainties, but also fundamentally dependent on conflicting interests and attitudes. So, when analysing benefits and costs of a transportation project or (alternative) networks we must always ask: to the benefit of whom? and at the cost of whom? Unfortunately, these questions have not generally been put in the public reports on the three European links in Denmark.

13.7 CONCLUSIONS ON THE DECISION-MAKING PROCESS

The case of the large bridges in Denmark is a textbook example of concrete confrontations between in, part, right-wing and left-wing politics, but primarily economic growth versus environmental attitudes.

The original political situation in Denmark was that a minority of conservatives and a small centre party were in favour of the large bridges, constructed as combined road and rail connections, while a conservative-liberal party with farmer roots was inclined to resist the bridges as being primarily a Copenhagen interest, and some small centre parties, social democrats and left-wing parties preferred a bored railway tunnel—or continued ferry services.

The Scandinavian Link lobbyists decided, in their strategy to establish a political majority in Denmark, in favour of combined road and rail connections crossing the Sound and Fehmarn Belt, using their efforts and money on propaganda for the Great Belt Bridge in which they had only a minor interest. They succeeded in persuading the conservative-liberal party by appeal to its positive interest in the effective functioning of the EC Single Market. The social democrats were persuaded to drop their resistance against the motorway part of the fixed connections, partly through pressure from trade union leaders claiming employment at large-scale construction works and by the alleged creation of extra economic growth by the existence of fixed connections, partly by arguing that pollution by motor vehicle traffic would be of minor importance and be outweighed by the disappearance of ferry pollution.

In this work Scandinavian Link was heavily supported by the Danish motor lobby, consisting of car and truck organizations, oil companies, motor vehicle importers and repair workshops and trade unions interested in employment effects. That left only two small centre parties, the political left wing and 'greens', as opponents to combined fixed connections, i.e. a rather small minority in the Danish Folketing (Parliament). Having won the Great Belt battle, the next step was to argue that the combined Sound bridge was a logical consequence of the first step. The Fehmarn Belt connection was still held back from public debate.

This politological and sociological process may be illustrated by Figures 13.4 and 13.5, indicating the usual law-making procedure in Denmark and the converted law-making process for the large combined bridges.

This usual procedure described in Figure 13.4 was, by the combined forces of the European Round Table, the EC Commission, Scandinavian Link, the Danish motor lobby and powerful trade union top leaders, converted into the procedure described in Figure 13.5.

Using the Sound Bridge plan as an example the practical methods used to manipulate the political decision to build the combined Sound Bridge were the following:

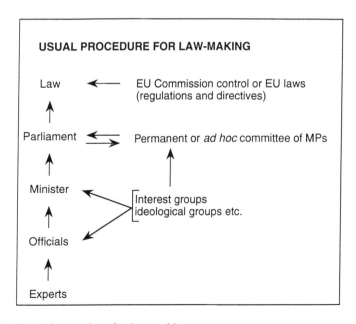

Figure 13.4. Usual procedure for law-making

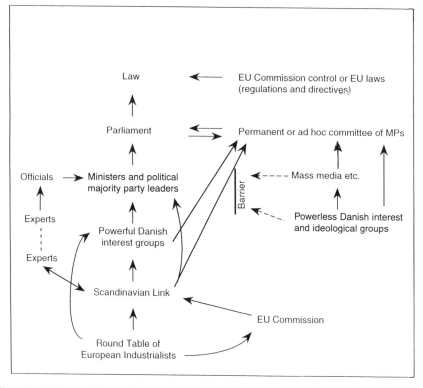

Figure 13.5. Actual law-making on trans-European links through Denmark

Methods used in the decision-making process in Denmark on trans-European links through Denmark

1. Lobbying leading politicians and trade union leaders.
2. Research-like reports with help from 'borrowed' public officials.
3. Appeal to national pride, presumed growth effects and competitiveness, employment effects, etc.
4. Party discipline in the Danish Parliament.
5. Under-evaluation of construction costs. Start with 'discount design', later introduce higher and more expensive qualities.
6. Political 'guarantee' that a combined bridge could be fully paid by the users, without any state subsidies.
7. 'Optimistic' traffic prognoses, later further increased when necessary to 'secure' economic feasibility.
8. Under-evaluation of environmental damages, later technical changes to reduce maritime damages.
9. Promises in general terms that 'no environmental damage will occur'.

10. In Denmark warnings that Sweden would never accept a railway connection without motorway.
11. Omission of full cost calculations for a bored railway tunnel for trains with motor vehicle wagons.
12. Time pressure to adopt law 'before political impetus fades away'.
13. Speeding up of expropriations and destruction of buildings necessary for the construction of landing strips and motorway routes.
14. Efforts to reach the 'point of no return', before further doubts or environmental claims were raised.
15. Denial of the existence of critical notes and corrections, alternative calculations, etc.
16. Reference to a secret Swedish Railways report, later destroyed (!), claiming a bored railway tunnel to be uneconomical.
17. Delay of information to organizations, the press and opposition politicians.

Result. A political majority for bored railway tunnels was converted into a majority for combined motorway/railway bridges.

During the prolonged period of public debate a series of official Danish and Swedish reports has been published (Miljøministeriet, Ministeriet for offentlige Arbejder, De danske og svenske Øresundsdelegationer, Storebælts-forbindelsen, several reports 1983 through 1993) constantly changing a number of presumptions and conditions. Later calculations have increased construction costs coast to coast and on land, new problems of maritime damages in the Baltic Sea and in the Sound have emerged, assumptions of the value of time savings for car and truck drivers have been attacked, and so has the underlying belief in a resulting extraordinary economic growth in the Greater Copenhagen region and South Scania in Sweden.

Consequently, the Sound Consortium and the Danish and Swedish authorities have made new, more 'optimistic' prognoses for car, bus and truckdriving; the firmly expressed principle of full user cost payment with no state subsidies has been modified, at first for construction costs, later more generally. The presupposed real rate of interest has gradually been reduced from 7 per cent to 5 per cent and further to 4 per cent over the 30 years' repayment period in spite of increasing rates in 1994. Also, the principle of motor vehicle payment corresponding to existing ferry fares Elsinore–Helsingborg minus expenses by own driving (= 160 DKK) seems now to be questioned. It will restrict car traffic to a fraction of the technical capacity of the bridge and make full cost coverage impossible, while a fare of 40 DKK would increase car traffic from a calculated 3 million (originally 2.3 million) to 11 million or more and secure a fine financial return. The aim of these changes has been to demonstrate a firm official belief in the profitability of the investment.

In May 1994, Danish and Swedish Greenpeace presented an elaborated project for a bored railway tunnel with trains for motor vehicles between Copenhagen–Malmoe and applied to the two states for permission to construct the tunnel.

Concerning the assumptions (1)–(5) mentioned in Section 13.2, the following conclusions can be drawn:

1. No proof or probability has been given that the Great Belt Bridge and the Sound Bridge constitute the most economical use of resources in transport policy.
2. The theory that large transport investments create considerable economic growth remains a postulate.
3. The public reports have not made it probable that economic calculations of a cost/benefit type constitute objective comparisons of alternative transport infrastructure investments.
4. National 'flagship' (prestige) investments like the Great Belt Bridge are given highest priority by politicians by 'choosing to ignore the economy' (citation of a former transport minister).
5. The real decisions are not made in the Danish Parliament, but by powerful inter-European companies, combining their efforts in their European Round Table with the Danish motor lobby and Danish trade union interests acting together as the dominant economic and political actors.

REFERENCES

Arbetsgruppen för utvidgat ekonomisk samarbete i Norden (Working Group for Increased Nordic Economic Cooperation) (1995) *Norden finns (The North exists)*.

Commission of the European Communities (CEC) (1988) *The Realisation of the Single Market*, Brussels (chaired by Paolo Cecchini).

De danske og svenske Øresundsdelegationer (Danish and Swedish Sound Delegations) (1985, 1987, 1989) *Faste Øresundsforbindelser (Fixed Sound Links)*.

European Communities (1992) *Treaty on the European Union*.

Group Transport 2000 Plus (EC Commission) (1990) *Transport in a Fast Changing Europe*.

Miljøministeriet (Ministry of the Environment) (1983) *Trafikforbindelser mellem Øst- og Vestdanmark—en landsplanmæssig vurdering (Traffic Connections between East and West Denmark—a National Planning Evaluation)*. *Øresundsregionen—en europæisk storby (The Sound Region—a European Metropol)*, Miljøministeriet (Ministry of the Environment) (1993) (In coop. with the Malmöhus County Council, Sweden).

Ministeriet for offentlige Arbejder (Ministry of Public Works, now Ministry of Transport) (1982–83) *Trafikforbindelser mellem Øst- og Vestdanmark* (3 volumes).

Ministeriet for offentlige Arbejder (Ministry of Public Works, now Ministry of Transport) (1985) *85-rapporten om Storebælt* (The '85 Report on Great Belt).

Ministeriet for offentlige Arbejder (Ministry of Public Works, now Ministry of Transport (1985) *En fast forbindelse. Ejerforhold, finansiering, beskæftigelse, beta-*

lingsbalance og rentabilitet (A Fixed Connection: Ownership, Financing, Employment, Balance of Foreign Payments, Profitability).

Round Table of European Industrialists (1984) *Missing Links.*

Round Table of European Industrialists(1988) *Needs for Renewing Transport Infrastructure in Europe.*

Scandinavian Link (1987) *Transport 1988–2000. Huvudrapport om Scandinavian Link (Main Report of S.L.)*

Storebæltsforbindelsen A/S (Great Belt Connection) (1990) *Fixed Links across Sound and the Fehmarn Belt—Traffic and Economy.*

Trafikministeriet og Miljøministeriet (1991, 1992, 1993) *Miljø Øresund (Environment).*

14 In Search of Sustainable Transport Systems

PETER NIJKAMP and JAAP VLEUGEL
Free University of Amsterdam, The Netherlands

14.1 INTRODUCTION

On a world-wide scale, transport by road and air have grown rapidly and are projected to continue this growth rate in the coming decades. There is also a broad recognition that further mobility growth would become unacceptable, as transport and mobility are among the major sources of environmental degradation. And it is therefore understandable that in recent years intense debates have started on the question of how to curb the environmental stress from our mobile network society (see also European Commission 1992). In principle, several options can be envisaged in order to alleviate the external costs of modern transport systems: moral conviction, strict regulations (and enforcement thereof), user charge principles (e.g. road pricing, Pigovian taxation), sophisticated environmentally benign technologies (e.g. route guidance, zero-emission cars) and alternative modes of physical planning (e.g. compact city design). Whatever the ultimate options chosen may be, it is clear that any reduction target in environmental stress has to be assessed from both an environmental sustainability viewpoint and from a cost-effectiveness viewpoint (cf. Tisdell 1991). Such an assessment may be based on evaluation criteria that are internal in the transport system or on criteria that mirror an overall systemic efficiency and sustainability. This provokes the question of the most appropriate level of reduction of environmental pollution by the transport sector vis-à-vis other economic sectors.

The nowadays popular notion of *sustainable development* (see World Commission on Environment and Development 1987) does not offer quantitative criteria which would guide policy-makers and planners in their efforts to reach a process of change in which resource use, productive investments, technological development and institutional changes are brought into harmony with one another. Consequently, a policy strategy aiming at a more sustainable transport system has to identify quantitative criteria which would offer guidelines on the maximum allowable contribution to environmental degradation by the transport sector (cf. Verhoef and Van den Bergh

European Transport and Communications Networks: Policy Evolution and Change. Edited by David Banister, Roberta Capello and Peter Nijkamp. © 1995 John Wiley & Sons Ltd.

1994; Vleugel 1994). This presupposes knowledge on the total permissible pollution in a given area and in a given time frame, as well as knowledge of the share of the transport system in this total volume of pollution (for different emittents). The aim of this chapter is to develop some thoughts on the question of identifying the maximum allowable pollution share by the transport sector, assuming a critical level of maximum resource use, a maximum carrying capacity, a maximum environmental utilization space, a maximum sustainable yield or some other critical threshold level for environmental decay. We will use here the notion of *maximum environmental capacity use* (MECU) to indicate the maximum resource use of a given environmental capital stock that, in a given time period, is compatible with both socio-economic objectives and environmental quality conditions now and in the future. The question we want to answer in this chapter is whether at the sectoral level, viz. that of the *transport system*, the idea of a transport-specific MECU, denoted here as MECUTS, can be operationalized. We will look in particular at alternative new possibilities, with particular emphasis on economic backgrounds, to control pollution emission in the transport sector. Before commencing this task, in the next section we will first offer some background information on the issue of sustainable development in the context of transport systems.

14.2   ENVIRONMENTAL STRESS: BACKGROUNDS

Most human activities (including transport and mobility) generate environmental stress. Often the nature or the size of this effect does not cause serious harm, but in our modern network economy the spatial interaction in terms of physical movement of people and goods has risen to an unprecedented degree, with the result that modern transport systems are causing severe environmental degradation. This holds for all modes of transport, although some modes—notably road transport—cause significantly more damage (see for an illustrative overview Tables 14.1 and 14.2).

It is clear from Tables 14.1 and 14.2 that each transport mode—be it private or public—contributes to environmental degradation. There is no environmentally benign transport activity, but some modes are less harmful than others.

A peculiar feature of transport is its *integrating* function in a network economy: transport is a derived activity which is dependent on and determined by all other economic activities. Besides, transport, and its necessary infrastructure, has also a structuring impact on the space-economy. Consequently, the assessment of environmental impacts of the transport sector is not an easy task, as transport is only a small part of a long chain of

Table 14.1. Important environmental effects by transport mode

Topic	Road	Rail	Air	Water
Climate change[1]	×	×	×	×
Squander[2]	×	×	×	×
Acidification[3]	×	×	×	×
Diffusion[4]	×			×
Spatial needs and intersection[5]	×	×		
External safety[6]	×	×		
Smell[6]	×		×	
Noise[6]	×	×	×	
Degradation of built environment[6]	×			
Aridification[7]	×			
Removal[8]	×			

Source: Van Wee et al (1992), Table 2.1.1 (modified).
[1]Contribution to the greenhouse effect because of energy consumption and the reduction of the ozone layer by CFCs in cooling and painting.
[2]E.g. inefficient use of fossil fuels and (rare) metals.
[3]Including emissions by power stations.
[4]Especially toxic chemicals like cadmium, asbestos and lead.
[5]For infrastructure, vehicle use and parking.
[6]Disturbance and damage to non-users.
[7]Infrastructure construction consumes large amounts of sand and other filling materials, which drastically changes the landscape, as large holes remain, which are increasingly turned into lakes.
[8]Vehicle bodies, tyres, etc.

subsequent activities. This chain can be broken down as follows:

1. production (including delivery of raw materials) and sales of goods;
2. ownership and use of transport vehicles:
   (a) individual modes
   (b) collective modes
   (c) integrated multimodal transport systems
   (d) logistic chains and structures;
3. depreciation and scrapping policy for old vehicles;
4. extraction of raw materials for physical infrastructure;
5. construction and maintenance of infrastructure.

Effective environmental policy in the transport sector presupposes sufficient insight into the driving forces of human activities, in terms of both nature and size in our 'post-Fordist' way of living. Two main categories are (Opschoor 1992):

1. *Material* processes, such as economic progress or technological innovations. These processes also include 'counter production' serving to alle-

Table 14.2. Contribution of the transport sector to environmental pollution in The Netherlands

Source Data 1985	Road	Mobile, other	Mobile, total	Other	Total
CO_2(Mtonne)	21	5	26	156	182
NO_x (ktonne)	262	79	341	211	552
SO_2 (ktonne)	11	23	34	237	271
NH_3 (ktonne)	0	0	0	285	285
H^+ (10^9 mol)	5.0	2.1	7.2	27.9	35
VOS (ktonne)	221	10	231	277	508
CO (ktonne)	923	42	965	356	1321
Lead (tonne)	1180	0	1180	223	1403
Particulates (tonne)	17	9	26	68	94
Waste (Mtonne)	0.6	0.02	0.6	48	49
Oil (ktonne)	35	8	43	59	102
Wrecks (ktonne)[1]	510[1]	p.m.	510 + p.m.	0	510 + p.m.
Tyres (ktonne)	61	12	73	0	73
Noise	59/19[2]				
Smell	61[3]				

Source: Van Wee et al (1992), Table 2.1.1a.
[1] 1986 data.
[2] Questionnaire data: percentage of people (severely) hindered.
[3] Percentage of people hindered (data 1989).

viate environmental degradation caused by the economy (e.g. catalytic converters, antinoise screens, etc.). Given this bias in our GNP composition and definition, the demand for a green GNP is conceivable.

2. *Structural* factors, such as institutional factors (e.g. the social security system) and ideological motives (e.g. belief in economic progress). These backgrounds often lead to a situation where a choice in favour of more polluting, rather than less polluting, activities is made, as is also witnessed by the popularity of individual transport modes.

In our modern society an in-built automatism to favour less sustainable activities, and hence also less sustainable modes of transport, appears to exist. This can be explained from three background factors in a network society which will now be concisely discussed: *separation, market failure* and *policy inertia*.

Separation in a network means a lack of direct behavioural connectivity among actors, so that causes and impacts show a distance in terms of *scale* (e.g. micro versus macro interests), *time* (e.g. current versus future generations) and *geographical space* (e.g. the NIMBY phenomenon). Separation in the transport sector takes place, since:

1. Individual contributions to environmental pollution (e.g. the use of a single car) are usually only marginal.
2. The link from a global (or macro) environmental problem to individual behavioural adjustments is not straightforward.
3. Behavioural changes by (a group of) individuals will discourage other (groups of) individuals to adopt the same strategy (the so-called *prisoner's dilemma*).

Consequently, most actors in the transport sector follow a *free rider* strategy. This means that voluntary behavioural changes are not likely to take place, so that rational individual behaviour appears to provoke the so-called *tragedy of the commons* (Hardin 1968). This intrinsic conflict between individual (or group) interest and collective (or macro) values is often referred to as the *social dilemma*.

Market failure refers to the lack of a proper price system for market transactions. Since environmental goods such as fresh air or quiet do not have an unambiguous value mirrored in a commonly accepted and paid for price level, a distortion in choice behaviour takes place. Such externalities lead to underpriced goods and cause a higher level of use than would be desirable from a social perspective. This phenomenon of market failure in the transport sector has provoked recent debates on 'user charge' and 'road pricing' principles. It has to be admitted that a policy for coping with market failures is a sound strategy from an economic viewpoint, but in practice many impediments appear to exist (e.g. lack of information, distributional effects, etc.). A survey of arguments can be found in Emmerink et al (1994).

Finally, *policy inertia* has to be mentioned. The existence of market failures has generated a wide array of policy initiatives in order to alleviate the associated external effects. In many cases however, such initiatives were uncoordinated or insufficiently based on reliable information on behavioural responses of travellers. Examples are incompatible combinations of regulations, subsidies and charges in the transport sector. As a result, an improvement did not take place, or did so unsufficiently, so that then a case of so-called *government or intervention* failures emerged (see e.g. Barde and Button 1990). In Table 14.3 a concise overview of various failure processes in the transport sector is given.

However, the problem is that, in view of government interests in public transport, a 'clean' economic framework for the transport sector as a whole is hard to identify, so that a frame of reference for effective environmental policies for the entire transport sector is badly missing. As a result, we observe a great deal of inertia in policy initiatives, so that environmental management in this sector is characterized by second-best or even third-best strategies. Sustainable transport policies are therefore extremely rare.

Table 14.3. Actors and failure processes in transportation

Process	Specific processes		
Government failure	Policy aims Instruments: legal, prices	Policy inertia Separation	Misjudgements, (awareness of) underestimation of environmental concerns vis-à-vis competing concerns
Market failure: public transport operators	Decisions regarding investments, cooperation, competition, product quality	Separation Market failure	Unwillingness or inability to supply adequate alternatives to the private car Externalisation of costs
Market failure: transport user	Choice of mode, mobility level, etc.	Separation Market failure	Few people choose environmentally more benign transport modes Externalization of costs

14.3 MAXIMUM ENVIRONMENTAL CAPACITY USE OF THE TRANSPORT SECTOR (MECUTS)

There is no doubt of the need to reduce environmental emissions in the transport sector. Table 14.4 contains—in summary form—the results of three distinct Dutch scenarios for the year 2010, followed by the required reduction in emissions in the same year. These figures illustrate clearly that the accepted environmental standards will hardly be met in the year 2010.

Next, in Tables 14.5 and 14.6 the necessary reductions in emissions by the transport sector, both passenger transport (Table 14.5) and goods transport (Table 14.6), implied by formulated policies are given. These figures cover transport by road, rail and air. It is clear from this information that the air pollution by road transport has to be reduced significantly, while at the same time personal mobility also has to decline. This target is hard to reach as car ownership is forecast to grow from 5.5 million to 7 or 8 million cars. Finally, substitution from road to rail is foreseen. Such goals can of course only be achieved if these policies are supported by progress in transport technology and new ways of physical planning, etc.

Before addressing in greater detail the issue of (the identification and measurement of) MECUTS, we will first pay some more attention to MECU in general. Environmental capacity has to be interpreted in terms of functions provided by environment and nature (cf. Costanza and Daly 1992, Turner 1993). Such functions concern in particular:

Table 14.4. Emission reduction scenarios for the Netherlands; All sources

	Reductions expected in 2010 (%)[1]			Reductions needed in 2010[1]
	Scenario I	Scenario II	Scenario III	
CO_2	+35	+35	−20 to 30	−20 to 30
SO_2	−50	−75	−80 to 90	−80 to 90
NO_x	−10	−60	−70 to 80	−80 to 90
NH_3	−33	−70	−80	−80 to 90
HC	−20	−50	−70 to 80	−80
CFCs	−100	−100	−100	−100
Effluents into river Rhine and North Sea	−50	−75	−75	−75 to 90
Waste dumping	0	−50	−70 to 80	−80 to 90
Noise severe[2]	+50	0	−15	−70 to 90
Smell[2]	+10	−50	−60	−70 to 90

Source: Ministerie VROM et al (1989), p. 107, Table 5.1.1.
[1]Percentage reduction with respect to 1985 levels.
Scenario I = current (so-called NMP-1) environmental policy package.
Scenario II = maximum use of all conceivable policy measures.
Scenario III = mix of measures aimed at emission reduction and source related meaures.
[2]Idem, in terms of number of people hindered.

Table 14.5. Emission reduction and mobility targets for passenger transport according to recent Dutch policy memoranda

	1986	1994 (NMP/+)	2000	2010
Levels				
NO_x[1]	163	100	40 (−75%)	40
HC[1]	136		35 (−75%)	35
CO_2[1]	23.000	26.400	23.000	20.700
Noise:				
dB(A)[2]	80		74	70
number of hindered severely) %[3]	61% (20%)		10–15%	0
Index				
Spatial needs[4]	100	100	[5]	[5]
Foreseen mobility level with no policy changes (index)	100	124	140	172
SVV-II policy		120	126	156
SVV-II + NMP-policy		117	120	148
NMP-2 policy			130	135

Source: Ministerie VROM et al (1989), pp. 194–196, 205; Ministerie VROM et al (1990), p. 50; Ministerie VROM et al (1993), p. 142 and 146; and RIVM (1993), p. 22 (note 5).
[1]NO_x, HC and CO_2 in kilotonnes per year.
[2]peak level dB(A) per car.
[3]1990 data.
[4]index 1986 = 100, no targets after the year 2000.
[5]no further intersection in rural areas and compensation elsewhere if new links are still needed. These targets are hard to reach where car ownership is forecasted to grow from 5.5 million to 7 or 8 million cars; 'standstill' seems therefore illusory.

Table 14.6. Emission reduction and mobility targets for freight transport according to recent Dutch policy memoranda

	1986	2000	2010
NO_x[1]	122	72 (−35%)	25 (−75%)
HC[1]	46	30 (−35%	12 (−75%)
Noise[2]	81–80	75–80	70
Mobility	100		140[3]

Source: Ministerie VROM et al (1989), pp. 195–196, 205; Ministerie VROM et al (1990), p. 50.
[1]NO_x and HC in kilotonnes per year.
[2]In dB(A) per bus/lorry.
[3]Wishful thinking, as in NMP (p. 194) the foreseen growth of mobility is 70–80%.

1. *economic* functions (e.g. a resource base, recreational opportunities);
2. *ecological* functions (e.g. regulation, absorption, signalling);
3. *cultural* functions (e.g. landscape beauty, educational values).

The question then is whether it is possible to establish a tolerance level of maximum use of such functions based on principles of environmental sustainability. It goes without saying that the identification of such critical threshold levels (e.g. in terms of carrying capacity, maximum load, maximum sustainable yield) is fraught with difficulties. Nevertheless, in policy practice we observe that governments are increasingly inclined— despite some degree of ambiguity and arbitrariness—to accept such critical threshold levels as signposts for environmental strategies, using the minimum regret principle (or any other meaningful risk principle) as a major justification. Consequently, notwithstanding some ranges of uncertainty, more and more countries define standards on carbon dioxide, nitrous oxides, methane and other greenhouse gases.

One may in general assume that such environmental target standards are fixed according to a level that keeps the existing stock of environmental resources more or less intact (see Pearce et al 1989). This then defines in operational terms the MECU in general, but it does not yet show the distributional consequences of such generic sustainability policies. Thus the sector-specific reduction of pollution is at stake here. In this framework various policy principles can be distinguished:

1. an equal percentage reduction of pollution in all sectors;
2. a sector-specific reduction in pollution that is proportional to the environmental stress caused by that sector;
3. a sector-specific reduction in pollution that is based on sectoral cost-effectiveness measures for pollution abatement;
4. a sector-specific reduction in pollution that is inversely proportional to the intensity of environmental measures taken by that sector in the past;

5. a sector-specific reduction in pollution that is proportional to the growth rate of its pollution in the past years.

Such distinctions are also loosely related to the concept of strong and weak sustainable development, depending on the question of whether overall environmental quality improvement is strived for or whether environmental quality decline in some sector or region can be compensated for by improvement elsewhere.

Clearly, in a dynamic economy technological progress may alleviate some of the problems inherent in the limits of a MECU, but this needs to be supported by behavioural and institutional responses (e.g. the potential offered by global computer networks).

The above observations apply to a large extent also to MECUTS, where two problems appear to emerge: (1) which is the share of the transport sector in overall reduction of pollutants; and (2) which is a share of subsectors within the transport system in terms of pollution reduction? The answer to such questions depends on various evaluation factors:

1. the nature and size of effluents by the transport sector;
2. the costs of emission reduction in the transport sector;
3. the modal split and composition of passengers and freight in the transport sector;
4. the spatial externalities caused by various transport activities;
5. the volume and distribution of renewable and non-renewable resources used by subsectors in the transport system.

In trying to assess a proper level of MECUTS, it has to be recognized that such levels may be site-specific due to lack of other choice (e.g. modal) opportunities in transport (Whitelegg 1993). For example, Newman and Kenworthy (1991) found in their comparative study on energy consumption and automobile dependence that land use intensity is in many cities the decisive factor. The more intensive urban land use, the shorter the travel distances and the higher the opportunities for public transit. Differences of 20–30 per cent in petrol consumption have been found between American and European cities. This suggests that urban land use intensity might also act as a principle for allocating a certain level of environmental utilization to segments of the transport sector. A related problem is of a distributional nature (see Sachs 1983): higher-income people appear to live in lower densities in general and hence are forced to use the car because of absence of profitable public transport in low-density areas. Thus, if quota systems based on the MECUTS notion were allocated to (segments of) the transport sector, it might mean that certain groups are disproportionately hit, which might lead to much public opposition. Clearly, the above remarks do not only apply to residential density, but also to industrial density, retail and service density, etc. (see also Kasanen and Savolainen 1993).

A specific problem is caused by the modal split of the transport sector. Once a level of MECUTS has been established and accepted, the next question is how much the share of rail systems vs. road systems vs. waterway systems, etc. would have to be. This would imply a confrontation of the performance of each transport mode (e.g. in terms of person kilometres or tonne kilometres) with the volume of effluents caused by it. Again, the same type of policy principles as mentioned above for distributing the emission constraints may in principle be used.

Clearly, at the very end one may even raise the question of how much individual vehicles would have to be affected in their operation. This brings us into the area of so-called *mobility quota*, which would mean that a maximum permissible distance to be driven by a vehicle in a certain period (or the maximum amount of fuel to be consumed by a car in a given period)—without any extra charge (in the form of a Pigovian tax)—would have to be established. Needless to say that theoretically such an option is comparable to quota systems in resource management, but this solution may lead to substantial problems regarding implementing and controlling such a system because of the 'large numbers' case in transport. However, in future this problem may be solved by smart technology.

Two types of approach may in principle be envisaged to solve the above choice problems. One would be to compose packages of policy measures (charges, technology solutions, prohibitions, etc.) and to evaluate the feasibility of each of them by means of a cost-effectiveness strategy, taking for granted a given level of MECUTS, established for example by a government or transport authority. Another approach would be to investigate different reduction scenarios in subsectors of the transport system and to evaluate such policy scenarios against the background of the above evaluation factors by means of multicriteria analysis. An intermediate approach might be to specify in advance achievement levels of environmental quality, so that a tension index can also be created for the discrepancy between actual and desirable performance. This index may then be used for assessing weight factors in a goals-achievement multicriteria analysis.

It is clear that the above issues require much more substantive research, notably (see also Vleugel 1994):

1. identification of relevant and proper MECUTS indicators;
2. fixation of acceptable critical thresholds levels;
3. measurement and monitoring of actual environmental performance of the transport (sub)sector(s);
4. assessment and evaluation of various MECUTS options (including spatial alternatives such as car-free cities, technology options such as telematics, and behavioural changes in mobility and life-style patterns);

5. cost and benefit assessment of alternative pollution abatement strategies in a MECUTS context.

In reality, policy decisions seem to be based on consensus formation incorporating compromise choices regarding ecologically sustainable development of various sectors, including the transport sector (cf. Nijkamp and Blaas 1994). This also means that effective policy strategies based on MECUTS guidelines are hard to find. This can also be explained by returning to the above-mentioned three impediments which caused the social dilemma.

The first problem, *separation*, can be coped with by voluntary individual internalization of social costs of environmental decay (e.g. less use of cars), by market internalization (e.g. Pigovian taxes), or by forced internalization (e.g. parking prohibitions). All such measures may have a substantial impact on sustainable development, but the public resistance and the policy inertia during several decades do not offer many hopeful perspectives, although recently the tides in Europe are changing. Examples are policies for car-free cities or even 'zero-emission cities'.

The second issue, *market failure*, may be dealt with by a system of standards that is compatible with MECUTS indicators. Here maximum load factors, maximum emission standards or quota for resource use may be introduced. In this context, advances in information and transport (systems) technology may provide new departures for more sustainable mobility (e.g. telematics, telecentres policies). However, in most cases neither mobility substitution nor reduction is achieved by means of telematics.

Finally, the problem of *policy reform* may be tackled by seeking institutional reforms, such as the organization of countervailing powers (e.g. public transport lobbies, international cooperation and agreements on environmentally benign transport modes).

In view of the social dilemma and the conflict between efficiency and equity, it seems that a 'packaging' of policies, with sufficient emphasis on financial incentives, but also with strict constraints whenever needed, offers the best opportunities.

14.4 CONCLUDING REMARKS

The above observations mean that in practice a fair balance has to be found between incentives and penalties, between economic and institutional (e.g. standards) measures, and between behavioural and technological responses. An interesting approach may be added here, in order to make sure that future generations would have sufficient options for environmental quality, it may be useful to think in terms of insurance strategies. The estimated annual insurance premiums that would have to be paid in order to guarantee such options would have to be attributed as additional social costs to be borne by the current economy (including the transport sector). It goes without saying that our paper on MECUTS has provoked more questions than answers, but it is clear that transportation science will need to intensify its research efforts to make sure that the transport sector will not continue to be a persistent unsustainable evil in a mobile network society.

REFERENCES

Barde, P. and K. Button (eds) (1990) *Transport Policy and the Environment, Six Case Studies*, Earthscan, London.
Costanza, R. and H. Daly (1992) Natural capital and sustainable development, *Conservation Biology* **6**(1), 47–63.
Emmerink, R., P. Nijkamp and P. Rietveld (1994) *How · Feasible is Congestion-Pricing?* Research paper, TI 94–62, Tinbergen Institute, Amsterdam.
European Commission (1992) *Green Paper on the Impact of Transport on the Environment*, Brussels.
Hardin, G. (1968) The tragedy of the commons, *Science*, **162**(13), 1243–1248.
Kasanen, P. and M. Sarolainen (1993) Changes in the structure of food retail and wholesale: effects on energy demand, *Papers in Regional Science*, **72**(4), 405–423.
Ministerie van Verkeer en Waterstaat (1990) *Tweede Structuurschema Verkeer en Vervoer, deel d: regeringsbeslissing*, TK 1989/1990, no. 20 922, nos. 15–16, Den Haag, SDU.
Ministerie VROM et al (1989) *Nationaal Milieubeleidsplan, Kiezen of verliezen*, TK 1988/89, no. 21 137, nos. 1–2, SDU Den Haag.
Ministerie VROM et al (1990) *Nationaal Milieubeleidsplan-plus*, TK 1989/1990, no. 21 137, nos. 20–21, SDU, Den Haag.
Ministerie VROM et al (1993) *Nationaal Milieubeleidsplan 2*, TK 1993/1994, no. 23 560, nos. 1–2, Den Haag SDU.
Newman, P. W. G. and J. R. Kenworthy (1991) *Cities and Automobile Dependence*, Gower, Aldershot.
Nijkamp, P. (ed.) (1993) *Europe on the Move*, Gower, Aldershot, UK.
Nijkamp, P. and E. Blaas (1994) *Impact Assessment and Evaluation in Transport Planning*, Kluwer, Boston/Dordrecht.
Opschoor, J. B. (1992) Sustainable development, the economic process and economic analysis, in J. B. Opschoor (ed.), *Environment, Economy and Sustainable Development*, Wolters-Noordhoff, pp. 25–52.
Pearce, D.W., A. Markandya and E. B. Barbier (1989) *Blueprint for a Green Economy*, London, Earthscan.
RIVM (1993) *Nationale Milieuverkenning 3, 1993–2015*, Bilthoven.
Sachs, W. (1983) Are energy-intensive life-images fading? *Journal of Economic Psychology*, **26**(3), 347–365.
Tisdell, C. (1991) *Economics of Environmental Conservation*, Elsevier, New York.
Turner, R. K. (1992) *Sustainable Environmental Economics and Management Principles and Practice*, Belhaven, London.
Verhoef, E. and J.C.J.M. van den Bergh (1994) *Transport and Environmental Sustainability*, Research paper, TI 94-63, Tinbergen Institute, Amsterdam.
Vleugel, J. M. (1994) *De Milieugebruiksruimte voor Verkeer en Vervoer, Verkenning*

(eerste fase), Dutch Railways, Division Corporate Development, Free University, ESI, Amsterdam.

Wee, G. P. van, and Waard, J. van der (eds) (1992) *Verkeer en Vervoer in de Nationale Milieuverkenning 2 1990–2010*, RIVM, Bilthoven.

Whitelegg, J. (1993) *Transport for a Sustainable Future*, Belhaven, London.

World Commission on Environment and Development (WCED) (1987) *Our Common Future* (Brundtland Report), Oxford Unviersity Press, Oxford.

15 European Telecommunications Policy

KENNETH BUTTON
Loughborough University, UK

15.1 INTRODUCTION

Liberalization and deregulation have been a recurrent theme of micro-economic policy since the passing of the Airline Deregulation Act 1978 in the USA (see Button and Swann 1989). This trend has gradually but inevitably caught up with the European telecommunications sector. An industry which has traditionally been characterized by state-owned monopolies is increasingly being privatized and processes of regulatory reform are producing quasi-competitive conditions. At one level this change is being brought about by the perceived need of many individual national governments to improve the efficiency of their own industries, but this process has also been spurred on by initiatives of the European Union. (To avoid confusion, the term European Union is used throughout, although many of the changes discussed occurred at times when other titles were relevant—for instance, European Community).

> The single market is a challenge. In the sphere of telecommunications, it will offer a new framework of regulation and management, on a higher level, favouring research and development and the establishment of high capacity networks which will be an essential dynamic element of our productive capacity (J. Delors in Ungerer and Costello 1988)

With the Single European market initiative completed and efforts at creating a single European currency stalled (if not aborted) there has been considerable interest in developing Europe's infrastructure. Transport and communications form a cornerstone of the strategy with, in the former case, the creation of concepts such as Trans-European Networks (TENs). In the telecommunications field there is also a perceived need for such a strategy, although with rather more emphasis on the regulation and control of the sector than with the strict provision of the hardware. The point was made in the Commission's 1993 White Paper:

> The diffusion of best practice aimed at business should be promoted and the development of Community-wide applications favoured. To this end, an appropriate regulatory and political environment should be created and the

European Transport and Communications Networks: Policy Evolution and Change. Edited by David Banister, Roberta Capello and Peter Nijkamp. © 1995 John Wiley & Sons Ltd.

implementation of trans-European telecommunication services stimulated. (Commission of the European Communities 1993)

The creation of such an appropriate operational environment raises a number of important questions. Change has already been forced on the sector as technological advances have put pressure on the long-standing institutional arrangements which have pertained in Europe. The matter, therefore, is really one of defining the need for further Union action to steer and modify developments which are already occurring and the extent to which such actions should be proactive rather than seeking to simply respond to national regulatory shifts which are moving against the broader interests of members. Underlying this must be some idea of both the preferred nature of telecommunications markets and the type of institutional arrangements which it is thought would bring this about. Tied to this is a more general question of whether telecommunications should be treated as an industry in its own right or whether it should be seen as an intermediate activity which can be manipulated to meet wider, social, spatial and industrial objectives.

The limited evidence that is emerging on this question is that, at least to date, telecommunications policy as deployed has not proved to be a potent tool in regional development. Capello (1994) provides an analysis of the implications of the STAR Programme for Southern Italy.

The subsequent discussion initially provides background information about the nature of telecommunications in Europe and highlights some of the national differences which exist. It then provides a brief examination of the types of market structure which recent commentators have thought to underlie the market for telecommunications services. It also offers a few thoughts on the difficulties of devising regulatory regimes which achieve the types of objectives which seem to be Union priorities. There is, finally, consideration of the actual policies which both individual Member States and the Union itself have been pursuing in recent years.

Since the main focus of the chapter is on the implications of telecommunications policy for economic development and spatial variation, some areas of policy are given very limited coverage. The protection of personal data and privacy and the interaction with industry policy are examples.

15.2 THE EUROPEAN TELECOMMUNICATIONS SYSTEM

The EU telecommunications industries amount to an annual market of about $84 billion (of a world market of $285 billion) with a further market of $26 billion for equipment. Current projections suggest an annual growth of about 8 per cent for services and 4 per cent for equipment to the year 2000. The sector, however, despite its perceived strategic importance for the future of the European Economy was until recently extraordinarily diverse in nature as between countries. Table 15.1 provides background data.

All systems involved a strong government presence but beyond that significant variations existed. Individual countries often had peculiar national standards and generally telecommunications were encompassed in monopolies embracing a range of other activities. In the cases of Germany, Spain, the UK and France, for example, there was a single supplier, but Italy had five suppliers divided along regional and functional lines. Countries such as Germany and France also used procurement policy to assist their telecommunications equipment industries, whereas Denmark had an internationally open procurement stance. Equally, in some instances, such as the UK and Germany, banking and postal services were combined with telecommunications and while most countries operated their telecommunications as a public enterprise, a small number (such as Spain) developed public–private corporations.

The different regulatory approaches found within Europe, and the accompanying variations in efficiency, technology and pricing which accompany them, have led to major differences in the services offered customers. In particular, the costs of telecommunication services can vary considerably between countries. By way of a broader international comparison, Rank Xerox Ltd's annual phone bill in Europe is triple the charges for a comparable volume in the US. Figure 15.1 provides some general indication of this with respect to industrial users but telecommunications is a multiproduct

Table 15.1. Some basic features of the major European telecommunications suppliers (1992)

Country	Exchange lines (million)	Productivity (employees per '000 lines)	Revenue (US$ billion)
Germany	35[1]	70[1]	30[1]
France	30	50	24
UK(BT)	26	61	24
Italy	24	59	17
Spain	14	57	12
Turkey	9[1]	96[1]	3[1]
Netherlands	8	51	6
Greece	4	64	1
Belgium	4[1]	63	4[1]
Denmark	3	59	3
Portugal	3	79[1]	2
Ireland	2	110	1

Source: Derived from Adonis (1993).
[1]Estimates.

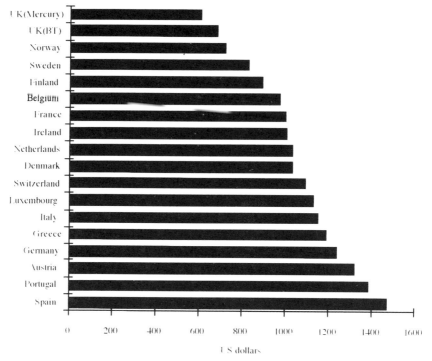

Figure 15.1. Monthly costs for a multinational company in the capital city (June 1994) (*Source*: Adonis 1994)

sector and the pattern of costs and the nature of the services offered is quite complex. Another indicator could well provide a somewhat different picture. There are signs, however, of considerably differing levels of productivity between the European suppliers (see again Table 15.1) and in the amount of investment going into the sector (Table 15.2). Perhaps the only solid fact which cannot easily be manipulated, however, is that overall the costs of telecommunications are falling in all European countries and for international telecommunication services.

15.3   BACKGROUND TO CHANGE

Major changes in policy seldom come about because of any single influence but are the result of a convergence of forces. In the context of European telecommunications this has certainly been the case. At one level there has been a significant shift in the way regulation is perceived. The traditional idea that regulation is to serve the public purpose and to counteract a range

Table 15.2. Percentage of GNP invested in telecommunications

Country	1981	1982	1983	1984	1985	1986	1987	1988
Italy	0.59	0.67	0.71	0.68	0.66	0.61	0.61	0.68
France	0.74	0.67	0.64	0.63	0.76	0.79	0.70	
Germany	0.73	0.75	0.72	0.78	0.85	0.82	0.82	
UK	0.57	0.54	0.51	0.58	0.56	0.57	0.58	

Source: Capello (1991)

of market failures (Kahn 1970) began to be seriously questioned in the 1970s. The notion of intervention failure and the upsurge of interest in free market economies in Chicago and the institutional insights offered by the public choice school of thinkers brought into doubt the universal desirability of regulation as a panacea to market imperfections and suggested that in many cases the cure was worse than the disease. In particular, asymmetric information theory began to cast doubt upon the ability of regulators to control markets where suppliers have control over information flows. Those who are supposed to be regulated in these circumstances have the ability to manipulate the system to their advantage. The objectivity of regulators may also be questioned in circumstances where their own future career may benefit from extending and making regulatory regimes more complex.

In itself these changes in the intellectual approach to regulation may not have brought about much change in a heavily regulated European tele-communications market but there were other factors involved. One of these was the development of contestability theory and the resurrection of nine-teenth century ideas, albeit in a much more rigorous framework, that poten-tial competition may be as potent as actual competition (Baumol et al 1982). Even in conditions where even potential entry might be seen as inadequate because of, for example, the existence of high sunk costs, new ideas on the way to regulate natural monopolies, especially in the context of rapid tech-nological change which posed serious problems for traditional rate-of-return regulation, began to emerge (Train 1991; Foster 1992).

Another, more practical, influence on the shaping of contemporary European telecommunications policy was what happened outside of Europe. North America, in particular, is often cited as providing one regulatory regime of interest. The US telecommunications market, while liberalized from the position pertaining until the early 1980s, has not, though, been fully deregulated, mainly, it is argued, because much of the market remains monopolistic. The actual *de facto* changes which have occurred, are however, relevant to the European situation. Until 1984 there was an effec-tive regulated quasi-monopoly situation in the US dominated by AT&T (Cowhey 1990). A similar story can be found with regard to Canada (Janisch 1989). The overriding concern had been with ensuring universal services by taking advantage of economies of scale and by adopting system-wide pricing which would maximize the availability of services (von Auw 1986). To this end, and after a period of regulation by the Interstate Commerce Commission, the Federal Communications Commission was established in 1934 to oversee rates, service facilities and construction certificates regulations. The long-standing private ownership-monopoly-regulation model was set in place.

The sea change in policy came in 1984 when it was decided that competition in non-local phone services was preferable to regulated telecommunications monopoly. How to introduce competition into local telecommunications in the US is still the subject of considerable debate—see, for example, Baumol and Sidak (1994) and Kahn and Taylor (1994). The practical result was the split of AT&T from the regional Bell operating companies (RBOCs) and the imposition on the latter of restrictions over the type of business open to them. Open access became a key element of the policy and AT&T, the RBOCs and others have to ensure equal access to all users of unbundled network features (e.g. billing information). Obligatory open network architecture ensures full disclosure to all service providers of the network design. The short-term impression gained from the US was that the reforms had not produced any diminution in service quality and that economic efficiency had been enhanced. As Horwitz (1989) has pointed out, however, this should have been taken in the context of the continuation of the old regulatory framework which in practice continued to operate. What also occurred, however, was a serious adverse imbalance of trade in telecommunications equipment, since other markets were still heavily regulated, and in the direction of calls (high prices elsewhere constraining calls into the US—see Table 15.3). The response of the US was to press for more open trade in equipment and services which, in turn, influenced EU policy via GATT negotiations.

15.4 EUROPEAN UNION POLICY

European Union telecommunications policy has been slow to evolve—for more detailed historical accounts of developments in EU policy see Bauer

Table 15.3. Telecommunications traffic to and from the US (1989)

Country	Traffic to US (A)	Traffic from US (B)	Ratio (A/B) (%)
Germany	192	419	46
Greece	20	52	38
Italy	71	146	49
Netherlands	48	62	77
UK	463	545	85

Source: Cited in Hurst (1991)

and Steinfeld (1994); Cowhey (1990); Mostershar (1993); Wilkinson (1989) and Noem (1992). Initially the procurement of telecommunications equipment was exempt from the general rules for public sector procurement, and throughout the 1970s the Commission attempted unsuccessfully to change this. The late 1970s, however, saw a change in policy as national governments, and especially that of France, became aware of the economic importance of telecommunications. The fragmented nature of the European market became more transparent and this weakened the national coalitions which existed between PTTs, domestic suppliers and users. The increased efficiency of non-European, and especially Asian, producers added pressure to this trend. More generally, the emerging experiences of the US in market liberalization (e.g. in aviation and public utilities) began to exert important demonstration effects at a time when 'stagflation' was a serious problem for European economies.

The Commission reacted by creating a Task Force in 1983 on information technology and telecommunications to develop an active policy approach and this was followed in the following year by the establishment of the Senior Officials Group on Telecommunications attended by representatives of telecommunications operators. In addition, a wide range of other forums was rapidly established to foster coordination, such as the Senior Officials Group on Information Technology Standards, the Senior Officials Advisory Group on the Information Market and special committees within the RACE and ESPRIT programmes. In the key area of standardization, the Technical Recommendations Applications Committee was set up in 1984 and the European Telecommunications Standards Institute in 1988.

In terms of policy, the Commission submitted its initial proposals to the Council in 1984. The pace over the following few years was rapid and by the end of 1988 over a dozen directives, decisions, regulations and recommendations had been adopted. Further, in 1986 the Task Force was merged with other departments in the European Commission to become the Directorate-General for Telecommunications, Information Industries and Innovation. Parallel to this, the Directorate General for Competition stepped up its efforts to introduce more competition into the sector. This action was reinforced by the European Court in its ruling that recommendations of the International Telegraph and Telephone Consultative Committee of the International Telecommunications Union were not legally binding.

In 1984, the Council of Ministers (Council of the European Communities 1984) issued recommendations concerning the implementation of harmonization in the field of telecommunications which formed the basis of the Green Paper (Commission of the European Communities 1988) on the development of telecommunications. By delineating the major characteristics of the future European telecommunications environment, the Green Paper identifies areas in which Union action is seen as being relevant. Member States are free to pursue their own policies in other areas as long as they remain compatible with the Treaty of Rome and the Single European Act. The main positions provide for the eventual attainment of the following:

1. A full liberalization of the supply of terminal equipment.
2. All restrictions on competition in value-added services are eliminated.
3. The Member States may specify exclusive or special rights for the provision of the network infrastructure by the telecommunications administrations for the exclusive provision of a restricted number of basic services as far as this is considered essential for the safeguarding of public service objectives.
4. To guarantee a level playing field in this mixed monopoly–competitive environment, there is a need for: Union-wide standards and inter-operability conditions; the development of common principles of open network provision; the separation of regulatory and operational activities of the telecommunications administrations; and provisions to monitor the competitive behaviour of public and private providers of telecommunications services.
5. The formulation of consistent Union positions in international negotiations as well as with third party countries.
6. Recommendations for the monitoring of the social impact of telecommunications developments.

A number of policy issues were to be at the discretion of national governments, e.g. the introduction of competitive conditions in the areas of network infrastructure and reserved services, the ownership form of the major telecommunications providers, and the specific regulatory arrangements to oversee the industry. Unlike the US situation, however, the Green Paper sees incumbent telecommunications authorities as potential competitors in every telecommunications market segment.

While the Green Paper is a legally non-binding communication by the Commission, it was followed in 1990 by the issuing by the Commission of its Services Directive which required the achievement by Member States of the following objectives:

1. Withdrawal of all special or exclusive rights for the supply of telecommunication services, other than voice telephony.
2. Authorizing simple resale services from 1993.
3. Licensing procedures for data services to be determined and notified to the Commission by 30 June 1992.
4. The establishment of objective and non-discriminatory access to networks.
5. The setting of conditions of use and charges for use of the public network to be non-discriminatory.
6. The regulatory function, including equipment type approval and allocation of frequencies, to be separated from the telecommunications organizations.

Some additional comments are justified on the implementation of the Green Paper. First, terminal equipment policy in the EU is based on the principle of open competitive provision subject to the fulfilment of essential requirements and the principle of mutual recognition of certificates stating the conformity of equipment with these essential requirements between Member States. Essential requirements here relate to: guaranteeing user safety, as well as the safety of public telecommunications network operators; electromagnetic compatibility requirements; protection of the public network from harm; effective use of the radiofrequency spectrum where appropriate; and the inter-working of terminal equipment with the public telecommunications network. The Council had issued a Directive on the initial stage of mutual recognition of type approval for telecommunications terminal equipment in 1986 which was repealed as of November 1992 on the approximation of the laws of the Member States concerning telecommunications equipment. In 1988, the Commission laid down the framework for the introduction of Europe-wide competition in the provision of terminal equipment. The Commission, assisted by a newly created advisory committee, the Approvals Committee for Terminal Equipment, identifies the type of terminal equipment for which common technical regulations are required. The respective technical standards are to be developed by the relevant standardization bodies and transferred into common technical regulations, which then become mandatory guidelines for all certification procedures. This simplifies the process of type approval. Whether it will fully open the terminal equipment market, though, will depend on the specific common technical requirements that are drafted.

Second, the policies of telecommunications service liberalization and Open Network Provision (ONP) are closely related. The ONP recognized that if infrastructure remained in the hands of monopoly providers then fair and non-discriminatory access would be needed if competitive non-reserved services other than those of the PTT were to exist. It can be seen as a necessary precondition to effectively liberalize the segment of non-reserved services and to provide a level playing field in an environment where the suppliers of the network infrastructure and reserved services are allowed to fully compete in the non-reserved services markets. Member States differ on such matters as the delimitation of reserved versus competitive services, the time schedule for service liberalization and the legal instrument for their implementation. The Commission argued in favour of open markets, but some members of the Telecommunications Council of Ministers placed more emphasis on public service goals.

The resulting compromise was the Services and Open Network Provision Directives of 1990 which confirmed the intention to fully open the value-added services markets for competition. For data communications services, however, it means that Member States with poorly developed data communication networks were granted the possibility of extending the deadline for the introduction of simple resale, until 1996. Member States were also allowed, under Commission scrutiny, to impose public service obligations such as quality and coverage requirements on providers of data communications services if they are in the general economic interest.

The solution to the services issue also enabled progress in the implementation of provisions for ONP, the basic principles of which are the opening and harmonization of conditions for access to the public network infrastructure and services for new service providers as well as for users. Harmonization covers the areas of technical interfaces, usage conditions and tariffs principles. The conditions of ONP must not restrict access to networks and services except for narrowly defined public interest reasons and must be based on objective criteria, be transparent, published in an appropriate manner, guarantee equality of access and be non-discriminatory. Based on these principles, more detailed legislation is gradually being drafted, e.g. the Council enacted a Directive on the application of ONP to leased lines in 1992.

Third, the widespread use of subsidies has led to a substantial divergence of telecommunications tariffs from the cost of service provision. In its resolution of 1988 on the development of the common market for telecommunications services and equipment up to 1992, the Council made explicit that telecommunications administrations would need to move toward cost-based tariffs. The Commission conducted a review of progress in 1992 and, in addition to consideration of national tariffs, it launched a formal investigation into international tariff-setting after preliminary studies revealed anti-competitive arrangements between telecommunications organizations.

Finally, the position of the Green Paper to separate regulatory and operational activities was included in both the terminal equipment and the services Directives. However, the specific approaches chosen in the member countries vary, with some (e.g. Germany and the Netherlands) relying on ministerial departments and others (e.g. the UK) creating formally independent regulatory bodies.

Another proposal, that for the creation of a European Telecommunications Standards Institute (ETSI), has resulted in a major reform of the standards-setting process in the sector. This is of particular importance both from the equipment development perspective and for fostering a more rapid uptake of new telecommunications facilities such as video conferencing (Button and Maggi 1994). The ETSI was founded in 1988 and has evolved to become the main body in the creation of European telecommunications standards. To date, however, standards are still evolving.

The Council, in 1990, adopted a Directive on the procurement procedures of entities operating in the water, energy, transport, and telecommunications sectors. The implementation of this Directive, although slow, is designed to open telecommunications procurement markets to bidders from other member states.

With the exception of a provision to liberalize receive-only Earth stations, the Green Paper on telecommunications has made the satellite issue contingent on further study. The background document, the Green Paper on Satellites, was published in 1990 and proposes four major changes of the existing regulatory environment in order to exploit fully the potential of satellite communications.

1. Liberalization of the Earth segment for both receive-only and two-way terminals.
2. Free access to space segment capacity, subject to licensing procedures in order to safeguard exclusive or special rights; with regulatory provisions set up by member states in accordance with Union law and based on the consensus achieved in Union telecommunications policy.
3. Full commercial freedom for space segment providers, including direct marketing of satellite capacity to service providers and users, subject to compliance with the mentioned licensing procedures and the EU competition rules.
4. Harmonization measures as far as required to facilitate the provision of Europe-wide services.

As compared to the *status quo*, these positions should lead to a significant opening of the satellite-based services market, although the need for compatibility with established exclusive or special rights in individual Member States will limit the overall effect. Legislation to implement these basic provisions of the Green Paper on satellites is currently being drafted.

The various legislative initiatives by the European Commission with regard to telecommunications are inseparable from the overall Union competition policy. Such Directives as those on terminal equipment and services, for example, are rooted in Treaty of Rome provisions espousing competition in goods and services. The Directives merely attempt to implement the provisions in a way that is consistent with the goals of the Treaty and are specifically focused on the equipment and services sectors.

Two additional aspects of the Treaty, though, are of particular importance. First, it specifically permits the European Commission to issue Directives directly to member states in order to ensure the application of competition policies to public undertakings or situations where Member States grant exclusive privileges. Second, it prohibits various forms of anti-competitive behaviour, including agreements that may affect trade and any abuses of a dominant position. These points have been instrumental in influencing the shape and pace of telecommunications reform beyond the general legislative processes.

15.5 NATIONAL STRATEGIES

Individual EU members have not adopted exactly the same stance regarding the way telecommunications policy should develop. Loose cooperation between the European PTTs did exist, e.g. in the European Conference of Postal and Telecommunications and also through the International Telecommunications Union. One of the main reasons for the somewhat differing emphasis found between national policies relates to general attitudes to regulation. The Anglo-Saxon-type approach of countries such as the UK emphasizes the simple internal efficiency of an industry while the continental philosophy of France looks at regulation in a broader context and, therefore, regulatory controls have traditionally been closer to government.

The Green Paper called only for a separation between the regulatory and operational functions of telecommunications administrations and contained no guidelines as to how this separation should be achieved. Regulatory functions may remain within the government ministry in charge of telecommunications, or they may be given over to a new, independent agency with no explicit link to the ministry. With the exception of the UK, which has created an independent regulatory agency (OFTEL), countries have kept their regulatory function within the relevant ministry.

The alternatives raise a number of issues. First, is a ministry capable of acting in an unbiased manner when dealing with competitors to an operator dependent on it? Following public choice theory, are ministers willing to make unpopular decisions that might have negative consequences in elections? An alternative perspective is that independent regulators are often less accountable.

The objectives of the regulation can include a variety of goals such as ensuring universality of services, availability of services, efficient management of scarce resources, and fair competition in those areas where permitted. The tools for meeting such objectives have traditionally embraced control of licences, tariffs, and, increasingly, interconnection requirements and usage conditions, but these tools may also vary between countries. The UK and France, for example, employ a price-cap scheme, which limits any price increases to the rate of inflation minus some percentage (although France's cap is less severe than the UK's). The (RPI–X) formulation in the UK is the classic example. Carsberg (1991) offers comments on its use in practice. The UK has also deployed other incentive regulation schemes, with financial penalties on poor performance, in an effort to provide the potential incentive under a price-cap regime for suppliers to maximize short-term returns by avoiding investment and allowing performance to deteriorate.

There are no specific EU guidelines, other than the separation of regulatory and operation functions, regarding the status of public telecommunications operators (PTOs). They can be administrations that remain a part of a ministry, state-owned companies that operate under the same legal framework as private firms, or privately owned companies. In several EU states, such as Germany, the PTO has retained the status of public administration tied to a ministry. The Belgium telecommunications operator has moved from administration to state-owned company status. This latter structure, however, because of the lack of earmarking, provides the operator with less control over investment strategy, because it is dependent on government budgets, whereas operating revenues are, in essence, state income that may be spent on other programmes. Employees also often enjoy civil servant status, reducing flexibility in dealing with the labour force. For these reasons, it may be difficult to reconcile administration status with competition in the service sector, as new competitors will not have the same constraints.

A more common situation has been to change the PTO's status from administration to state-owned corporation; such bodies are often capable of financing investment out of their own revenues. They can, therefore, respond quite quickly to market shifts. Some state-owned companies, such as the Netherlands PTT, operate according to private business laws and have stock, even though the government owns 100 per cent of the shares. France Telecom has a more *ad hoc* status and is not fully subject to private business laws. Corporatization is often viewed as a pre-privatization step, which can help enhance the value of a private share offering by making the operator more efficient while the internal ethos moves away from that of a government bureaucracy.

Moving on from this, one feature of many national strategies is the priority given to privatization. There is no single reason for this. Rather, privatization is variously seen as a way of attracting private finance into an industry in need of major infrastructure investment, as a device to enhance internal efficiency by reducing bureaucratic management, to reduce public expenditure commitments or as a way of developing cross-border activities within Europe. There may also be broader objectives, often of a political nature such as the UK concept of the share-owning democracy.

The pace and form of privatization has varied significantly across Europe. In part this is explained by quite important variations in national attitudes to public and private ownership *per se* but equally there were initially a variety of models of public ownership in place, and change is in part being dictated by perceptions of just how well any national public system met national objectives. The way in which the early privatizations have been viewed has also provided a learning and demonstration effect. Nevertheless, the fact is that major privatizations have taken place (most notably in the UK) and a number of major players are planning to follow a similar path in the near future (Table 15.4).

It would also be a mistake to view all privately held operators as alike. One consideration, for example, is the degree to which the government maintains some stake in a private telecommunications operator. The UK

Table 15.4. Planned privatizations

Company	Country	Estimated value (£ billion)	State holding (%)	Year
KPN	Netherlands	8.6	100	1994
Telefónica	Spain	8.1	31	1994
STET	Italy	7.5	53	1995
France Telecom	France	21.6	100	1996 +
Telekom	Germany	14.8	100	1996 +

Source: derived from Bowden (1994)

government, for example, initially retained 51 per cent of the shares of BT following its privatization in 1984. Other governments hold stakes in their partially private operators (e.g. see again Table 15.4), including the Spanish government with a significant stake in Telefónica and the Italian government with a majority ownership of SIP (the local telephone service provider) and Italcable (the intercontinental service provider).

Privatization of telecommunications operators relates to processes of liberalization and regulation in key ways (Vickers and Yarrow 1988; Foster 1992). A private firm is usually assumed to have a greater incentive to be efficient in its use of resources. Econometric evidence from the energy sector indicates, however, that it is the reform in regulation inevitably accompanying privatization that is the dominant impact on performance levels (Button and Weyman-Jones 1994). The unconstrained pursuit of profits can, however, conflict with social objectives and, hence, some form of regulation generally remains. Even with privatization, however, there are few incentives for a private monopoly to improve quality and innovate. To achieve these objectives, some form of competition is generally introduced. For example, in the UK, a second private operator, Mercury, was introduced to compete with the privatized BT. The exclusivity provision is frequently introduced to deal with newly privatized operators and new facilities-based entrants, so as to ensure investor confidence. The mere presence of a competitor has often proved to be inadequate to control the power of the former monopoly suppliers and regulations are gradually being introduced.

Although the European Commission's policies allow member states to maintain the exclusive monopoly provision of basic telephony and telex, there is no restriction against authorizing competition in these services if a country so desires. The UK has been willing to permit competition in these so-called reserved services, beginning first with the creation of duopoly between Mercury and BT, but since 1991 other network operators will be authorized to offer voice telephony services and resale of leased capacity is now allowed. More recently, in 1993, new decisions have laid the groundwork for opening voice resale to competition with complete deregulation of market entry coming in 1998.

As in the above case, a monopoly on network provision is allowed but not mandated by the EU, and an individual country may authorize competitive operators. Likewise, as noted above, within Europe only the UK permits such facilities-based competition. A decision to authorize facilities-based competition introduces the need for new regulatory initiatives aimed at nurturing new entrants and protecting them from the overwhelming market power of established PTOs.

15.6 CONCLUSIONS

Telecommunications is a very dynamic sector. It is also one which needs a clearly defined legal and institutional framework in which to operate if provision is to be efficient. In the past decade, however, it has been undergoing fundamental change in terms of the regulatory framework under which it operates in Europe. Much of the focus of this shift in the regulatory framework has concerned efforts to enhance the competitive nature of supply in some spheres of telecommunications services, such as terminal equipment and specialized services. This has been accompanied by rather less forceful moves to introduce competition into infrastructure development where the role of network externalities and other forces are less certain.

Whether these developments will be sufficient to bring about the rapid transformation of the system to meet the needs of the next century given the commitments of countries such as the US to develop sophisticated information highways is less certain. The necessary changes will be painful, perhaps involving the loss of over a million jobs in the sector by the end of the century (Levine et al 1993). This will be difficult to achieve when many national systems are still protected and the attitudes of the old PTTs have been changing only slowly. The changes will be needed, however, not only to ensure that current relative levels of service are maintained, but also to enable the full benefits of the greatly enhanced potential which will be afforded if the infrastructure investments planned in the EU come to fruition.

REFERENCES

Adonis, A. (1993) Whose line is it anyway?, *Financial Times*, October 11, p. 15.
Adonis, A. (1994) A crucial role in companies large and small, *Financial Times*, June 15, pp. 1–11.
Bauer, J. M. and C. Steinfield (1994) Telecommunications initiatives of the European Communities, in C. Steinfield, J. M. Bauer and L. Caby (eds), *Telecommunications in Transition*, Sage, London.

Baumol, W. J. and J. G. Sidak (1994) *Toward Competition in Local Telephony*, MIT Press, Cambridge, MA.

Baumol, W. J., J. C. Panzar and R. D. Willig (1982) *Contestable Markets and the Theory of Industry Structure*, Harcourt Brace Jovanovich, New York.

Bowden, A. (1994) Getting ready to face the public, *Investors Chronicle*, April 22, p. 86.

Button, K. J. and R. Maggi (1995) Videoconferencing and its implications for transport: an Anglo-Swiss perspective, *Transport Reviews*, 15(1), 59–75.

Button, K. J. and D. Swann (eds) (1989) *The Age of Regulatory Reform*, Clarendon Press, Oxford.

Button, K. J. and T. Weyman-Jones (1994) The impacts of privatisation policy in Europe, *Contemporary Economic Policy*, 12, 23–33.

Capello, R. (1991) L'assetto istituzionale nel settore delle telecomunicazioni, in R. Camagni (ed.), *Computer Networks: Mercati e Prospettive delle Tecnologie di Telecomunicazione*, Etas Libri, Milan.

Capello, R. (1994) *Spatial Economic Analysis of Telecommunications Network Externalities*, Avebury, Aldershot.

Carsberg, B. (1991) Office of Telecommunications: competition and the duopoly review, in C. Veljanovski (ed.), *Regulators and the Market: An Assessment of the Growth of Regulation in the UK*, Institute for Economic Affairs, London.

Commission of the European Communities (1987) *Towards a Dynamic European Economy: Green Paper on the Development of the Common Market for Telecommunications Services and Equipment*, COM(87) 290.

Commission of the European Communities (1988) *Towards a Competitive Community-wide Telecommunications Market in 1992: Implementing the Green Paper on the Development of the Common Market for Telecommunications Services and Equipment*, COM(88) 48.

Commission of the European Communities (1993) *Growth, Competitiveness, Employment – The Challenges and Ways Forward into the 21st Century. (White Paper)*, Luxembourg, Office for Official Publications of the European Communities.

Council of the European Communities (1984), Council Recommendation of 12 November 1984 Concerning the Implementation of Harmonisation in the Field of Telecommunications (84/549/EEC), OJL 298/49.

Cowhey, P. F. (1990) Telecommunications, in G. C. Hufbauer, (ed.) *Europe 1992: An American Experience*, Brookings Institution, Washington.

Foster, C. D. (1992) *Privatisation, Public Ownership and the Regulation of Natural Monopoly*, Blackwell, Oxford.

Horwitz, R. B. (1989) *The Irony of Regulatory Reform: The Deregulation of American Telecommunications*, Oxford University Press, Oxford.

Hurst, C. (1991) The market for international telecommunications services: some arguments for a competitive environment, *EIB Papers*, No. 17, pp. 27–53.

Ireland, J. (1994) *The Importance of Telecommunications to London as an International Financial Centre*, London Business School, London.

Kahn, A. E. (1970) *The Economics of Regulation: Principles and Institutions*, MIT Press, Cambridge, MA.

Kahn, A. E. and W. E. Taylor (1994) The pricing of inputs sold to competitors: a comment, *Yale Journal on Regulation*, 11, 225–240.

Levine, J. B. with G. E. Schares, J. Flynn, P. Dwyer and J. Rossant (1993) Wake-up call, *Business Week*, December 20, pp. 38–40.

Mosteshar, S. (1993) *European Community Telecommunications Regulation*, Graham and Trotman, London.

Muller, J. (1988) *The Benefits of Completing the Internal Market for Telecommunica-*

tion – Equipment and Services in the Community, Office for Official Publications of the European Communities, Luxembourg.

Noem, E. M. (1992) *Telecommunications in Europe*, Oxford University Press, Oxford.

Swann, D. (1992) *The Single European Market and Beyond – A Study of the Wider Implications of the Single European Market*, Routledge, London.

Train, K. E. (1991) *Optimal Regulation: The Economic Theory of Natural Monopoly*, MIT Press, Cambridge, MA.

Ungerer, H. with N. Costello (1988) *Telecommunications in Europe*, Office for Official Publications of the European Communities, Luxembourg.

Vickers, J. and G. Yarrow (1988) *Privatisation: an Economic Analysis*, MIT Press, Cambridge, MA.

Von Auw, A. (1986) *Heritage and Destiny: Reflections on the Bell System in Transition*, Praeger, New York.

Wilkinson, C. (1989) Completing the internal market for information technology and telecommunications, in D. Shorrock, (ed.), *European Communications: Technologies and Regulations of the Single Market*, Blenheim, London.

16 Network Diversity or Network Fragmentation? The Evolution of European Telecommunications in Competitive Environments

ANDREW GILLESPIE and JAMES CORNFORD
University of Newcastle upon Tyne, UK

16.1 INTRODUCTION

This chapter is concerned with the nature of the evolution which is occurring in telecommunications provision in Europe, and, in particular, with its geographical outcomes and implications. 'Evolution' is perhaps misleading here, for what we are witnessing is rather a fundamental discontinuity or rupture between two quite different paradigms of telecommunications provision, a rupture which of course is paralleled by other major shifts in the 'regime of accumulation' and its regulation (Lüthje 1993). Within the domain of telecommunications provision within Europe, we are seeing the demise of the unitary network, provided under monopoly conditions within individual nation states, and the rise of multiple networks, competitively supplied, and displaying new forms of articulation between globalization and localization.

Setting aside for the moment the technological underpinnings of this shift from unitary to multiple networks, which are considered below, its implications can be interpreted in two quite different ways; one which accentuates the positive aspects of network diversity, the other which stresses rather the risks of network fragmentation.

A particularly useful interpretative framework has been developed by Mansell (1993) in her study of the political economy of network evolution. She provides two contrasting paradigms for understanding the relationship between technical and institutional change in the development of telecommunications provision. The first she terms the *idealist* model, which is derived from theories which envisage the emergence of mature and fully articulated competitive markets, and which assumes that technology automatically provides solutions to current problems and imperfections. In contrast, what Mansell terms the *strategic* model is rooted in theories of

European Transport and Communications Networks: Policy Evolution and Change. Edited by David Banister, Roberta Capello and Peter Nijkamp. © 1995 John Wiley & Sons Ltd.

Table 16.1. Mansell's idealist and strategic models compared

The idealist model	The strategic model
Permeable seamless networks	Fragmented networks
Ubiquity (universal service diffusion)	Reduced ubiquity in service diffusion
Demand-led telecommunications industry	Supply-led industry, multinational user pressure
Open-systems, common interface standards	Weak stimuli for competition in most submarkets
Co-operative partnerships, transparent network access	Monopolisation and rivalry, non-transparent network access
Minimal regulation to achieve efficiency and equity	Increasing regulation

Source: Mansell (1993) p. 193 and 195

imperfect competition and oligopolistic rivalry, and emphasizes the way in which new market distortions become embedded in the design of technical artefacts such as telecommunications networks. The main characteristics of network evolution in these contrasting interpretations are indicated in Table 16.1.

In the later stages of this chapter, we shall focus on the geographical aspects of service provision. Under an idealist, or network diversity, interpretation, we would expect to see competition delivering benefits to all places, with ubiquity of access to the seamlessly interconnected networks. Under the strategic, realist or network fragmentation interpretation, in contrast, we would expect to see geographically differentiated levels of competition and network access emerging. Based upon an analysis of the evolution of telecommunications provision in the UK, which, in the European context at least, has gone furthest in facilitating the shift from unitary to multiple networks, we shall argue that the network fragmentation interpretation has greater empirical validity; and further, that in consequence, new forms of regulation are needed in Europe to ensure that emerging telecommunications environments do not lead to a sharp differentiation between places in the range and quality of services provided.

Firstly, however, the historical context for the current shift from unitary to multiple networks is outlined, followed by an examination of the reasons for the shift, with a particular focus on the balance between technological changes and changes in the structure of markets, institutions and regulatory frameworks.

16.2 FROM MANY TO ONE, AND ONE TO MANY

For much of this century, telecommunications services have been delivered within the context of a particular institutional form, that of regulated monopoly. It is important to recognize that there is nothing in the nature of telecommunications systems that dictates that monopoly is the only, or only rational, form of delivering telecommunications services. Even the long accepted so-called 'natural monopoly' in the connection of individual subscribers to the local telephone exchange (usually termed the 'local loop'), needs to be understood as a historically specific, socially—rather than technologically—determined institutional form.

In the United States, for example, the Bell monopoly emerged only after a 25-year period of vigorous competition between Bell and a large number of independents which sprung up after the expiration of the original patents (Mueller 1993). This competition took the form of rivalry between quite separate, unconnected systems. Whereas Bell had been pursuing a strategy modelled on Western Union's telegraph network, that of a national network connecting large cities, the independents targeted firstly the suburbs and smaller towns that the Bell network had initially chosen not to serve. Bell's response to the competitive threat was, firstly, to greatly extend its own network coverage in these areas, leading to 'dual service', or what Mueller (1993) terms 'access competition'.

The second part of the Bell company's competitive response was to develop a vision of a single, fully interconnected national system, for which it attempted to get political backing. This vision was articulated by Theodore Vail, the President of AT&T, in the company's 1909 Annual Report, as follows:

> The Bell system was founded on the broad lines of 'One System', 'One Policy', 'Universal Service', on the idea that no aggregation of isolated independent systems not under common control, however well built or equipped, could give the public the service that the inter-dependent, intercommunicating, universal service could give (AT&T 1909, p. 18).

Political backing was gradually obtained for the elimination of the fragmentation caused by access competition. It was these diseconomies of fragmentation—in the sense of network externalities foregone—rather than any notion of supply-side economies of scale, that led, by the early 1920s, to the establishment of regulated monopoly provision of telecommunications in the United States (Mueller 1993).

What is it, seventy years on, that has led, or in Europe is leading to, the undermining of the monopoly-provided unitary networks? Undoubtedly, technological changes, coupled with increasing pressure from large users for cheaper and more specialized services, have played a major role in the shift. Let us consider each in turn.

16.3   TECHNOLOGICAL CHANGE AND THE 'GEODESIC' NETWORK

New wireless transmission technologies, including satellite, microwave radio and cellular radio, have come to provide viable, and in some cases, highly desirable alternatives to copper wire, coaxial cable or fibre-based networks. Radio technologies in particular have put the final nail in the coffin of natural monopoly arguments; even the local loop, regarded as the last bastion of natural monopoly, is now wide open to alternative delivery systems (Cave and Sharma 1993).

Further, within the 'conventional' wire-based telephone network, the convergence of computing with telecommunications has posed a profound challenge to the unitary network. This challenge finds expression in the architecture and degree of centralized control within telecommunications systems. Mulgan (1991) describes how telephone networks were traditionally configured in switching hierarchies, which reached from the home through local to central switches, organized in the form of a pyramid or tree and branch structure under the control of the PTT. In one sense, the huge economies of scale embodied in new digital switches fostered this type of centralized control over the network. Simultaneously, however,

> cheap processing power ... has served to erode traditional structures of control. ... As costs of processing and switching fell more rapidly than the cost of transmitting signals, fundamentally different network topologies became viable, more like a mesh or grid than the traditional pyramid, with multiple points of access and interconnection (Mulgan 1991, p. 43).

With computerization, intelligence has been increasingly dispersed not only throughout the network but also into customer premises equipment, such as terminals, PBXs and local area networks (LANs); LANs indeed represent the ultimately dispersed intelligence, in which every node is a switch. An influential study by Peter Huber in the late 1980s compared the emerging architecture of telecommunications networks to Mandelbrot's fractals, in which the whole consists of a multitude of parts based on similar topologies. He termed this the 'geodesic' network, after Buckminster Fuller's geodesic domes (Huber 1987).

Undoubtedly, then, technological change has played an important role in the shift from the unitary to multiple networks. This does, however, rather beg the question of where the technology comes from, how and why it has come into being in the form that it has. As Mansell (1993, p. 196) reminds us, a self-contained account of technical progress,

> is of limited value unless it is coupled with the economic and political factors that contribute to the biases operating within the innovation and technical selection process. These factors shape the design and implementation of opera-

tional telecommunication systems. They are tightly enmeshed within the political economy of institutional change and in the strategies adopted by suppliers and some segments of the telecommunications user community.

16.4   USER PRESSURES AND THE POLITICAL ECONOMY OF NETWORK FRAGMENTATION

It is important to recognize that much of the technological change which is driving the emergence of more flexible and decentralized architectures comes not from the traditionally constituted telecommunications industry but from the computer industry and, particularly, from computer users. As telematics applications have become increasingly central and strategic to major users, often becoming integral to their core competitive strategies, so their demands for more specialized and optimized networks have proliferated. The specialized and segmented nature of these demands means that they cannot be accommodated in a unitary network.

A good current example is provided by the rapid deployment of a new generation of data services, known as switched multi-megabit data services (SMDS). These are metropolitan-area and wide-area public packet data networks which support data rates from 1.5 to 34 Mbit/s. Developed in the late 1980s by Bellcore, the research organization for the US Regional Bell Operating Companies, the service has been commercially launched in the US, Germany (as Datex-M) and the UK, with trials taking place in many other countries. The emergence of SMDS is explained by user pressures to interconnect their LANs, and to run LAN applications over wide areas. The pre-existing solution to this demand need, leased digital circuits, was both prohibitively expensive (due to the high bandwidth required for short bursts) and inflexible (due to the lack of switching capability).

SMDS is designed then for a specific purpose, to provide a *public* network service for connecting geographically distributed *private* token rings and Ethernets. Technological advances such as SMDS, therefore, have not occurred in a vacuum; they can be interpreted as the telecommunications industry's response to user pressures. As such, both technical advance and the proliferation of specialist networks reflect and embody the underlying power struggle between different institutions and interest groups, particularly that between the telecommunications industry and its users.

According to Eli Noam, the shift from a unitary to multiple networks is an inevitable outcome not of the failure of the monopoly provision model but rather of its success, resulting in what he terms the 'tipping of network coalitions' (Noam 1993). His argument is that monopoly provision has succeeded in making telephone use universal and essential; as the system expands, however, political group dynamics take place which lead to redistribution and over-expansion. This provides increasing incentives for certain users to exit from the network, and to the eventual 'tipping' of the network from a stable, single coalition to a system of separate subcoalitions. Noam terms this 'the tragedy of the network commons', for 'as in a Greek tragedy, their [the PTTs] preventative actions only assure their doom' (Noam 1993, p. 46).

If, as has happened in the present period (though not in the competitive pre-monopoly era in the United States outlined earlier), the exiting users can force the regulators to allow the interconnection of alternative networks with the existing unitary public network, then the exiters can have the best of both worlds; customized, optimized, privatized networks, plus the network externalities deriving from interconnection with the public network.

As has already been suggested, explicit within the shift towards multiple competitively provided networks is a shift in the balance of power between institutions. The centralized, unitary network had a very well-established coalition of institutions and power structures supporting it, a coalition which Noam (1987) has termed the 'postal-industrial complex' and comprising the PTTs themselves, equipment manufacturers, trade unions, and political parties of the centre and left, coalescing around a political programme based on regulated monopoly provision, universal service, cross-subsidies and *de facto* domestic protectionism.

Power has clearly shifted to a new coalition which is dominated by large users, in which the considerable power of the PTTs and telecommunications equipment suppliers has been partly neutralized by the counterbalance of the computer industry. Large user pressures are leading not only to the de-regulation of telecommunications within national boundaries, but also to the globalization of telecommunications networks and the telematics applications which run across them.

The idealist interpretation of this shift in the balance of power towards users, which is associated with the demise of the monopoly-provided unitary network, stresses the greater degree of user choice which network diversity brings. However, the strategic model interpretation, which focuses on the political economy of network evolution, emphasizes the limitations to this 'choice' for those users that are not part of the new telematics coalition:

> in spite of the claim that the public network is becoming more open and accessible to users in general, in fact, it is becoming more closely attuned to the needs of a specific segment of the user community; namely, the globally operating firms. The risk, in the absence of effective policy and regulation, is that access to the electronic communication environment for some users will become increasingly limited at the same time that it is being extended for large corporate users (Mansell 1993, p. 197).

In Section 16.5, we examine the nature of this risk, concentrating on the geography of service provision in one of the world's most liberalized telecommunications environments, the United Kingdom. The proposition to be explored is that the diverse, competing networks which have been fostered by this regime are geographically selective in terms of the user communities which they are serving.

16.5   FRAGMENTED GEOGRAPHIES: THE EXAMPLE OF TELECOMMUNICATIONS SUPPLY IN THE UNITED KINGDOM

Liberalizing the telecommunications sector is intended to provide a number of generalized benefits to users, including service innovation and lower prices. In the case of the customer equipment market, these benefits are location-independent. The inherent spatiality of network infrastructures and network services, however, means that the benefits associated with new networks can be spatially differentiated, due to the interplay between the location of particular types of demand and the competitive strategies of the suppliers. The experience of introducing competition into telecommunications networks in the UK over the 1984–1994 period provides some indications of how this interplay works out in practice.

16.5.1   THE FIXED LINK DUOPOLY

Following the 1984 Telecommunications Act, Mercury Communications Ltd was licensed to provide competition for British Telecom in the field of 'fixed network' services. Mercury was required, by the terms of its licence, to build a national network, with a list of specified cities which had to be connected to the network within a specified number of years. Mercury exceeded the terms of its licence, with the initial 'figure of eight' network connecting London with Birmingham and Manchester soon being extended to the north-east of England and Scotland.

Although the main competitor network to BT can thus claim to be national, it is far from being universally available. Mercury's marketing strategy has been to concentrate on providing an alternative long-distance trunk network for the largest business customers. Direct access to the company's network is obtained by fibre in over forty city centres and by microwave radio from locations with line-of-sight access to one of Mercury's trunk distribution nodes. These distribution nodes are located to serve large cities and other major concentrations of demand, such as business parks.

For non-metropolitan areas, and for SMEs regardless of their location, the so-called fixed-link duopoly has had very limited expression, leaving a *de facto* monopoly in most areas. Through the selective targeting of major business customers, Mercury now provides 11 per cent of total business lines in the UK, but fewer than 2 per cent of residential lines (see Table 16.2).

Table 16.2. Development of fixed-link competition in Britain, 1987–1993

	Business lines (numbers, 000s)				Business lines (per cent)		
	BT	Mercury	Cable	Total	BT	Mercury	Cable
1987	4 359	7	0	4 366	99,84	0.16	0.00
1988	4 558	38	0	4 596	99.17	0.83	0.00
1989	5 307	110	0	5 417	97.97	2.03	0.00
1990	5 551	226	0	5 777	96.09	3.91	0.00
1991	5 795	391	0	6 186	93.68	6.32	0.00
1992	5 866	582	0	6 448	90.97	9.03	0.00
1993	5 970	725	20	6 715	88.91	10.80	0.30

	Residential lines (numbers, 000s)				Residential lines (per cent)		
	BT	Mercury	Cable	Total	BT	Mercury	Cable
1987	17 549	0	0	17 549	100.00	0.00	0.00
1988	18 106	6	0	18 112	99.97	0.03	0.00
1989	18 703	15	0	18 718	99.92	0.08	0.00
1990	19 246	35	0	19 281	99.82	0.18	0.00
1991	19 573	90	4	19 667	99.52	0.46	0.02
1992	19 729	200	31	19 960	98.84	1.00	0.16
1993	20 114	336	124	20 574	97.76	1.63	0.60

Sources: OFTEL, ITC Data
Mercury Business = 2100 (direct) and 2200 (indirect) services;
Mercury Residential = 2300 (indirect) services; Mercury figures for March;
Cable figures for April; Cable figures not split between business and residential until 1993.

Recently, however, pressure from the regulator, OFTEL, concerned at the 'cherry-picking' strategy which Mercury have being pursuing, is beginning to lead to a more widely marketed service, albeit a service which is dependent on BT exchange lines being used to gain access to Mercury's long-distance trunk network.

16.5.2   CABLE TELECOMMUNICATIONS

In order to attempt to introduce competition into the 'local loop', an objective which the fixed-link duopoly had conspicuously failed to achieve, the Government's 1991 review of telecommunications policy sought to encourage the slowly developing cable TV industry to provide local telecommunications services. The raising of restrictions on non-EC ownership of UK cable operations at the end of the 1980s, combined with the new opportunity to provide telecommunications services as well as television over their networks, has led to substantial interest from US and Canadian cable and telecommunications companies, who have come to dominate the UK cable industry (Cornford and Gillespie 1993).

Cable networks are, however, very much confined to urban areas, with pronounced regional variations in cable coverage apparent as a result of differing degrees of metropolitanization. On a fourfold geographical sub-division of Britain, for example, the percentage of homes within cable franchise areas varies from 95 per cent in the core region of the Greater South East and 81 per cent in the largely metropolitan 'Heartland' region (basically the west Midlands, the north-west and Yorkshire), but reaches only 59 per cent and 58 per cent respectively in the less urbanized 'emergent southern' and 'periphery' regions (Cornford and Gillespie 1992). Competitive local fixed-link telecommunications networks will, therefore, be largely confined to cities and core regions, with much lower investment in alternative infrastructures in the UK's peripheral regions with their more rural settlement structures.

16.5.3 MOBILE CELLULAR RADIO SERVICES

At the same time as the British Telecom/Mercury duopoly of fixed-link networks was created in the mid-1980s, a second, and much more effective, duopoly was created for mobile cellular radio. Whereas competition proved difficult to establish in fixed-link services, due to the strength of the monopoly created by British Telecom's fully developed existing network, in the context of a completely new service the establishment of a duopoly regime proved successful in stimulating the rapid roll-out of the competing Cellnet and Vodafone networks. Rapid network roll-out was encouraged further by the nature of the mobile telecommunications market, in which demand is spatially generalized through the travel patterns of mobile subscribers, rather than being dependent on locally generated demand. As a result, the requirement imposed in the licences of the two mobile operators for their networks to be able to reach 95 per cent of the UK population by the year 2000 has been easily exceeded, with 98 per cent of the population able to receive mobile services. Although the services are thus spatially generalized, they remain premium services in terms of pricing, and are aimed primarily at business users.

16.5.4 PERSONAL COMMUNICATIONS NETWORKS

Radio-based systems offer the possibility of competitive telecommunications networks which do not require substantial investment in a duplicate wire-based 'last mile'. In the UK three personal communications networks (PCNs) were licensed in the late 1980s to provide such services; delays have been experienced in launching these services, and only two licence holders remain. The first, Mercury One-to-One, launched its service in 1993, but only within the 'M25 ring' around London. The second PCN service, Orange, operated by Hutchison Telecom, was launched in 1994 with a more extensive coverage; it is expected that 60 per cent of the population will be covered by the year end.

It is, however, not yet clear how far PCNs will penetrate into rural areas. Certainly the roll-out of the Orange network displays a clear strategy of focusing on metropolitan areas and the major transport corridors connecting them. For rural areas which do not lie along these corridors, the prospects for future supply are uncertain. One study which attempted to estimate the likely extent of PCN coverage in Wales, for example, based on the levels of demand required for profitable operation, suggested that only the industrialized south-east and a small area in north-east Wales would be covered, leaving most of the country without a PCN service (National Economic Research Associates 1992).

16.5.5 THE OVERALL PICTURE

The emerging picture of telecommunications provision in the UK is thus one of dynamic service innovation coupled with marked geographical variation in the pattern of advanced service availability. Particular concentrations of demand, such as the City of London, are receiving huge amounts of investment by multiple competing telecommunications network operators— in the City, sophisticated users can choose between at least five suppliers of switched broadband services (BT, Mercury, COLT, Energis and the cable company). Surrounding these peaks of the telecommunications demand/ supply profile, are larger, predominantly urban and suburban hinterlands which are now beginning to see the emergence of competitive telephony through both cable and PCNs and, perhaps in future, other telecommunications services. Finally there are the areas of *de facto* monopoly of narrowband switched services, including rural areas and small, free-standing towns, as well as some deprived inner city areas which are effectively being 'red-lined' by the cable companies (Cornford and Gillespie 1994).

As significant as the differentiated pattern of competitive network supply is BT's competitive response, which, unsurprisingly, tends to mirror the strategies of its competitors by focusing investment and resources on those market segments, with their associated geographies, in which the risk of losing market share is greatest. The result threatens, in the longer term, to produce an increasingly uneven pattern of telecommunications provision in the UK, with substantial differentiation becoming apparent between what might be termed 'hot spots', 'warm haloes' and 'cold shadows'. In the former, a virtuous circle of expressed demand brings forward investment in new networks, while in the latter a vicious circle of low demand and/or high cost of network provision stifles infrastructure investment.

16.6 MULTIPLE NETWORKS AND THE ROLE OF REGULATION

In the era of the unitary network, the primary role of regulation was to prevent the abuse of monopoly power. This role is still necessary, because although monopoly power has been eroded, in certain market segments it remains strong, as the previous section has demonstrated. Although under competitive markets the established PTTs inevitably lose market share, their existing unitary networks still give them considerable power. Hence in the UK BT has been a vigorous advocate of complete deregulation, knowing that in the absence of regulatory controls its market power would enable it to destroy new entrants through predatory pricing and other anti-competitive strategies.

Further, the established PTTs have a clear interest in preventing the fragmentation of their existing markets through attempts to re-establish a technical vision of the all-embracing unitary network. Current debates over Broadband-ISDN and the 'intelligent network' can clearly be interpreted as the PTT's competitive response to the threat from rival network suppliers. Mansell (1993), for example, argues that the design of the intelligent network, including the creation of interface standards and control over the access to intelligence within the network, 'can be used to maintain or extend the scope of their markets and to strengthen the strategic oligopoly which is emerging from the monopolistic structures of the past' (Mansell 1993, p. 166).

In the new telecommunications environment, therefore, a key role for regulation is to ensure that the multiple networks can be re-integrated, in order to prevent network externalities being lost. Without regulation, there would be a very real risk of competition leading to the fragmentation of telecommunications supply between competing but unconnected networks, just as existed in the early years of competitive telecommunications markets in the US. In such an environment, the regulator often has to force inter-connection on the dominant supplier, who would otherwise prevent it. Inter-connection is a very complex technical and political issue, which requires competitors to gain access to network intelligence, to the central switching facilities of the PTTs. In the UK, for example, inter-connection between the BT and Mercury networks was a tortuous, drawn-out and highly contested process, which could only be achieved by the regulator, OFTEL, having the authority to arbitrate between the two parties and to force a settlement.

Beyond the two roles for regulation defined above—that of preventing the abuse of monopoly/oligopoly power and that of ensuring the interconnection of multiple networks—a third role concerns the geographical aspects of the risk of fragmentation in network provision in a competitive telecommunications environment. During the monopoly regime, 'universal service' within national territories became one of the dominant, albeit not necessarily very clearly defined, assumptions upon which monopoly provision rested and became politically legitimated (Hills 1993). The question now being posed is whether the commitment to universal service can be maintained in multiple network, competitive telecommunications environments.

16.7   REDEFINING UNIVERSAL SERVICE

A commitment to universal service exists in the UK with respect to basic telephony in the terms of BT's licence, with the cost of servicing 'uneconomic' customers shared between established operators through the mechanism of Access Deficit Contributions. However, no such commitment is imposed, or mechanism established, with respect to 'advanced' services. At the time of the 1990–91 so-called 'duopoly review' of telecommunications policy in the UK, various bodies, such as the Scottish Development Agency and OFTEL's Scottish Advisory Committee on Telecommunications, argued that ISDN should be brought within BT's universal service obligation. The argument was resisted by BT, however, and was not accepted by the Government, and ISDN continues to be supplied by BT on purely commercial grounds, taking into account the cost of providing the service and the revenues it is likely to generate.

At the European level, the need to define a workable and sustainable conception of universal service is particularly important, given the considerable disparities which exist between the Union's core and peripheral regions in terms of existing levels of provision of telecommunications services. It has been estimated, for example, that to upgrade the public infrastructure of the Community's designated less favoured regions to the level available in the rest of the Single Market would need some 40 billion ECU of investment (Commission of the European Communities 1992).

Within the context of the impending liberalization of telecommunications markets in Europe, there is then a pressing need to provide and implement a new definition of 'universal service', one which is both sufficiently flexible to cope with the realities of competing, specialized (as opposed to universalized) services, and sufficiently robust to ensure that Europe's regions and rural areas share in the benefits of service innovation, cost reduction and institutional reform. Unless this balance can be struck, the Union's proposals for a Common European Information Space will come into conflict with existing objectives concerning regional convergence and European cohesion.

Progress has been made towards developing the basis of the EU's policy towards universal service in a competitive environment. In the April 1993 communication to the Council on telecommunications (Commission of the European Communities 1993a), the political goal of universal service was defined as 'making available a defined minimum service of specified quality to all users at an affordable price'. Although it is recognized that the Community's policy towards developing universal service will need to go beyond basic telephony to encompass 'certain advanced services' (Commission of the European Communities 1993b, p. 4), existing data services and ISDN are likely to be incorporated within the definition only by the 'non-mandatory application of ONP [Open Network Provision] principles' (Commission of the European Communities 1993b).

It thus seems inevitable that in an environment of competing, specialized services, not all services will be made available everywhere. Evidence from the UK suggests that while the preservation of universal service (defined in terms of geographical access at least) is relatively easy to achieve for voice telephony, and while the commercial logic of mobile cellular radio services leads towards reasonably 'universalized' networks in geographical terms, for more specialized services, such as the SMDS high-speed data service considered in an earlier section, widespread geographical coverage is extremely unlikely.

The shift from a unitary, universal network to multiple, competing, specialized networks will thus inevitably lead to the increasing geographical differentiation in levels and qualities of telecommunications service provision. Given the growing importance of telecommunications to firm competitiveness, we can thus anticipate that this differentiation will have significant implications for the geography of economic development and locational decision-making.

REFERENCES

AT&T (1909) *Annual Report*.

Cave, M. and Y. Sharma (1993) Competitive developments in local telecommunications in the UK, Department of Economics, Brunel University, Middlesex (mimeo).

Commission of the European Communities (1992) *1992 Review of the Situation in the Telecommunications Services Sector*, Commission of the European Communities, Brussels, 21 October.

Commission of the European Communities (1993a) *Communication to the Council and the European Parliament on the Consultation on the Review of the Situation in the Telecommunications Services Sector*, COM(93)159, 28 April.

Commission of the European Communities (1993b) *Developing Universal Service for Telecommunications in a Competitive Environment*, Communications to the Council and European Parliament (draft).

Cornford, J. and A. Gillespie (1992) The coming of the wired city? The recent development of cable in Britain, *Town Planning Review*, **63**(3), 243–264.

Cornford, J. and A. Gillespie (1993) Cable systems, telephone and local economic development, *Telecommunications Policy*, **17**(6), pp. 589–602.

Cornford, J. and A. Gillespie (1994) Competition in the local loop: the development and implications of cable telephony in Britain, paper presented at the *International Telecommunications Society European Regional Conference*, Chania, Crete, 2–3 September.

Hills, J. (1993) Universal service: a social and technological construct, *Communications and Strategies*, **10**(2), 61–83.

Huber, P. (1987) *The Geodesic Network: 1987 Report on Competition in the Telephone Industry*, US Department of Justice, Antitrust Division, Washington DC.

Lüthje, B. (1993) On the political economy of 'post-fordist' telecommunications: the US experience, *Capital and Class*, **51**, 81–117.

Mansell, R. (1993) *The New Telecommunications: A Political Economy of Network Evolution*, Sage, London.

Mueller, M. (1993) Universal service in telephone history: a reconstruction, *Telecommunications Policy*, **17**(5), 352–370.

Mulgan, G. J. (1991) *Communication and Control: Networks and the New Economies of Communication*, Polity, Cambridge.

National Economic Research Associates (1992) *Study of Telecommunications Provision in Wales and its Impact on Regional Economic Development*, Final Report to the Welsh Development Agency, the Development Board for Rural Wales and the Office of Telecommunications.

Noam, E. (1987) The public telecommunications network: a concept in transition, *Journal of Communications*, **37**, 30–48.

Noam, E. M. (1993) Network tipping and the tragedy of the common network: a theory for the formation and breakdown of public telecommunications systems, *Communications and Strategies*, **10**, 42–72.

17 European Transport and Communications: Lessons for the Future

DAVID BANISTER
University College London, UK

ROBERTA CAPELLO
Politecnico di Milan, Italy

PETER NIJKAMP
Free University of Amsterdam, The Netherlands

17.1 PROLOGUE

Europe is rapidly changing in terms of both socio-political and economic features; especially in the last decade, old political barriers are removed and a new 'map' of Europe is gradually drawn. Traditional boundaries disappear and traditional patterns of competition—within national borders—are increasingly replaced by vigorous competition on a multinational scale. Over the past decades a succession of agreements (e.g. the General Agreement on Tariffs and Trade—GATT) and trading unions (the European Community, the European Economic Area and the European Free Trade Association), have developed which have attempted to remove (or at least to reduce) the effects of various tariff, quota and subsidy systems and other protectionist measures which have been introduced at various times by national governments to protect their individual economic growth or interests.

All these socio-political and economic changes drive the European economy towards an international 'network economy', in which all European countries tend to become part of the same 'trade block' and where 'intra-country' competition is increasingly replaced by a higher-order 'inter-trade-block' competition (Nijkamp 1993). New communications and transport networks may hence be regarded as the new 'carriers'—in both a literal and symbolic sense—of these new patterns of trade and interaction. Transport and communications networks are nowadays seen as the new weapons, upon which the competitiveness of firms and comparative advantages of regions and nations will increasingly depend. This new 'network

European Transport and Communications Networks: Policy Evolution and Change. Edited by David Banister, Roberta Capello and Peter Nijkamp. © 1995 John Wiley & Sons Ltd.

economy' is based on large flows of people, commodities and information transmitted among countries and regions; the globalization trend of the economy has inevitably caused an increase in transportation and communications channels and use. At the same time, large technological changes are apparent in both the transport and the telecommunications sectors, enabling the implementation of large and efficient networks throughout Europe. The technological revolution opens many new opportunities to implement advanced and sophisticated telecommunications networks, as well as efficient transport systems, but at the same time it also creates new threats in terms of environmental problems and organizational and financial bottlenecks and these have to be overcome in order to exploit the new potential of these networks.

This book provides ample empirical evidence on the transition phase through which the transport and telecommunications sectors are going. The transition phase is not limited to technological transformation; it deals at the same time with institutional changes (especially in the telecommunications sector) and with a deep reorganization and restructuring of the whole economy. Interesting and strategic *policy implications* emerge from the broad spectrum of analyses presented in this book. These are worth mentioning here and can be interpreted as *lessons for the future*. These lessons relate to three different issues, namely:

1. the need to overcome nationally oriented transport and telecommunications network planning;
2. the need for a better coordination among transport and telecommunications policy-makers;
3. the need for a linkage between territorial planning (including physical planning) and the new network planning.

17.2 A EUROPEAN NETWORK PLANNING

Historically, major transitions in European economic systems were always accompanied (or sometimes induced) by major changes in transport and communications networks. Various types of drastic restructuring or transformation phenomena have occurred in the past centuries. For instance, Andersson and Strömqvist (1988) distinguished four major transformations (so-called 'logistical revolutions') in the past millennium, each characterized by the emergence, acceptance and adoption of fundamentally new types of infrastructure. The following transitional phases were distinguished:

1. The *Hanseatic Period* (from the thirteenth century onwards) based on an integration of sea and land transport, through which northern Italy was linked to the European Hanseatic League.

2. The *Golden Age* (from the beginning of the sixteenth century) based on the improvement in ship building and navigation techniques, through which Europe was connected to other continents.

3. The *Industrial Revolution* (from the beginning of the nineteenth century onwards) based on new transport and industrial systems technology, through which new world markets could be created.

4. The *Information Revolution* (from the 1970s onwards) based on sophisticated interaction and communications channels, through which knowledge and information transfer was possible on a worldwide scale. In this framework just-in-time (JIT) systems and material requirements planning (MRP) are evolving as new management principles. The rapid developments in the area of new information technology have also led to the emergence of integral logistics. This may mark the beginning of a 'new area' (the 'fourth logistic revolution').

Transport and telecommunications networks are increasingly becoming strategic carriers of the rapid and deep transformation of Europe's socio-political features. However, some severe bottlenecks still exist which hamper the implementation of these networks at a European level. The most drastic bottlenecks that may be foreseen are (1) the still prevailing national planning orientation of these networks, and (2) the missing links which are still in place all over Europe. Fortunately, the notion of trans-European networks is a first policy response to the above issue.

As far as the planning of these networks is concerned, the Internal European Market requires for its efficient functioning the existence of advanced and sophisticated means of transport and telecommunications. What is needed in this context is European, and not national, planning and action in infrastructure policy, based on knowledge of past successes and failures in infrastructure planning and of the future needs of the economy. Only then can the value added of European integration be reaped. Unfortunately, interests in the European scale of networks has until recently not yet been very successful and significant, as transport and telecommunications policy is seldom performed from this perspective. In the telecommunications networks, for example, interests of national carriers are still too strong to induce a European planning policy. In this field, to avoid different (national) technical standards in networks, which would otherwise have hampered the harmonization of European advanced telecommunications networks, the Community has taken charge of the problem and has imposed via its Green Paper the implementation of an open network provision (ONP). This example demonstrates that national interests in these sectors are still stronger than European intentions, and that we are still far away from spontaneous policy initiatives from a European perspective.

This still nationalistic behaviour results from the development of Europe as a partly integrated and partly competitive network system of countries, regions and metropolitan areas. It is evident that a delay in the entry to a unique European network will imply a loss of many opportunities and hence cause significant costs. This also means that network infrastructure policy is of critical importance for the future competitive position of Europe as a whole, and also of its constituent regions and cities. In this context, both transport and telecommunications sectors play a crucial role, as they are the vehicles for both European integration and intra-European competitiveness.

The second drastic barrier to the achievement of a European transport and telecommunications network is the existence of 'missing networks', not only in terms of physical missing networks, but also in terms of different organizational, institutional and financial settings which still differ a lot among countries and which inevitably hamper the achievement of a unified European strategy in both transport and telecommunications network planning. As is witnessed in a recent study on 'missing networks' in Europe (see Maggi et al 1992; Nijkamp and Vleugel 1993), this aspect presents a serious concern which will have to be addressed in the future to overcome the present nationalistic view on the development of these networks.

17.3   A COORDINATION BETWEEN TRANSPORT AND TELECOMMUNICATIONS POLICY-MAKERS

Another crucial condition for a strategic implementation of transport and telecommunications networks in the future is a coordinated initiative between policy-makers in these sectors. It is in fact becoming more and more evident that the strategic importance of the future organization of society at large depends on the simultaneous use of both physical and virtual means of transport.

Transport and telecommunications sectors have often been regarded as two distinct and separate fields, growing at different growth rates, according to different technological advances, and facing a different growth in demand. The technological revolution in the telecommunications sector at the beginning of the 1980s had a particular influence on the corporate organization, by allowing new forms of external contacts with suppliers and customers (e.g. via the use of Electronic Data Interchange). At the same time, these new external linkages could only be put in place with a different transport structure. A recent study on this aspect (Capello and Gillespie 1993) has concluded that the most likely future scenario of the industrial organization is the 'network firm' scenario, built on the assumption that the inadequacies of Fordist mass production are overcome, but that the oppositional outcomes suggested by the 'flexible specialization' school are more verbally advocated than firmly based on empirical evidence. The 'network firm' scenario will be attracted towards intermediate forms of 'quasi-organization' that are assuming an ever more important role as an alternative to full vertically integrated or vertically disintegrated production systems. The 'make-together' organization—à la Williamson (1975)—among firms seems to be one dominant organizational structure which will emerge in the future.

If this is true, the 'make-together' form of organization implies a high volume of information transmitted between firms, in the form of horizontal intercorporate information flows. At the same time, high volumes of intermediate products need to be transported between cooperative firms. This new industrial organization has a peculiarity in respect to previous organizations; it needs at the same time both advanced transport and telecommunications systems, and these systems have to be integrated in terms of organization, planning, technological aspects and spatial development.

The JIT system and the new logistics systems are an evident example of the new organizational structure taking place. However, in order to have an efficient JIT system, or an efficient logistic chain, advanced telecommunications and transport systems need to be integrated; both networks have to be developed with the same geographical patterns, with the same technological advances and with the same carrying capacity. It is difficult to envisage an efficient organization in a region, when advanced transport systems are put in place but where telecommunications networks are still very old. The strength of these industrial organizations is related to the implementation of technological and geographical integration of strategic factors in both the transportation and communications systems.

Unfortunately, policy-makers in these two fields are not used to developing such networks which need to take the actions of other policy-makers into account. Traditionally, their way of thinking has been dictated by problems of individual sectors. At present an integrated policy strategy has to be advocated for the sake of efficiency of these two sectors in the future.

The need for a less pronounced sectoral planning orientation is also instigated in view of different transport means. The flexibility of routes and modes of transport, envisaged by some experts as the solution to congestion, requires a common policy strategy among different modes of transport, in terms of both organizational and geographical development. Even in this context, a lack of coordination among policy-makers has to be recognized.

17.4   A COORDINATION BETWEEN NETWORKS AND TERRITORIAL PLANNING

There is ample evidence of the fact that transport and telecommunications networks may be seen as the 'carriers' in both a literal and symbolic sense of the spatial configuration of the economy. Their spatial development influences the competitiveness of geographical areas, at the urban, the regional and the national level. As already mentioned, many authors have argued that these networks are becoming the strategic weapons upon which the competitiveness of industrial, regional and national systems will depend.

There is one aspect which is in general overlooked, but which is important for the success in the exploitation of these networks; they have to be implemented on the basis of the socio-economic specificity of the area. In other words, the full exploitation of these networks takes place when they are developed on the basis of the needs of the local (regional or national) economy. This implies the need for integrated transport and communication systems to be developed in conjunction with broader spatial (urban and regional) planning, since only in this way will transport and telecommunications networks be developed on the basis of the real needs and necessity of the emerging 'network economy'.

There have been many attempts to exploit transport and telecommunications networks for economic development. In terms of economic policy, the European Community itself has interpreted telecommunications systems as a way to decrease the gap between advanced and backward regions of the Community. A large Community programme, called STAR, was launched in 1987 in order to enhance economic development in 'Objective 1' regions of the Community. The success of the programme in increasing economic growth in these regions depends very much on various specific local conditions which have to be present. One of the most important conditions is the use of the new telecommunications systems for the industrial and economic specificity of the area in which they are implemented and for its future development plans. Only in this way can these technologies become strategic weapons for local potential users, and this in turn will allow the development of economic, urban and social specificities of the area (Capello 1994).

This statement refers to the idea that transport and telecommunications networks in themselves are not sufficient forces for generating indigenous local economic development. On the contrary, they have to be thought of as strategic instruments to be exploited with reference to broader spatial economic planning. In this way, supply-driven transport and communications projects with little or no connection to real demand requirements and needs can be avoided, and the future development of these leading technological infrastructures would rather be evaluated in terms of their contribution to the creation of an integrated economy for a unified Europe.

17.5   A RESEARCH AGENDA FOR THE FUTURE

As we have seen above, the socio-political changes at the European level have strong implications in terms of policy strategies for the future. At the same time, however, these changes reinforce the scientific interest in these two strategic fields, and especially strengthen the importance of studies in specific research areas.

This book provides new background information on research areas. However, some major fields of work have still some uncovered scientific aspects, due to both the changing socio-political situation and the increasing importance that these infrastructure technologies are now assuming. In this context, a research policy can be set out for the future. This would include important areas such as:

1. the impact of political changes on transport systems;
2. the increasing demand due to the increasing mobility patterns, and its consequence at the environmental level;
3. the declining friction of distance;
4. the role of new actors in a liberalizing transport and communication market.

Numerous research questions are raised by the demise of communism in central and eastern Europe and the former Soviet Union and the rapid move away from centrally planned economies towards more free-market economies based on consumers' choice. One crucial question which emerges is the way in which a more deregulated supply of these technologies could be useful to develop transport and telecommunications infrastructure in advance of this fragmented geopolitical framework.

Eastern economies represent a large market opportunity, where a re-development of old infrastructure and the development of new and more advanced transport and telecommunications networks represent large market opportunities. However, the best market structure through which these technologies are developed is a strategic problem; a more monopolistic structure might bear a higher risk of failure of the development of these infra-structures and could at the same time guarantee an even geographical development of these strategic infrastructures. Alternatively, a competitive structure could foster the development of more advanced technologies in a shorter time period.

In relation to the changing geography of Europe, a research field which becomes increasingly strategic is the efficiency, productivity and competitiveness in transportation and telecommunications networks. These become strategic weapons in the development of metropolises and main points in the world economy. To an ever greater extent, manufacturing, marketing and distribution functions of firms are becoming internationalized. The efficiency and productivity of international transportation operations may determine the sources of raw materials and the location of production of goods (who is the importer and who is the exporter) and operational efficiency will surely influence the distribution function (who is the carrier). The ability to have a leading role in technological innovation will not only define the competitive advantage in transportation operations, but it will also determine the share of world markets in transportation products (Schofer and Boyce 1985).

The increasing transport demand, stemming from the new mobility patterns associated with the new geography of Europe, faces severe environmental constraints. In the framework of environmental aspects, there are three distinct and at the same time interrelated issues which are becoming of high scientific interest, namely (1) congestion, (2) pollution and (3) energy use. At the urban scale, traffic congestion is becoming a very large negative externality which is strongly affecting agglomeration economies associated with urban life. This growing urban road congestion has stimulated renewed worldwide interest in urban rail-based transit systems which provide fast, congestion-free movement through urban areas and especially to and from the central business district. Associated with this aspect are the issues of economic regeneration, reduction in pollution and accidents, and traffic decongestion.

Pollution is another environmental aspect associated with transport demand increase whose importance is rising and for which our current insights are not yet satisfactory. This problem holds at both the urban, regional and national level. The increase in the carrying capacity of airports, for example, finds a clear limit in the associated noise and air pollution that airports generate. Best 'sustainable policies' for meeting travel demand requirements and environmental constraints are necessary.

The impact of transport on the local and global environment and the use by transport of increasingly scarce and expensive non-renewable energy resources cause growing concern and pose many research questions (Knowles 1993). At the urban scale, traffic noise, atmospheric pollution and traffic accidents all diminish the quality of life, especially in densely populated urban areas. Traffic policies can reduce traffic flows and pollution, and help residents to regain some control over their local environment. Efforts to reduce the effects of exhaust emissions and to increase the energy efficiency of transport systems seem likely to be overtaken by increases in traffic volumes. Appropriate fiscal measures are often advocated, but are still to be found and implemented.

Another strategic research field associated with transport and telecommunications network development is the declining friction of distance. Location remains important as technological innovations in transport and telecommunications continue to reduce time and space differentially, and as the scale of economic activity increases and land use and the environment are affected (Knowles 1993). Transportation will increase the speed of movement, and thus enhance interaction along development corridors between the main centres to the detriment of smaller centres and peripheries. A major research question related to this aspect is how much these development patterns exacerbate and increase the still existing strong difference between centre and periphery.

Another interesting question concerns the role of new actors (e.g. distributors, forwarders, carriers) in a multimodal European network, where competition is increasing. This means that new transport and distribution systems may emerge, which serve to fulfil the needs of clients at minimal costs. Especially in freight transport this will have far-reaching consequences, as transport, distribution and internal logistics will become a more integrated chain.

The above-mentioned research issues are only a few of the great many interesting research areas which are related to transport and telecommunications networks and services. This book offers some suggestions on how to address some of the main research issues. It has the aim of both covering some strategic research issues, and of stimulating future research work in this field for the development of a more properly functioning common Europe.

REFERENCES

Andersson, A. E. and U. Strömqvist (1988) The emerging C-society, in D. Batten and R. Thord (eds), *Transportation for the Future*, Springer-Verlag, Berlin, pp. 64–82.

Capello, R. (1994) *Spatial Economic Analysis of Telecommunications Network Externalities*, Avebury, Aldershot.

Capello, R. and A. Gillespie (1993) Transport, communications and spatial organisation: conceptual framework and future trends', in P. Nijkamp (ed.), *Europe on the Move*, Avebury, Aldershot, pp. 43–65.

Knowles, R. D. (1993) Research agendas in transport geography for the 1990s, *Journal of Transport Geography*, **1**(1), 3–11.

Maggi, R., I. Masser and P. Nijkamp (1992) Missing networks in Europe, *Transport Reviews*, **12**(4), 311–321.

Nijkamp, P. (1993) Challenges to European transport policy analysis', in P. Nijkamp (ed.), *Europe on the Move*, Avebury, Aldershot, pp. 1–16.

Nijkamp, P. and J. Vleugel (1993) Missing networks in European telecom systems, in H. Bakis, R. Abler and M. Roche (eds), *Corporate Networks, International Telecommunications and Interdependence*, Belhaven Press, London, pp. 77–98.

Schofer, J. L. and D. E. Boyce (1985) Conference summary and conclusions, *Transportation Research*, **5/6**, 351–354.

Williamson, O. (1975) *Markets and Hierarchies: Analysis and Antitrust Implications*, Free Press, New York.

Index